# Learning Science in Informal Environments

## People, Places, and Pursuits

**Committee on Learning Science in Informal Environments**

Philip Bell, Bruce Lewenstein, Andrew W. Shouse, and Michael A. Feder, *Editors*

Board on Science Education

Center for Education

Division of Behavioral and Social Sciences and Education

NATIONAL RESEARCH COUNCIL
*OF THE NATIONAL ACADEMIES*

THE NATIONAL ACADEMIES PRESS
Washington, D.C.
**www.nap.edu**

**THE NATIONAL ACADEMIES PRESS 500 Fifth Street, N.W. Washington, DC 20001**

NOTICE: The project that is the subject of this report was approved by the Governing Board of the National Research Council, whose members are drawn from the councils of the National Academy of Sciences, the National Academy of Engineering, and the Institute of Medicine. The members of the committee responsible for the report were chosen for their special competences and with regard for appropriate balance.

This study was supported by Grant No. ESI-0348841 between the National Academy of Sciences and the National Science Foundation. Any opinions, findings, conclusions, or recommendations expressed in this publication are those of the author(s) and do not necessarily reflect the views of the organizations or agencies that provided support for the project.

**Library of Congress Cataloging-in-Publication Data**

Learning science in informal environments : people, places, and pursuits / Philip Bell ... [et al.], editors ; Committee on Learning Science in Informal Environments, Board on Science Education, Center for Education, Division of Behavioral and Social Sciences and Education.
     p. cm.
  Includes bibliographical references and index.
  ISBN 978-0-309-11955-9 (hardcover) — ISBN 978-0-309-11956-6 (pdf)  1. Science—Study and teaching. 2. Learning.  I. Bell, Philip, 1966- II. National Research Council (U.S.). Committee on Learning Science in Informal Environments.
  Q181.L49585 2009
  507.1—dc22

                              2009007899

Additional copies of this report are available from the National Academies Press, 500 Fifth Street, N.W., Lockbox 285, Washington, DC 20055; (800) 624-6242 or (202) 334-3313 (in the Washington metropolitan area); Internet, http://www.nap.edu.

Suggested citation: National Research Council. (2009). *Learning Science in Informal Environments: People, Places, and Pursuits.* Committee on Learning Science in Informal Environments. Philip Bell, Bruce Lewenstein, Andrew W. Shouse, and Michael A. Feder, Editors. Board on Science Education, Center for Education. Division of Behavioral and Social Sciences and Education. Washington, DC: The National Academies Press.

# THE NATIONAL ACADEMIES
*Advisers to the Nation on Science, Engineering, and Medicine*

The **National Academy of Sciences** is a private, nonprofit, self-perpetuating society of distinguished scholars engaged in scientific and engineering research, dedicated to the furtherance of science and technology and to their use for the general welfare. Upon the authority of the charter granted to it by the Congress in 1863, the Academy has a mandate that requires it to advise the federal government on scientific and technical matters. Dr. Ralph J. Cicerone is president of the National Academy of Sciences.

The **National Academy of Engineering** was established in 1964, under the charter of the National Academy of Sciences, as a parallel organization of outstanding engineers. It is autonomous in its administration and in the selection of its members, sharing with the National Academy of Sciences the responsibility for advising the federal government. The National Academy of Engineering also sponsors engineering programs aimed at meeting national needs, encourages education and research, and recognizes the superior achievements of engineers. Dr. Charles M. Vest is president of the National Academy of Engineering.

The **Institute of Medicine** was established in 1970 by the National Academy of Sciences to secure the services of eminent members of appropriate professions in the examination of policy matters pertaining to the health of the public. The Institute acts under the responsibility given to the National Academy of Sciences by its congressional charter to be an adviser to the federal government and, upon its own initiative, to identify issues of medical care, research, and education. Dr. Harvey V. Fineberg is president of the Institute of Medicine.

The **National Research Council** was organized by the National Academy of Sciences in 1916 to associate the broad community of science and technology with the Academy's purposes of furthering knowledge and advising the federal government. Functioning in accordance with general policies determined by the Academy, the Council has become the principal operating agency of both the National Academy of Sciences and the National Academy of Engineering in providing services to the government, the public, and the scientific and engineering communities. The Council is administered jointly by both Academies and the Institute of Medicine. Dr. Ralph J. Cicerone and Dr. Charles M. Vest are chair and vice chair, respectively, of the National Research Council.

**www.national-academies.org**

# COMMITTEE ON LEARNING SCIENCE IN INFORMAL ENVIRONMENTS

# Acknowledgments

This report would not have been possible without the important contributions of National Research Council (NRC) leadership and staff, and many other organizations.

First, we acknowledge the support and sponsorship of the National Science Foundation (NSF). We particularly thank David Ucko, deputy division director of the Division of Research on Learning in Formal and Informal Settings, whose initial and continuing engagement with the committee supported and encouraged the development of the report.

We also acknowledge the contributions of participants in the planning process. In particular, a number of people participated in a planning meeting to define the scope of the study. We thank Alan Friedman, New York Hall of Science for chairing that meeting. We also thank Lynn Dierking and John Falk, Oregon State University; Kathleen McLean, Independent Exhibitions; and Martin Storksdieck, Institute for Learning Innovation, for preparing papers to elicit discussion at the planning meeting. The success of the meeting was largely due to the insights provided by the meeting participants, including Sue Allen, The Exploratorium; Dennis Bartels, TERC; Rick Bonney, Cornell Lab of Ornithology; Kevin Crowley, University of Pittsburgh; Zahava Doering, Smithsonian Institution; Sally Duensing, King's College London; John Durant, at-Bristol; Kirsten Ellenbogen, Science Museum of Minnesota; Patrice Legro, Koshland Museum of Science; Bruce Lewenstein, Cornell University; Mary Ellen Munley, Visitor Studies Association; Wendy Pollock, Association for Science-Technology Centers; Dennis Schatz, Pacific Science Center; Leona Schauble, Vanderbilt University; Marsha Semmel, Institute of Museum and Library Services; Cary I. Sneider, Boston Museum of Science; Elizabeth Stage, Lawrence Hall of Science, University of California, Berkeley; David Ucko,

NSF; and Ellen Wahl, Liberty Science Center. Following the planning meeting Julie Johnson, Science Museum of Minnesota, consulted with the project to help assemble the committee.

Over the course of the study, members of the committee benefited from discussion and presentations by the many individuals who participated in our four fact-finding meetings. In particular, our initial framing of the domain of science learning in informal environments underwent significant revisions and refinements as a result of the scholarly and thoughtful contributions made by the background paper writers, presenters, and responders. At our first meeting, Lynn Dierking, Oregon State University, gave an overview of the informal learning field in science, technology, engineering, and mathematics. Shalom Fisch, MediaKidz Research and Consulting, discussed the effects of educational media. Sheila Grinell, Strategic Designs for Cultural Institutions, spoke about the recent evolution of practice in informal science. George Hein, Lesley University and TERC, discussed the need for the field to be both cautious and bold. Jon Miller, Northwestern University, described a framework for understanding the processes through which children and adults learn about science, technology, and other complex subjects.

The second meeting included a diverse set of presenters. Maureen Callanan, University of California, Santa Cruz, described the sociocultural and constructivist theories of learning. Kevin Dunbar, University of Toronto, summarized the cognitive and neurocognitive mechanisms of science learning and how they play out in informal environments. Margaret Eisenhart, University of Colorado, Boulder, discussed the aspects of informal learning environments that afford opportunities to underserved or underrepresented populations. Leslie Goodyear, Education Development Center, Inc., and Vera Michalchik, SRI International, presented methods and findings from evaluations of informal programs that serve underrepresented or underserved populations. Kris Gutiérrez, University of California, Los Angeles, gave specific examples of how informal learning environments serve diverse populations. Karen Knutson, UPCLOSE, University of Pittsburgh, discussed views of learning science in informal environments inherent in programs and evaluations. K. Ann Renninger, Swarthmore College, gave an overview of theories of motivation and how they map to learning in informal environments.

At the third meeting, the committee heard evidence about the science learning that takes place in various informal venues and pressing policy issues in the field. Bronwyn Bevan, The Exploratorium; Christine Klein, an independent consultant; and Elizabeth Reisner, Policy Study Associates, participated in a panel discussion of current policy issues in informal learning environments. Deborah Perry, Selinda Research Associates, Inc., described how exhibits and designed spaces are constructed for learning science. Saul Rockman, Rockman Et Al, discussed the evidence of science learning from traditional forms of media. Bonnie Sachatello-Sawyer, Hopa Mountain, Inc., gave an overview of the design and impact of adult science learning programs.

At the fourth meeting, the public session was concerned primarily with the status of the papers prepared to support the committee's work and the organizational structure being implemented in NSF as it relates to this project. David Ucko provided an overview of the new organizational structure and focus of the education program offices at NSF.

At our final meeting, the committee discussed the planned practitioner volume on science learning in informal environments that the Board on Science Education is developing as a resource for practitioners based on the evidence, findings, and conclusions of this consensus study. Two of the current study members are also members of the oversight group for the practitioner volume: Sue Allen, The Exploratorium, and Gil Noam, Harvard University. The five other members of the practitioner volume oversight group also attended our final meeting: Myles Gordon, consultant; Leslie Rupert Herrenkohl, University of Washington; Natalie Rusk, MIT Media Lab; Bonnie Sachatello-Sawyer; and Dennis Schatz, Pacific Science Center. We are grateful to each member of the group for providing us with excellent feedback. The practitioner volume, sponsored by NSF's Division of Research on Learning in Formal and Informal Settings, the Institute for Museum and Library Services, and the Burroughs Wellcome Fund, will be released following publication of this report.

We also acknowledge the efforts of the eight authors who prepared background papers. Arthur Bangert and Michael Brody, Montana State University, along with Justin Dillon, King's College London, were asked to review the literature on assessment outcomes. Laura Carstensen and Casey Lindberg, Stanford University, along with Edwin Carstensen, University of Rochester, were asked to synthesize the literature on older adult learning in informal environments. Shirley Brice Heath, Stanford University and Brown University, was asked to describe how issues of diversity influence individual conceptions of science. The Institute for Learning Innovation was asked to review the evidence in evaluation studies of the impact of designed spaces. Bryan McKinley Jones Brayboy, University of Utah, and Angelina E. Castagno, Northern Arizona University, were asked to review and synthesize the literature on native science. K. Ann Renninger, Swarthmore College, was asked to review research on interest and motivation in the context of learning science in informal environments. Rockman Et Al was asked to provide a review of the evidence of the impact of traditional media (e.g., television, radio, print). Sarah Schwartz, Harvard University, was asked to provide a synopsis of the scope and institutional investments in after-school and out-of-school-time programs.

Many individuals at the NRC assisted the committee. The study would not have been possible without the efforts and guidance of Jean Moon, Patricia Morison, and Heidi Schweingruber. Each was an active participant in the deliberations of the committee, helping us to focus on our key messages and conclusions. In addition, they made profound contributions to the development of the report through periodic leadership meetings with the committee co-chairs and the NRC staff. We are grateful to Victoria Ward and Kemi Yai, who arranged logistics for our meetings and facilitated the proceedings of the meetings themselves. We would also like to thank Rebecca Krone for assisting with the construction of the reference lists in each chapter of the report. The synthesis of the diverse literatures reviewed in this report would not have been possible without the efforts of Matthew Von Hendy, who conducted multiple literature searches and acquired copies of studies essential to our review.

This report has been reviewed in draft form by individuals chosen for their diverse perspectives and technical expertise, in accordance with procedures approved by the Report Review Committee of the NRC. The purpose of this independent review is to provide candid and critical comments that will assist the institution in making its published report as sound as possible and to ensure that the report meets institutional standards for objectivity, evidence, and responsiveness to the charge. The review comments and draft manuscript remain confidential to protect the integrity of the deliberative process.

We wish to thank the following individuals for their review of this report: David Anderson, Department of Curriculum Studies, University of British Columbia; Bronwyn Bevan, Informal Learning and Schools, The Exploratorium, San Francisco, CA; Ilan Chabay, Public Learning and Understanding of Science (PLUS), University of Gothenburg, Sweden; Lynn D. Dierking, Free-Choice Learning and Department of Science and Mathematics Education, Oregon State University; Shalom Fisch, Office of the President, MediaKidz Research and Consulting, Teaneck, NJ; Shirley Brice Heath, Anthropology and Education, Stanford University and Department of Education, Brown University; Bonnie L. Kaiser, Office of the Dean of Graduate and Postgraduate Studies, The Rockefeller University; Frank C. Keil, Department of Psychology, Yale University; Leona Schauble, Peabody College, Vanderbilt University; and Cary I. Sneider, Boston Museum of Science, Portland, OR.

Although the reviewers listed above provided many constructive comments and suggestions, they were not asked to endorse the conclusions and recommendations nor did they see the final draft of the report before its release. The review of this report was overseen by Adam Gamoran, Center for Education Research, University of Wisconsin–Madison, and May Berenbaum, Department of Entomology, University of Illinois, Urbana–Champaign. Appointed by the NRC, they were responsible for making certain that an independent examination of this report was carried out in accordance with institutional procedures and that all review comments were carefully considered. Responsibility for the final content of this report, however, rests entirely with the authoring committee and the institution.

Philip Bell and Bruce Lewenstein, *Co-chairs*
Andrew W. Shouse, *Senior Program Officer*
Michael A. Feder, *Senior Program Officer*

# Contents

**Appendixes**

# Summary

Science is shaping people's lives in fundamental ways. Individuals, groups, and nations increasingly seek to bolster scientific capacity in the hope of promoting social, material, and personal well-being. Efforts to enhance scientific capacity typically target schools and focus on such strategies as improving science curriculum and teacher training and strengthening the science pipeline. What is often overlooked or underestimated is the potential for science learning in nonschool settings, where people actually spend the majority of their time.

Beyond the schoolhouse door, opportunities for science learning abound. Each year, tens of millions of Americans, young and old, explore and learn about science by visiting informal learning institutions, participating in programs, and using media to pursue their interests. Thousands of organizations dedicate themselves to developing, documenting, and improving science learning in informal environments for learners of *all* ages and backgrounds. They include informal learning and community-based organizations, libraries, schools, think tanks, institutions of higher education, government agencies, private companies, and philanthropic foundations. Informal environments include a broad array of settings, such as family discussions at home, visits to museums, nature centers, or other designed settings, and everyday activities like gardening, as well as recreational activities like hiking and fishing, and participation in clubs. Virtually all people of all ages and backgrounds engage in activities that can support science learning in the course of daily life.

The Committee on Learning Science in Informal Environments was established to examine the potential of nonschool settings for science learning. The committee, comprised of 14 experts in science, education, psychology, media, and informal education, conducted a broad review of the literatures that inform learning science in informal environments. Our charge specifically included assessing the evidence of science learning across settings, learner age groups, and over varied spans of time; identifying the qualities of learning experiences that are special to informal environments and those that are shared (e.g., with schools); and developing an agenda for research and development.

The committee organized its analysis by looking at the places where science learning occurs as well as cross-cutting features of informal learning environments. The "places" include everyday experiences—like hunting, walking in the park, watching a sunrise—designed settings—such as visiting a science center, zoo, aquarium, botanical garden, planetarium—and programs—such as after-school science, or environmental monitoring through a local organization. Cross-cutting features that shape informal environments include the role of media as a context and tool for learning and the opportunities these environments provide for inclusion of culturally, socially, and linguistically diverse communities.

We summarize key aspects of the committee's conclusions here, beginning with evidence that informal environments can promote science learning. We then describe appropriate learning goals for these settings and how to broaden participation in science learning. Finally, we present the committee's recommendations for practice.

## PROMOTING LEARNING

Do people learn science in nonschool settings? This is a critical question for policy makers, practitioners, and researchers alike—and the answer is yes. The committee found abundant evidence that across all venues—everyday experiences, designed settings, and programs—individuals of all ages learn science. The committee concludes that:

- Everyday experiences can support science learning for virtually all people. Informal learning practices of all cultures can be conducive to learning systematic and reliable knowledge about the natural world. Across the life span, from infancy to late adulthood, individuals learn about the natural world and develop important skills for science learning.
- Designed spaces—including museums, science centers, zoos, aquariums, and environmental centers—can also support science learning. Rich with real-world phenomena, these are places where people can pursue and develop science interests, engage in science inquiry, and reflect on their experiences through sense-making conversations.
- Programs for science learning take place in schools and community-based and science-rich organizations and include sustained, self-organized activities of science enthusiasts. There is mounting evidence

that structured, nonschool science programs can feed or stimulate the science-specific interests of adults and children, may positively influence academic achievement for students, and may expand participants' sense of future science career options.

- Science media, in the form of radio, television, the Internet, and hand-held devices, are pervasive and make science information increasingly available to people across venues for science learning. Science media are qualitatively shaping people's relationship with science and are new means of supporting science learning. Although the evidence is strong for the impact of educational television on science learning, substantially less evidence exists on the impact of other media—digital media, gaming, radio—on science learning.

## DEFINING APPROPRIATE OUTCOMES

To understand whether, how, or when learning occurs, good outcome measures are necessary, yet efforts to define outcomes for science learning in informal settings have often been controversial. At times, researchers and practitioners have adopted the same tools and measures of achievement used in school settings. In some instances, public and private funding for informal education has even required such academic achievement measures. Yet traditional academic achievement outcomes are limited. Although they may facilitate coordination between informal environments and schools, they fail to reflect the defining characteristics of informal environments in three ways. Many academic achievement outcomes (1) do not encompass the range of capabilities that informal settings can promote; (2) violate critical assumptions about these settings, such as their focus on leisure-based or voluntary experiences and nonstandardized curriculum; and (3) are not designed for the breadth of participants, many of whom are not K-12 students.

The challenge of developing clear and reasonable goals for learning science in informal environments is compounded by the real or perceived encroachment of a school agenda on such settings. This has led some to eschew formalized outcomes altogether and to embrace learner-defined outcomes instead. The committee's view is that it is unproductive to blindly adopt either purely academic goals or purely subjective learning goals. Instead, the committee prefers a third course that combines a variety of specialized science learning goals used in research and practice.

### Strands of Science Learning

We propose a "strands of science learning" framework that articulates science-specific capabilities supported by informal environments. It builds on the framework developed for K-8 science learning in *Taking Science to School* (National Research Council, 2007). That four-strand framework aligns tightly with our Strands 2 through 5. We have added two additional strands—Strands 1 and 6—which are of special value in informal learning environments. The six strands illustrate how schools and informal environments can pursue complementary goals and serve as a conceptual tool for organizing and assessing science learning. The six interrelated aspects of science learning covered by the strands reflect the field's commitment to participation—in fact, they describe what participants do cognitively, socially, developmentally, and emotionally in these settings.

Learners in informal environments:

Strand 1: Experience excitement, interest, and motivation to learn about phenomena in the natural and physical world.

Strand 2: Come to generate, understand, remember, and use concepts, explanations, arguments, models, and facts related to science.

Strand 3: Manipulate, test, explore, predict, question, observe, and make sense of the natural and physical world.

Strand 4: Reflect on science as a way of knowing; on processes, concepts, and institutions of science; and on their own process of learning about phenomena.

Strand 5: Participate in scientific activities and learning practices with others, using scientific language and tools.

Strand 6: Think about themselves as science learners and develop an identity as someone who knows about, uses, and sometimes contributes to science.

The strands are distinct from, but overlap with, the science-specific knowledge, skills, attitudes, and dispositions that are ideally developed in schools. Two strands, 1 and 6, are particularly relevant to informal learning environments. Strand 1 focuses on generating excitement, interest, and motivation—a foundation for other forms of science learning. Strand 1, while important for learning in any setting, is particularly relevant to informal learning environments, which are rich with everyday science phenomena and organized to tap prior experience and interest. Strand 6 addresses how learners view themselves with respect to science. This strand speaks to the process by which individuals become comfortable with, knowledgeable about, or interested in science. Informal learning environments can play a special role in stimulating and building on initial interest, supporting science learning identities over time as learners navigate informal environments and science in school.

The strands serve as an important resource from which to develop tools for practice and research. They should play a central role in refining assessments for evaluating science learning in informal environments.

## BROADENING PARTICIPATION

There is a clear and strong commitment among researchers and practitioners to broadening participation in science learning. Efforts to improve inclusion of individuals from diverse groups are under way at all levels and include educators and designers, as well as learners themselves. However, it is also clear that laudable efforts for inclusion often fall short. Research has turned up several valuable insights into how to organize and compel broad, inclusive participation in science learning. The committee concludes:

- Informal settings provide space for all learners to engage with ideas, bringing their prior knowledge and experience to bear.
- Learners thrive in environments that acknowledge their needs and experiences, which vary across the life span. Increased memory capacity, reasoning, and metacognitive skills, which come with maturation, enable adult learners to explore science in new ways. Senior citizens retain many of these capabilities. Despite certain declines in sensory capabilities, such as hearing and vision, the cognitive capacity to reason, recall, and interpret events remains intact for most older adults.
- Learning experiences should reflect a view of science as influenced by individual experience as well as social and historical contexts. They should highlight forms of participation in science that are also familiar to nonscientist learners—question asking, various modes of communication, drawing analogies, etc.
- Adult caregivers, peers, teachers, facilitators, and mentors play a critical role in supporting science learning. The means they use to do this range from simple, discrete acts of assistance to long-term, sustained relationships, collaborations, and apprenticeships.
- Partnerships between science-rich institutions and local communities show great promise for structuring inclusive science learning across settings, especially when partnerships are rooted in ongoing input from community partners that inform the entire process, beginning with setting goals.
- Programs, especially during out-of-school time, afford a special opportunity to expand science learning experiences for millions of children. These programs, many of which are based in schools, are increasingly folding in disciplinary and subject matter content, but by means of informal education.

# RECOMMENDATIONS

The committee makes specific recommendations about how to organize, design, and support science learning. These recommendations provide a research and development agenda to be explored, tested, and refined. They have broad reach and application for a range of actors, including funders and leaders in practice and research; institution-based staff who are responsible for the design, evaluation, and enactment of practice; and those who provide direct service to learners—scout leaders, club organizers, front-line staff in science centers. Here we make recommendations to specific actors who can influence science learning in practice. Additional recommendations for research appear in Chapter 9.

## Exhibit and Program Designers

Exhibit and program designers play an important role in determining what aspects of science are reflected in learning experiences, how learners engage with science and with one another, and the type and quality of educational materials that learners use.

**Recommendation 1:** Exhibit and program designers should create informal environments for science learning according to the following principles. Informal environments should

- be designed with specific learning goals in mind (e.g., the strands of science learning)
- be interactive
- provide multiple ways for learners to engage with concepts, practices, and phenomena within a particular setting
- facilitate science learning across multiple settings
- prompt and support participants to interpret their learning experiences in light of relevant prior knowledge, experiences, and interests
- support and encourage learners to extend their learning over time

**Recommendation 2:** From their inception, informal environments for science learning should be developed through community-educator partnerships and whenever possible should be rooted in scientific problems and ideas that are consequential for community members.

**Recommendation 3:** Educational tools and materials should be developed through iterative processes involving learners, educators, designers, and experts in science, including the sciences of human learning and development.

## Front-Line Educators

Front-line educators include the professional and volunteer staff of institutions and programs that offer and support science learning experiences. In some ways, even parents and other care providers who interact with learners in these settings are front-line educators. Front-line educators may model desirable science learning behaviors, helping learners develop and expand scientific explanations and practice and in turn shaping how learners interact with science, with one another, and with educational materials. They may also serve as the interface between informal institutions and programs and schools, communities, and groups of professional educators. Given the diversity of community members who do (or could) participate in informal environments, front-line educators should embrace diversity and work thoughtfully with diverse groups.

**Recommendation 4:** Front-line staff should actively integrate questions, everyday language, ideas, concerns, worldviews, and histories, both their own and those of diverse learners. To do so they will need support opportunities to develop cultural competence, and to learn with and about the groups they want to serve.

## REFERENCE

National Research Council. (2007). *Taking science to school: Learning and teaching science in grades K-8.* Committee on Science Learning, Kindergarten Through Eighth Grade. R.A. Duschl, H.A. Schweingruber, and A.W. Shouse (Eds.). Washington, DC: The National Academies Press.

# Part I

# Learning Science in Informal Environments

# 1

# Introduction

Humans are inherently curious beings, always seeking new knowledge and skills. That quest for knowledge often involves science: from a child's "Why is the sky blue?" to a teenager's inquiry into the dyes for a new t-shirt; from a new homeowner's concern about radon in the basement to a grandparent's search for educational toys for a grandchild. Each of these situations involves some facet of science learning in a nonschool, informal setting.

Experiences in informal environments for science learning are typically characterized as learner-motivated, guided by learner interests, voluntary, personal, ongoing, contextually relevant, collaborative, nonlinear, and open-ended (Griffin, 1998; Falk and Dierking, 2000). Informal science learning experiences are believed to lead to further inquiry, enjoyment, and a sense that science learning can be personally relevant and rewarding. Participants in them are diverse and include learners of all ages, cultural and socioeconomic backgrounds, and abilities. They include hobbyists, tourists, preservice teachers, members of online student communities, student groups, and families, who may explore experiences in the home, at work, in community organizations, or just about anywhere. Ideally these experiences enable learners to connect with their own interests, provide an interactive space for learning, and allow in-depth exploration of current or relevant topics "on demand." Box 1-1 provides several examples of informal science learning environments.

While drawing on and feeding human curiosity is a valuable end in its own right, informal environments for science learning may also make important practical contributions to society. Serious scientific concerns are ubiquitous in modern life—global warming, alternative fuels, stem cell research, the place of evolution in K-12 schools, to name just a few. Many

---

**BOX 1-1 Experiences in Informal Science Learning Environments**

- Visitors to whyville.net, a large social networking site on the Internet targeted at teenagers, find their chat sessions interrupted by the unexpected appearance of the word "Achoo!" Over a few days, the virus spreads through the community. Using resources from the Centers for Disease Control and Prevention made available on the site, visitors learn to identify how the virus spreads and how to prevent further infection (Neulight, Kafai, Kao, Foley, and Galas, 2007).

- A retired doctor and his wife travel several times a year, often with Elderhostel programs. During one trip, the program explores the history and culture of Montreal. On another trip, they learn to express themselves through art. Many of their trips involve the natural world: learning to conduct marine research in the Louisiana wetlands, observing elk in Colorado, and counting manatees in Florida (Hopp, 1998).

- A teenager with a collection of stuffed elephants gathered since the age of one receives a calendar with pictures of elephants as a gift. Bored, the teen browses the Wikipedia page about elephants. Excited by what he reads, he recalls years before attending a lecture on elephants given by a local university researcher. He contacts the researcher and joins her research group as an unpaid intern, analyzing sound recordings of elephants in African jungles. When he applies to colleges, his interest has shifted from international politics to biology and conservation.

---

people and scientific organizations have argued that, to successfully navigate these issues, society will have to draw creatively on all available resources to improve science literacy (American Association for the Advancement of Science, 1993; National Research Council, 1996).

Contrary to the pervasive idea that schools are responsible for addressing the scientific knowledge needs of society, the reality is that schools cannot act alone, and society must better understand and draw on the full range of science learning experiences to improve science education broadly. Schools serve a school-age population, whereas people of all ages need to understand science as they grapple with science-related issues in their everyday lives. It is also true that individuals spend as little as 9 percent of their lives in schools (Jackson, 1968; Sosniak, 2001). Furthermore, science in K-12 schools is often marginalized by traditional emphases on mathematics and literacy. This is quite evident under current federal education policy, which creates incentives for mathematics and literacy instruction and which appears to be reducing instructional time in science and other subject matters, especially in the early grades (e.g., Center on Education Policy, 2008). Finally—though it needn't be and isn't always so—much of science instruction in schools focuses narrowly on received knowledge and simplistic notions of scientific practice (Lemke, 1992; Newton, Driver, and Osborne, 1999; National Research Council, 2007; Rudolph, 2002). Clearly, informal environments can and should play an important role in science education now more than ever.

Learning science in informal environments has the potential to bolster science education broadly on a national scale. This is evident in reports from national initiatives to improve education in science, technology, engineering, and mathematics (STEM) in the United States. For example, both the Academic Competitiveness Council and the National Science Board were charged with reviewing the effectiveness of all federally funded STEM education programs, as well as recommending ways to coordinate and integrate the programs. The council's report cites informal education as one of three integral pieces of the U.S. education system (the other two being K-12 education and higher education) needed to ensure "U.S. economic competitiveness, particularly the future ability of the nation's education institutions to produce citizens literate in STEM concepts and to produce future scientists, engineers, mathematicians, and technologists" (U.S. Department of Education, 2007, p. 5). Federal interest in informal environments is also reflected in the National Science Board's report on the critical needs in STEM education (National Science Board, 2007). The National Science Board report stresses the need for coherence in this kind of learning and an adequate supply of well-prepared and effective STEM teachers. It calls for coordination of formal and informal environments to enhance curriculum and teacher development. Informal education is described as an essential conduit to increase public interest in and understanding and appreciation of science, technology, engineering, and mathematics. Furthermore, the report calls for the informal education community to be represented on a nonfederal national council for STEM education that would coordinate education efforts in this area.

This report echoes the need for greater coherence and integration of informal environments and K-12 functions and classrooms, and it urges a careful analysis of the goals and objectives of learning science in informal environments. While often complementary and sometimes overlapping with the goals of schools, the goals of informal environments are not identical to them. Differences may stem from the populations that participate in school and nonschool settings, the fact that participation is compulsory in K-12 settings (but is typically not in nonschool settings), and the relative emphasis placed on affective and emotional engagement across these settings. Yet, despite these differences schools and informal settings share a common interest in enriching the scientific knowledge, interest, and capacity of students and the broader public.

The emerging sense that informal environments can make substantial contributions to science education on a broad scale motivated the National Science Foundation's (NSF's) interest in requesting the study that resulted in this report. NSF is the leading sponsor for research and development in science education in informal settings. Its portfolio of sponsored activities includes program and materials development, research, and evaluation across a broad range of informal settings and areas of STEM education throughout the nation. This report describes numerous NSF-sponsored projects as well as projects sponsored through other public and private sources.

This report provides a broad description of science learning in informal environments and a detailed review of the evidence of their impact on science learning. It synthesizes literature across multiple disciplines and fields to identify a common framework of educational goals and outcomes, insights into educational practices, and a research agenda. The remainder of this chapter provides a brief historical overview of the literatures, a discussion of current issues driving research and practice, and a description of the characteristics of informal environments for science learning; it also describes the scope of the study and provides an orientation to the remainder of the volume.

## Emergence and Growth of Science Learning in Informal Environments

The early roots of America's education system developed in the late 18th century when informal learning institutions, such as libraries, churches, and museums, were seen as the main institutions concerned with public education. They were viewed as places that encouraged exploration, dialogue, and conversation among the public (Conn, 1998). The American Lyceum movement, which began in the 1820s, supported the growing movement of public education in the United States (Ray, 2005). Lyceums, modeled after the early Greek halls of learning, brought the public together with experts in science and philosophy for lectures, debates, and scientific experiments. In the late 1800s, the Chautauqua movement, a successor to the Lyceum movement, grew out of the social and geographic isolation of America's farming and ranching communities. Chautauquas, a type of educational family summer camp, brought notable lecturers and entertainers of the day to rural communities, where there was a strong hunger for both entertainment and education. These movements were driven by the notion that in a democratic nation, an educated populace is needed to inform public policy. They provided a conduit for bringing the science knowledge and practices of the day to an American public with limited access to information. At the same time, people often developed an intuitive sense of the natural world and scientific principles through activities like farming, gardening, and brewing alcohol—processes that were closely connected to daily life in an agrarian society.

Beginning in the mid-19th century the world's fairs or expositions brought people from around the world together to learn about developments in commerce, technology, science, and cultural affairs. World's fairs have been the site for initial broad dissemination of scientific and technological developments, especially during the period of industrialization, when developments like telephone communication were unveiled to vast publics. Recently, individuals' personal recollections of these events have been used as the basis for exploring what people attend to, learn, and recall from learning experiences in informal settings (e.g., Anderson, 2003; Anderson, Storksdieck, and Spock, 2007; Anderson and Shimizu, 2007).

The role and structure of informal learning in this country have evolved over the past 200 years. Today, technological advances have distanced people from traditional agrarian experiences. In some respects, members of this highly urbanized and technological society have fewer opportunities to explore the natural world than did their ancestors, who raised livestock and farmed. Science education has evolved in a new social context. News and entertainment media merge with natural history museums and science centers, after-school programs, and computer games and gaming communities to reshape the world and people's exposure to science.

Although many people are quick to point out a large and persistent resource gap between schools and nonschool settings, in recent years public and private funders have made significant investments to support informal environments for science learning. A 1993 report of the Federal Coordinating Council for Science, Engineering and Technology showed that the federal government spent about $67 million on "public understanding of science" activities and that the federal portion was probably only 10 percent of the total outlay for such activities (Lewenstein, 1994). Since 1993 the federal investment in informal science education has more than doubled, totaling $137.4 million in fiscal year (FY) 2006 (U.S. Department of Education, 2007). Increases in funding have also occurred in federal programs that provide informal environments for learning in general (not science specific), such as the 21st Century Community Learning Centers, an after- and out-of-school program. Originally, in FY 1995, $750,000 was allocated to the 21st Century Centers, and since then their funding has expanded to just under $1 billion in FY 2006 (Learning Point Associates, 2006). Additional funding for informal science learning comes from national foundations, nonprofit research organizations, and advocacy groups that are interested in supporting opportunities for underserved populations.

Organizations, consortiums, affinity groups, and publications concerned with learning science in informal environments have also proliferated over the past 50 years (Lewenstein, 1992; Schiele, 1994), as shown in Box 1-2. The post–World War II Soviet Sputnik Program, which in 1957 launched the

---

**BOX 1-2 50 Years of Major Events in Informal Science Learning (with primary focus on the United States)**

1957 – National Science Foundation (NSF) conducts first studies of public knowledge of science; repeated in 1979 and thereafter biennially.

1958 – NSF creates program on "Public Understanding of Science" (continues to 1981).

1961 – American Association for the Advancement of Science (AAAS) begins newsletter on "Understanding," linking science journalists, Hollywood film and television producers, mass communication researchers, adult educators, and museum staff (continues to 1967).

1962 – Founding of Pacific Science Center in Seattle.

1968 – Founding of the Lawrence Hall of Science at the University of California, Berkeley.

1969 – Founding of The Exploratorium in San Francisco.

1973 – AAAS creates a (short-lived) NSF-funded National Center for Public Understanding of Science, linking radio, television, schools, youth activities, and science kits.

1983 – NSF recreates Public Understanding of Science Program as Informal Science Education.

1985 – Royal Society's "Bodmer Report" on public understanding of science (UK) leads to sustained interest in research on related topics (Ziman, 1991; Irwin and Wynne, 1996).

1988 – Founding of Visitor Studies Association.

1989 – A grant awarded to the Association for Science-Technology Centers by the Institute for Museum and Library Services results in a series of articles called "What Research Says About Learning Science in Museums" in the association newsletter and two subsequent volumes with the same title.

1990 – First chair in the public understanding of science is established, at Imperial College, London.

---

first satellite into orbit, captured the attention of the U.S. public and galvanized support for domestic science education. For the first time, the federal government participated in K-12 and undergraduate curriculum development though its newly formed NSF, and a critical mass of top academics made a concerted push to improve science education. This began an era of wide-

1991 – An *International Journal of Science Education* special issue on informal science learning is published.

1992 – The journal *Public Understanding of Science* is established.

1994 – A conference funded by NSF results in publication of *Public Institutions for Personal Learning: Establishing a Research Agenda* (Falk and Dierking, 1995).

1996 – The first major informal learning research grant is awarded to the Museum Learning Collaborative, funded by a consortium of federal agencies.

1997 – A *Science Education* special issue on informal science learning is published.

1998 – NSF-funded conference results in publication of *Free-Choice Science Education: How We Learn Science Outside of School* (Falk, 2001).

2000 – NSF-funded conference results in publication of *Perspectives on Object-Centered Learning in Museums* (Paris, 2002).

2001 – Founding of the Center for Informal Learning and Schools.

2002 – A *Journal of Research on Science Teaching* special issue on informal science learning is published.

2002 – Free-choice/informal learning is added as a strand of graduate study in science and mathematics education in the College of Science at Oregon State University.

2004 – A conference called "In Principle, In Practice: A Learning Innovation Initiative" resulted in a preconference supplemental issue of *Science Education*, a postconference online publication called *Insights*, and a postconference edited book on informal science learning.

2004 – Founding of the Learning in Informal and Formal Environments Center.

2005 – Informalscience (http://www.informalscience.org) is launched to share evaluation and research on informal science learning environments.

2008 – NSF publishes *Framework for Evaluating Impacts of Informal Science Education Projects*.

spread interest in science centers, and, over the next decade, several of the leading institutions in informal science education were established.

More recently several education research organizations, which focus primarily on schools, have added special-interest groups devoted to informal learning and informal science. Numerous peer-reviewed journals have included special editions on informal science learning, and the journal *Science Education* added an informal learning section. New journals, such as *Public Understanding of Science* and *Science Communication*, have arisen as well. Furthermore, research and evaluations of informal science learning environments have become more available through websites, such as informalscience.org; research agenda-setting events have transpired in an attempt to explore and coordinate the research and evaluations (Royal Society, 1985; Irwin and Wynne, 1996); and NSF has published a framework for assessing their impact (Friedman, 2008).

## Need for Common Frameworks

With the growth of interest in science learning in informal environments and the diversification of venues, practitioners, and researchers, the literature has developed in a fractured and uneven manner. Several factors appear to contribute to the divergent trajectories of the research. First, the relationship between schools and informal environments for science learning has been unclear and contested, serving as an impediment to integration of what is understood about learning across these settings. In other words, research on schools rarely builds on findings from research in informal settings and vice versa.

Second, the goals of informal environments for science learning are multiple. Designed environments have historically focused on what attracts an audience and keeps it engaged, but experiences are not often framed in terms of learning (Commission on Museums for a New Century, 1984; Rockman Et Al, 2007). After-school programs were traditionally designed with goals that often focused on providing a safe and healthy environment for young children during the hours after school. The goals of these programs have been driven by the institutions that have traditionally supported them, and only recently has large support come from sources that are increasingly concerned with learning outcomes.

Third, since many fields of inquiry are invested in this work, the research base reflects a diversity of interests, questions, and methods from several loosely related fields. Historical sociological studies of the relationship between science and the public have largely focused on institutional issues, again without attention to learning. Anthropology and psychology tend to explore learning, but not educational design. Much of the empirical evidence on museums, zoos, libraries, media, and programs has emerged from visitor studies and may include learning outcomes, demographic profiles, and analysis of visitor behavior. Evaluations typically illustrate how a specific program, broadcast, or exhibit supports learning. The theoretical underpinnings of this work may not be explicit, and general implications for informal science education are often hard to discern. In addition educators, researchers, and policy makers who are accustomed to research on classroom settings may tend to rely on measures of learning that are not appropriate for informal settings. Education researchers, psychologists, anthropologists, practitioners, and evaluators all have interest in informal science learning, yet they tend to explore those interests in distinct ways and participate in distinct and often disconnected communities of inquiry.

Fourth, as funding for informal environments for science learning grows, so do questions about the responsible stewardship of investments and resources and its appropriate role in the educational infrastructure. Greater investment in an era of widespread accountability has brought greater scrutiny of whether and how science learning experiences in informal settings reach their goals. Designed spaces, after-school programs, and media developed to serve informal science learning ends are now faced with questions of how to prove they are having the impact many have long presumed.

Furthermore, this area of inquiry must navigate the uncertain relationship between research and practice. This is perhaps most evident in research on everyday learning for which linkages to an infrastructure for science learning may be unclear. Everyday learning—the things people learn by engaging in the everyday activities of life—has no institutional home, yet it is fundamental to learning science.

Fifth, media and information technology add a host of additional exciting dynamics with which researchers and practitioners must grapple. Advances in wireless technology, the expansion of the Internet, the advent of blogs and wikis, and the growth of games and simulations have changed the ways in which people access or are exposed to information related to science. New media may enhance dissemination of scientific knowledge, but they also raise questions about how and when media should be harnessed for science learning. Consider online gaming: it is a two-way medium (users are both receivers and senders of information), it allows for multimodal engagement (i.e., games can engage people in their preferred way, whatever it is), and as a networked environment it can leverage the small efforts of many users. In important ways these design features of gaming resonate with the philosophy of informal learning and call for greater analytic attention.

At the same time, while new media forms make it easier for nonscientists to get access to scientific information—for example, through university websites and government documentation centers—they also provide platforms for unverified information, incorrect explanations, speculative theories, and sometimes outright fraudulent claims. In many cases, information seekers may not have the tools to distinguish among the available information sources. The possibilities of media are exciting, yet the ability of researchers and practitioners in informal learning environments to keep pace with media and technological developments remains uncertain.

Diversity of perspectives, research approaches, and questions is necessary for the healthy development of research in any field of study. Yet a common language and common constructs that characterize the settings, goals, practices, and technologies that are central to the work are needed as building blocks for research and practice. Researchers can benefit from common constructs and language because they make it possible to clearly connect to and build on the work of their peers and predecessors to guide their work. Practitioners can benefit from common language and constructs because they facilitate clear communication, which is central to developing strong, dynamic professional cultures. The field itself benefits from common constructs that identify the commitments, core practices, and knowledge of the field for outsiders and newcomers to the field. Many individuals and organizations, including philanthropies, government agencies, and volunteers, are interested in science learning in informal environments. They need to understand the field well enough to engage with the work, support high-quality efforts, and assess its overall value to society.

Can clear, common constructs and language be identified? What are the goals of learning science in informal environments? What is known about leverage points for learning across the diverse settings involved? What are the possible relationships between schools and nonschool settings for science learning? What strategies allow educators to serve diverse audiences? How should one construe the influence of everyday learning, and how might it inform educational practice? Can the digital media age be harnessed to improve science learning? These are the kinds of questions that prompted this report.

## ABOUT THIS REPORT

With support from the National Science Foundation (NSF), the National Research Council established the Committee on Learning Science in Informal Environments to undertake this study. Selected to reflect a diversity of perspectives and a broad range of expertise, the 14 committee members include experts in research and evaluation, exhibit design, life-span development, everyday learning, science education, cognition and learning, and public understanding of science. In addition, the committee membership reflects a balance of experience in and knowledge of the range of venues for informal science museums, after-school programs, science and technology centers, libraries, media enterprises, aquariums, zoos, and botanical gardens.

### Committee Charge

This study was designed to describe the status of knowledge about science learning in informal environments, illustrate which claims are supported by evidence, articulate a common framework for the next generation of research, and provide guidance to the community of practice. The report covers issues of interest to museums, after-school programs, community organizations, evaluators, researchers, and parents. The committee's work was directed toward the following goals:

- Synthesize and extract key insights from the multiple sources of information that now shape the field, including evaluation studies, research activities, visitor studies, and survey mechanisms.
- Provide evaluators, practitioners, and researchers with an analysis of this synthesized research to begin to identify where common definitions and guiding epistemologies on science learning exist.
- Provide policy makers, scientific societies, academics, and others interested in informal education with a clear, credible, and research-based overview of the research.
- Guide future research, evaluation, and education needs by identifying what a future research agenda might look like, given the state and application of current knowledge-based frameworks.

The committee's charge was to respond to seven specific questions, which appear in Box 1-3. Following this report, a separate book is planned for publication by the National Academies Press. Based on the conclusions and recommendations of this report, the book that follows will interpret the research base for a practitioner audience. More information on that book is available at http://www7.nationalacademies.org/bose/LSIEP_Homepage.html.

## Approach and Scope

The committee conducted its work through an iterative process of gathering information, deliberating on it, identifying gaps and questions, gathering further information to fill these gaps, and holding further discussions. In our search for relevant information, we held four public fact-finding meetings, reviewed published and unpublished research reports and evaluations, and asked nine experts to prepare and present papers. At the fifth meeting, the committee intensely analyzed the findings and discussed our conclusions. We were particularly concerned to identify bodies of research that are characterized by systematic collection and interpretation of evidence and to show the ways in which these research literatures connect to each other. Some of the literatures drawn on include

- out-of-school and free-choice learning programs,
- diversity and learning,
- learning from media,
- learning in museums and other designed environments,
- the nature of learning, and
- everyday learning and families.

The committee has also drawn extensively on evidence that does not appear in traditional, peer-reviewed scholarly publications, although many of the

---

### BOX 1-3 Committee Charge

1. What is the range of theoretical perspectives, assumptions, and outcomes that characterize research on informal science?

2. What assumptions, epistemologies, or modes of learning science are shared between the formal and informal science education environments? How do informal science understanding and practice vary in diverse communities?

3. What evidence is there that people who participate in informal science activities learn concepts, ways of thinking, practices, attitudes, and aesthetic appreciation in these settings? What kinds of informal learning environments best support the learning of current scientific issues and concerns (e.g., global warming)? What are the organizational, social, and affective features of effective informal science learning environments vis-à-vis a range of learned competencies/outcomes?

4. Are some learning outcomes unique to informal environments? For example, is there evidence that informal learning environments support the learning of populations who have been poorly served by school science?

5. What is known about the cumulative effects of science learning across time and contexts? How do learners (young, middle-aged, adolescent, older adults) utilize informal science learning opportunities? How do these opportunities influence learners? Are informal learning experiences designed to suit the developmental trajectories of individuals?

6. What information is needed by practitioners in the field? What information is needed by academics seeking to build and enlarge relevant areas of advanced or graduate study? What information is needed by policy makers to affect policies that include informal environments within the scope of education-directed legislation?

7. What are promising directions for future research? Can common frameworks that link the diverse literatures be developed? If so, what would they look like?

---

projects and programs devoted to informal science learning have been the subject of formal evaluations, often conducted in rigorous and informative ways. When appropriate, and with sufficient detail to demonstrate the evidentiary value of the material, we have drawn on this evaluation literature.

At the first meeting, the committee discussed the charge with representatives of NSF and heard from a panel of researchers on the status of the field of informal science learning. This meeting was largely intended to provide committee members with a chance to internalize the charge and to obtain input from senior informal science educators, researchers, and evaluators as they began the study.

The second meeting was largely intended to provide the committee with information on the learning perspectives that guide or inform informal learning environments and how these environments can serve underrepresented or underserved populations. At the third meeting, the committee heard evidence about science learning that takes place in various informal venues and pressing policy issues. During this meeting, the committee identified seven topics for which they required a focused literature review from a range of experts with research interests in learning science in informal environments. These topics became the focus of commissioned papers. At the fourth meeting, the public session was concerned primarily with the status of the papers prepared to support the committee's work and the organizational structure being implemented in NSF as it relates to this project. The fifth meeting was taken up with the committee's deliberations.

This report is primarily concerned with characterizing the state of evidence about how and what people learn about science in informal environments throughout their lives. However, this broad scope, the divergent nature of the relevant literature, and the quantity of unpublished information prevented us from doing an in-depth analysis of all of the literature on the topic. Consequently, there are relevant literatures that this study does not consider or only touches on. They include adult workforce learning, the classroom as a site for informal learning, and media-based health interventions (e.g., in public health campaigns and international development).

## Focus of the Report

This report is an effort to develop a common framework for the broad and diverse fields of inquiry about informal environments for science learning. Prior efforts to synthesize these literatures and discern what is known and what is not have been minimal. Synthesis is a crucial step toward leveraging research to enhance practice and making strategic choices about which research questions to prioritize. One complicating factor in efforts to synthesize this work is that the evidence base reflects the diversity of the evidence and is informed by a range of disciplines and perspectives, including field-based research, evaluations, visitor studies, design studies, and traditional experimental psychological studies of learning. The purposes of these studies, their conceptions of learning and goals, and the methods and measures they employ vary tremendously. Consequently, there is no basis for targeted, systematic, and efficient knowledge accumulation, and it is difficult to leverage research to guide policy and practice.

A necessary step in developing a framework is to clearly define what learning is in informal environments. Informal learning institutions typically operate without the authority to compel participation, and they are not solely concerned with improving the science proficiency of children. The model of science learning the committee presents places special emphasis on providing entrée to and sustained engagement with science—reflecting the purview of informal learning—while maintaining an eye on the potential of informal science learning environments to support a broad range of science-specific learning outcomes and intersect with related institutional players. We use this broad definition of learning to build a coherent set of shared goals, to articulate particular strengths of the varied research bases involved, and to acknowledge the ways in which informal learning environments and K-12 schooling can complement one another.

As noted, this report reviews an extremely broad and diverse literature, and the committee needed to make decisions about how to focus and limit the fact-finding process in order to complete the project with the resources available. Finding ways to constrain fact-finding was particularly important in this study because there are currently few synthesized works, such as handbook chapters and commissioned research reviews, to draw on. Accordingly, the literature reflected in this volume is drawn primarily from North American publications. The committee acknowledges that there are clearly high-quality research literatures that are developed in other regions of the world, but given limitations on time and resources, we chose to use those that are most familiar to the U.S. audience. When international reports are included, these are works that are either seminal in the North American context or speak to important issues that the committee was unable to address otherwise. The report also reflects an emphasis on research from the past 20 years, a period during which sociocultural and cognitive accounts of learning are most prevalent.

## Organization of the Report

The report has four parts. Part I sets the stage, beginning with this introductory chapter. Chapter 2 provides a description of the theoretical frameworks that guide practice and research in science learning in informal environments. In Chapter 3 we illustrate the expected outcomes of engagement in these settings, what we call the strands of science learning. Also, this chapter includes guidance on appropriate methods and techniques for studying these outcomes and their development in informal settings.

Part II provides a detailed description of venues for learning science in informal environments. The individual chapters focus on everyday learning environments (Chapter 4), designed learning environments (Chapter 5), and programs for science learning (Chapter 6). Each includes a discussion of their defining features, how they support science learning, and the impact they have on the strands of science learning. Part III explores themes that emerge across the venues and configurations, focusing on diversity and equity in Chapter 7 and media in Chapter 8. Part IV contains a final chapter with the committee's broad conclusions about learning science in informal environments as well as recommendations for practice and future research.

## REFERENCES

American Association for the Advancement of Science. (1993). *Benchmarks for science literacy*. New York: Oxford University Press.

Anderson, D. (2003). Visitors' long-term memories of world expositions. *Curator, 46*(4), 400-420.

Anderson, D., and Shimizu, H. (2007). Factors shaping vividness of memory episodes: Visitors' long-term memories of the 1970 Japan world exposition. *Memory, 15*(2), 177-191.

Anderson, D., Storksdieck, M., and Spock, M. (2007). The long-term impacts of museum experiences. In J. Falk, L. Dierking, and S. Foutz (Eds.), *In principle, in practice: New perspectives on museums as learning institutions* (pp. 197-215). Walnut Creek, CA: AltaMira Press.

Center on Education Policy. (2008). *Instructional time in elementary schools: A closer look at changes for specific subjects*. Washington, DC: Author.

Commission on Museums for a New Century. (1984). *Museums for a new century*. Washington, DC: American Association of Museums.

Conn, S. (1998). *Museums and American intellectual life, 1876-1926*. Chicago: University of Chicago Press.

Falk, J.H. (2001). *Free-choice science education: How we learn science outside of school*. New York: Teachers College Press.

Falk, J.H., and Dierking, L.D. (Eds.). (1995). *Public institutions for personal learning: Establishing a research agenda*. Washington, DC: American Association of Museums.

Falk, J.H., and Dierking, L.D. (2000). *Learning from museums: Visitor experiences and the making of meaning*. Walnut Creek, CA: AltaMira Press.

Friedman, A. (Ed.). (2008). *Framework for evaluating impacts of informal science education projects*. Washington, DC: National Science Foundation.

Griffin, J. (1998). *School-museum integrated learning experiences in science: A learning journey*. Unpublished doctoral dissertation, University of Technology, Sydney.

Hopp, R. (1998). Experiencing Elderhostel as lifelong learners. *Journal of Physical Education, Recreation & Dance, 69*(4), 27, 31.

Irwin, A., and Wynne, B. (Eds.). (1996). *Misunderstanding science? The public reconstruction of science and technology*. New York: Cambridge University Press.

Jackson, P.W. (1968). Life in classrooms. In A. Pollard and J. Bourne (Eds.), *Teaching and learning in the primary school*. New York: RoutledgeFalmer.

Learning Point Associates. (2006). *21st century community learning centers (21st CCLC) analytic support for evaluation and program monitoring: An overview of the 21st CCLC program: 2004-05*. Naperville, IL: Author.

Lemke, J.L. (1992). *The missing context in science education: Science*. Paper presented at the Annual Meeting of the American Education Research Association Conference, San Francisco.

Lewenstein, B.V. (1992). The meaning of "public understanding of science" in the United States after World War II. *Public Understanding of Science, 1*(1), 45-68.

Lewenstein, B.V. (1994). A survey of public communication of science and technology activities in the United States. In B. Schiele (Ed.), *When science becomes culture: World survey of scientific culture* (pp. 119-178). Boucherville, Quebec: University of Ottawa Press.

National Research Council. (1996). *National science education standards.* National Committee on Science Education Standards and Assessment. Washington, DC: National Academy Press.

National Science Board. (2007). Science, technology, engineering, and mathematics (STEM) education issues and legislative options. In R. Nata (Ed.), *Progress in education* (vol. 14, pp. 161-189). Washington, DC: Author.

Neulight, N., Kafai, Y.B., Kao, L., Foley, B., and Galas, C. (2007). Children's participation in a virtual epidemic in the science classroom: Making connections to natural infectious diseases. *Journal of Science Education and Technology, 16*(1), 47-58.

Newton, P., Driver, R., and Osborne, J. (1999). The place of argumentation in the pedagogy of school science. *International Journal of Science Education, 21*(5), 553-576.

Paris, S. (2002). *Perspectives on object-centered learning in museums.* Mahwah, NJ: Lawrence Earlbaum Associates.

Ray, A.G. (2005). *The lyceum and public culture in the nineteenth century United States.* East Lansing: Michigan State University Press.

Rockman Et Al. (2007). *Media-based learning science in informal environments.* Paper commissioned for the National Research Council's Committee on Learning Science in Informal Environments, Washington, DC. Available: http://www7. nationalacademies.org/bose/Learning_Science_in_Informal_Environments_ Commissioned_Papers.html [accessed November 18, 2008].

Royal Society. (1985). *The public understanding of science.* London: Author.

Rudolph, J.L. (2002). *Scientists in the classroom: The cold war reconstruction of American science education.* New York: Pelgrave.

Schiele, B. (Ed.). (1994). *When science becomes culture: World survey of scientific culture* (Proceedings I). Boucherville, Quebec: University of Ottawa Press.

Sosniak, L. (2001). The 9% challenge: Education in school and society. *Teachers College Record, 103,* 15.

U.S. Department of Education. (2007). *Report of the Academic Competitiveness Council.* Washington, DC: Author.

Ziman, J. (1991). Public understanding of science. *Science, Technology and Human Values, 16*(1 Winter), 99-105.

# 2

# Theoretical Perspectives

Public discussions of learning usually focus on the experiences and outcomes associated with schooling. Yet a narrow focus on traditional academic activities and learning outcomes is fundamentally at odds with the ways in which individuals learn across various social settings: in the home, in activities with friends, on trips to museums, in potentially all the places they experience and pursuits they take on. The time that children spend pursuing hobbies of their own choosing—in such activities as building, exploring, and gaming—often provides them with experiences and skills relevant to scientific processes and understanding. Adults faced with medical conditions typically learn what they can do to manage them from a wide variety of information sources. Families spend leisure time at science centers, zoos, and museums engaged in exploration and sense-making. Communities defined by linguistic and cultural ties maintain science-related practices and socialize their children into their routines, skills, attitudes, knowledge, and value systems as a part of their daily activities and rituals.

For all these pursuits, the range of learning outcomes far exceeds the typical academic emphasis on conceptual knowledge. Across informal settings, learners may develop awareness, interest, motivation, social competencies, and practices. They may develop incremental knowledge, habits of mind, and identities that set them on a trajectory to learn more.

The ongoing connections among experiences, capabilities, dispositions, and new opportunities to learn continue throughout a person's life. The fundamental influence of early childhood experiences is increasingly recognized as providing the foundation for discipline-specific learning (National Research Council, 2007). As the population ages, demographic shifts heighten the need to understand the ongoing role that science learning has in the lives of adults, including the elderly.

The informal education community pursues a range of learning outcomes. The idea of lifelong, life-wide, and life-deep learning has been influential in efforts to develop a broad notion of learning, incorporating how people learn over the life course, across social settings, and in relation to prevailing cultural influences (Banks et al., 2007).

*Lifelong learning* is a familiar notion. It refers to the acquisition of fundamental competencies and attitudes and a facility with effectively using information over the life course, recognizing that developmental needs and interests vary at different life stages. Generally, learners prefer to seek out information and acquire ways of doing things because they are motivated to do so by their interests, needs, curiosity, pleasure, and sense that they have talents that align with certain kinds of tasks and challenges.

*Life-wide learning* refers to the learning that takes place as people routinely circulate across a range of social settings and activities—classrooms, after-school programs, informal educational institutions, online venues, homes, and other community locales. Learning derives, in both opportunistic and patterned ways, from this breadth of human experience and the related supports and occasions for learning that are available to an individual or group. Learners need to learn how to navigate the different underlying assumptions and goals associated with education and development across the settings and pursuits they encounter.

*Life-deep learning* refers to beliefs, ideologies, and values associated with living life and participating in the cultural workings of both communities and the broader society. Such learning reflects the moral, ethical, religious, and social values that guide what people believe, how they act, and how they judge themselves and others. This focus on life-deep learning emphasizes how learning is never a culture-free endeavor.

Taken together, these concepts of lifelong, life-wide, and life-deep learning help bring into view the breadth of human learning and emphasize the broad reach of informal settings. Figure 2-1 is a conceptual diagram that depicts the prevalence of lifelong and life-wide learning in formal and informal learning environments. Although there is significant variation for individuals, the diagram gives a rough estimation of the amount of time people routinely spend in informal (nonschool) learning environments over the life course. In addition to focusing on how learning is accomplished in specific informal settings, we consider how learning is accomplished across multiple settings—across shifting material and social resources, in the variety of ways people participate in and make use of their knowledge, their various social groupings, and their evolving purposes and expectations.

The idea of lifelong, life-wide, life-deep science learning informs the committee's approach to the charge. Thus, we explore a wide variety of places and social settings, which we refer to as venues and configurations. We defined a broad set of valued learning outcomes and examined the evidence related to each. Finally, we examined research on learners of all ages from very young children to the elderly.

## LIFELONG AND LIFE-WIDE LEARNING

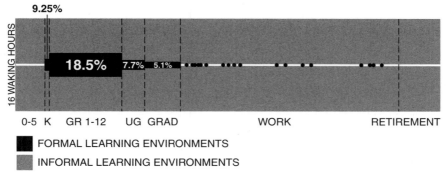

FORMAL LEARNING ENVIRONMENTS

INFORMAL LEARNING ENVIRONMENTS

FIGURE 2-1    *Estimated time spent in school and informal learning environments.*
*NOTE: This diagram shows the relative percentage of their waking hours that people across the life span spend in formal educational environments and other activities. The calculations were made on the best available statistics on how much time people at different points across the life span spend in formal instructional environments. This diagram was originally conceived by Reed Stevens and John Bransford to represent the range of learning environments being studied at the Learning in Informal and Formal Environments (LIFE) Center. Graphic design, documentation, and calculations were conducted by Reed Stevens, with key assistance from Anne Stevens (graphic design) and Nathan Parham (calculations).*
*SOURCE: Stevens (no date).*

In this chapter we begin by discussing some general theoretical perspectives of learning and exploring how some prominent frameworks used in research on learning in informal environments build on them. We then describe an ecological model of learning that provides multiple lenses for synthesizing how people learn science across informal environments. Building from the ecological perspective, we define the venues and configurations for learning and science learning strands that frame the remainder of this volume.

## INTEGRATING VIEWS OF KNOWLEDGE AND LEARNING

Research on learning science in informal environments reflects the diversity of theoretical perspectives on learning that have guided research. Over a century ago, scientists began to study thinking and learning in a more systematic way, taking early steps toward what are now called the cognitive sciences. During the first few decades of the 20th century, researchers focused on such matters as the nature of general intellectual ability and its distribution in the population. In the 1930s, they started emphasizing such issues as the laws governing stimulus-response associations in learning. Beginning in the 1960s, advances in fields as diverse as linguistics, computer science, and neuroscience offered provocative new perspectives on human development and powerful new technologies for observing behavior and brain functions.

The result during the past 40 years has been an outpouring of scientific research on the mind and the brain—a "cognitive revolution," as some have termed it. With richer and more varied evidence in hand, researchers have refined earlier theories or developed new ones to explain the nature of knowing and learning. Three theoretical perspectives of the nature of the human mind have been particularly influential in the study of learning and consequently in education: *behaviorist*, *cognitive*, and *sociocultural*. The relative influence of these perspectives over time has changed. Each emphasizes different aspects of knowing and learning with differing implications for educational practice and research (see, e.g., Greeno, Collins, and Resnick, in press).

Behaviorism describes knowledge as the organized accumulation of stimulus-response associations that serve as components of skills (Thorndike, 1931). People learn by acquiring simple skills which combine to produce more complex behaviors. Rewards, punishments, and other (mainly extrinsic) factors orient people to attend to relevant aspects of a situation and support the formation of new associations and skills. Cognitive theories, in contrast, focus on how people develop, transform, and apply structures of knowledge in relation to lived experience, including the concepts associated with a subject matter discipline (or domain of knowledge) and procedures for reasoning and solving problems. One major tenet of cognitive theory is that learners actively construct their understanding by trying to connect new information with their prior knowledge. This theoretical approach generally focuses on individual thinking and learning. Sociocultural theory builds on cognitive perspectives, but emphasizes the cultural origins of human development and explores how individuals develop through their involvement in cultural practices (e.g., Cole, 1996; Heath, 1983; Rogoff, 2003). In this view, individuals develop specific skills, commitments, knowledge, and identity as they become proficient in practices that are valued in specific communities.

From the perspective of educational practice, there are complementarities between cognitive and sociocultural accounts. The cognitive perspective, with its interest in characterizing an individual's knowledge structures, can help educators identify what a learner understands about a particular domain. This can be important for gauging where and how to initially engage a learner and what aspects of understanding require instructional support. Meanwhile, the sociocultural perspective can orient educators to patterns of participation and associated value systems that are important to learning. These may include analyses of expert practice in a particular domain such as how scientists communicate ideas to one another or forms of participation that are comfortable or culturally important to learners (e.g., how learners tend to communicate with one another). Identifying common ground between learners' practices and practices in the domains of interest may be a productive route to experiences that move learners toward deeper understanding and capability in the domain. For example, individuals learn to reason in science by crafting and using forms of notation or inscription that help represent the natural world. Crafting these forms of inscription can be viewed as being situated within a particular (and even peculiar) form of practice—creating representations and models—into which students need to be initiated.

All three theoretical perspectives have had some influence on the design of informal environments that support science learning. As a result, a number of theoretical views are in play in the research and they are not particularly well integrated. This limits the degree to which the study of learning science in informal environments functions as a field. In Box 2-1 we describe a few examples of perspectives on learning science in informal environments. We note that most draw on the cognitive and sociocultural traditions rather than behaviorism. Also, the list in Box 2-1 is intended to illustrate the range of perspectives and is not exhaustive.

## An Ecological Framework for Understanding Learning Across Places and Pursuits

A broad theory, or set of complementary perspectives, which could be refined through empirical testing, could help integrate the range of theories and frames currently in use (as represented in Box 2-1) and help generate core questions. To move in that direction, we propose an "ecological framework for learning in places and pursuits" intended to highlight the cognitive, social, and cultural learning processes and outcomes that are shaped by distinctive features of particular settings, learner motivations and backgrounds, and associated learning expectations. The term "ecological" here refers to the relations between individuals and their physical and social environments with particular attention to relations that support learning. The framework draws mainly from cognitive and sociocultural theories.

Our proposal is consonant with other calls for using an ecological perspective for accounts of human development and learning that can accommodate a range of disciplinary perspectives as well as the diversity of life experiences in a global society (Barron, 2006; Lee, 2008). It builds on a tradition of scholarship on the ecological nature of human development. This tradition has long recognized and taken into account the compound set of influences on learning and development originating from a person's experiences across myriad institutional contexts and social niches (family, school, playground, peers, neighbors, media, etc.) (Bronfenbrenner, 1977).

Within the ecological framework, we describe three cross-cutting aspects of learning that are evident in all learning processes: people, places, and

**BOX 2-1 Perspectives on Informal Environments for Science Learning**

A variety of perspectives have been developed to understand, define, or evaluate science learning in informal settings. Most of these perspectives have attempted to provide a broader frame for learning outcomes yet are compatible with the nature of learning in informal environments. These frameworks are based on or framed in terms of cognitive and sociocultural theories.

- The Contextual Model of Learning (Falk and Dierking, 2000) is a general framework for understanding informal or free-choice learning (see also Falk and Storksdieck, 2005, for an application and quantitative validation of the model). The model focuses on 12 key personal, sociocultural, and physical dimensions of learning. The model stresses visitor agenda, personal motivation, the sociocultural nature of learning, the importance of physical context, and long-term outcomes.

- The Multiple Identities Framework, grounded in situated cognition, explores factors associated with deciding what kind of person one wants to be or fears becoming and engaging in activities that make one part of the communities associated with a particular identity. It has been used to examine women negotiating the worlds of science and engineering, as well as race and gender in workplace settings (Tate and Linn, 2005; Packard, 2003).

- Third Spaces is a theoretical construct that lends itself to nonschool learning (e.g., Gutiérrez, 2008; Eisenhart and Edwards, 2004). Third spaces are outside the two typical spheres of existence: home and work or home and school for children. For telecommuters, for example, a coffee shop where they spend the work day could be construed as a third space. Third spaces are places where participants' everyday and technical (or scientific) language and experiences intersect and can be the site for fascinating accounts of informal learning.

- Situated/Enacted Identity (Falk, 2006; Rounds, 2006) focuses on audience expectation and audience agenda in terms of true, underlying interests that are intimately linked to the audience's enacted identity during a visit or free-choice learning experience. This framework is based on a large body of literature that considers the entry narrative of the visitor as a key factor in understanding motivation and learning from an informal learning experience.

- Family learning, though not a theoretical framework per se, has been an important way of reframing informal learning experiences, changing the focus from any single individual in a learning group, such as the child, to the entire family (Bell, Bricker, Lee, Reeve, and Zimmerman, 2006; Ellenbogen, Luke, and Dierking, 2004; Astor-Jack, Whaley, Dierking, Perry, and Garibay, 2007; Ash, 2003; Crowley and Galco, 2001; Ellenbogen, 2002, 2003; Borun et al., 1998). In this context, learning is defined as "a joint collaborative effort within an intergenerational group of children and significant adults." Outcomes include learning science concepts, attitudes, and behaviors and also learning about one another and the members of the group, as well as shaping and reinforcing individual and group identity. Family learning approaches are grounded in sociocultural theories and are currently transforming the way some museums and science centers are reorienting their missions, educational strategies, and experiences.

Other perspectives have been used to inform evaluation studies of learning in informal environments.

- Community of Practice (see Lave and Wenger, 1991) is a framework used to guide development and assessment of community-based efforts and professional development projects. This framework offers insight into participants' trajectories from science novices (peripheral members of the science community) to more active and core members, engaging in authentic science and sometimes even participating in apprentice-like activities with scientists, engineers, and technicians.

- Positive Youth Development and Possible Selves frameworks have been used primarily in assessing youth programs (Koke and Dierking, 2007; Luke, Stein, Kessler, and Dierking, 2007). They are grounded in sociocultural theory and address the broader developmental needs of youth, in contrast to traditional deficit-based models that focus solely on youth problems, such as substance abuse, conduct disorders, delinquent and antisocial behavior, academic failure, and teenage pregnancy. Positive Youth Development describes six characteristics of positively developing young people that successful youth programs foster: cognitive and behavioral competence, confidence, positive social connections, character, caring (or compassion) and contribution, to self, family, community, and ultimately, civil society. Possible Selves (Stake and Mares, 2005) proposes that individuals' perceptions of their current and imagined future opportunities serve as a motivator and organizer for their current task-related thoughts, attitudes, and behaviors, thus "linking current specific plans and actions to future desired goals."

cultures. Using each as a lens to examine learning environments enables us to tease out various factors at play in the learning process and better identify potential leverage points for improving learning.

### People-Centered Lens

This lens sheds light on the intrapsychological phenomena that are relevant to the purposes and outcomes of science learning in informal environments including: the development of interests and motives, knowledge, affective responses, and identity. Some of the relevant principles for the people-centered frame are encapsulated in *How People Learn* (National Research Council, 1999). These principles include the influence of prior knowledge on learning, how experts differ from novices, and the importance of metacognition. Other principles highlight the learning benefits of having experiences that provide one with a positive affect and that help identify personal interests, motives, and identities that can be pursued.

From early childhood onward, humans develop intuitive ideas about the world, bringing prior knowledge to nearly all learning endeavors. Children and adults explain and hear explanations from others about why the moon is sometimes invisible, how the seasons work, why things fall, bounce, break, or bend. Interestingly, these ideas develop without tutoring and are often tacit (individuals may remain unaware of their own ideas). Yet these ideas often influence behavior and come into play during intentional acts of learning and education. Thus, a major implication for thinking about informal science learning is that what learners understand about the world is perhaps as important as what we wish for them to learn through a particular experience. Accordingly, efforts to teach should not merely be about abstractions derived in knowledge systems like science, but should also focus on helping learners become aware of and express their own ideas, giving them new information and models that can build on or challenge their intuitive ideas.

Experts in a particular domain are people who have deep, richly interconnected ideas about the world. They are not just good thinkers or really smart. Nor are novices poor thinkers or not smart. Rather, experts have knowledge in a specific domain—be it chess, waiting tables, chemistry, or tennis—and are not generalists. Their ability to identify problems and generate solutions is closely connected to the things that they know, much more so than once believed (National Research Council, 2007). At the same time, expertise is not just a "bunch of facts"; the knowledge of experts resides in organized, differentiated constructs with which the expert works and applies fluidly. Research has documented how expertise development can begin in childhood through informal interaction with family members, media sources, and unique educational experiences (Crowley and Jacobs, 2002; Reeve and Bell, in press).

One way that experts work with their knowledge is through metacognition or monitoring their own thinking. Much of this work is done in the head and is not naturally accessible to others, although researchers have found it useful to ask knowledgeable people to talk aloud about their thinking while they engage in tasks. Metacognition, like expertise, is domain-specific. That is, a particular metacognitive strategy that works in a particular activity (e.g., predicting outcomes, taking notes) may not work in others. However, metacognition is not exclusive to experts; it can be supported and taught. Thus, even for young children and older novices engaged in a new domain or topic of interest, metacognition can be an important means of controlling their own learning (National Research Council, 1999). Accordingly, as a means of controlling learning, metacognition may have special salience in informal settings, in which learning is self-paced and frequently not facilitated by an expert teacher or facilitator.

At the individual unit of analysis, people-centered analyses might focus on the details of mental processes and evidence of acquiring knowledge, affective responses, or interest development. It may also attend to changes in the individual as a result of broader social and cultural processes.

It is important to note that a people-centered analysis is not the same as a cognitive perspective. Although both tend to examine individuals as the unit of analysis, a cognitive perspective is concerned with mentation, whereas people-centered analysis could also explore people's social actions, practices, and emotional worlds. Thus, within a people-centered analysis, shades of sociocultural and cognitive perspectives are evident.

Many approaches to designing informal science learning experiences reflect a people- or individual-centered approach to learning. For example, many museum experiences are designed to juxtapose museum goers' prior knowledge with the formal disciplinary ideas that can explain the natural phenomena they engage with in an exhibit or activity. This approach to design, focused on stimulating cognitive dissonance, is presumed to help learners question their own knowledge and more deeply reconstruct that knowledge, so that it comes to resemble that of the discipline in question.

One example of a framework that could be considered people-centered was developed by George Hein (1998). It allows for classification of museum-based and similar learning experiences along dimensions of the thinking they support or promote for participants. Hein's framework can be represented in a diagram depicting two orthogonal lines on a plane (see Figure 2-2). One plane represents the theory of knowledge (epistemology) embodied in an exhibit or museum. This ranges from realism (the world exists independently of human knowledge about it) to idealism (knowledge of the world exists only in minds and doesn't imply anything about the world "out there"). The second plane represents a theory of learning, which moves from a transmission model to a constructed model. This reflects a range from behaviorist commitments (e.g., knowledge is transmitted) to the variability in cognitive perspectives with respect to the *extent* to which knowledge is learner-constructed.

Hein's simple diagram can be used to classify the pedagogical approach

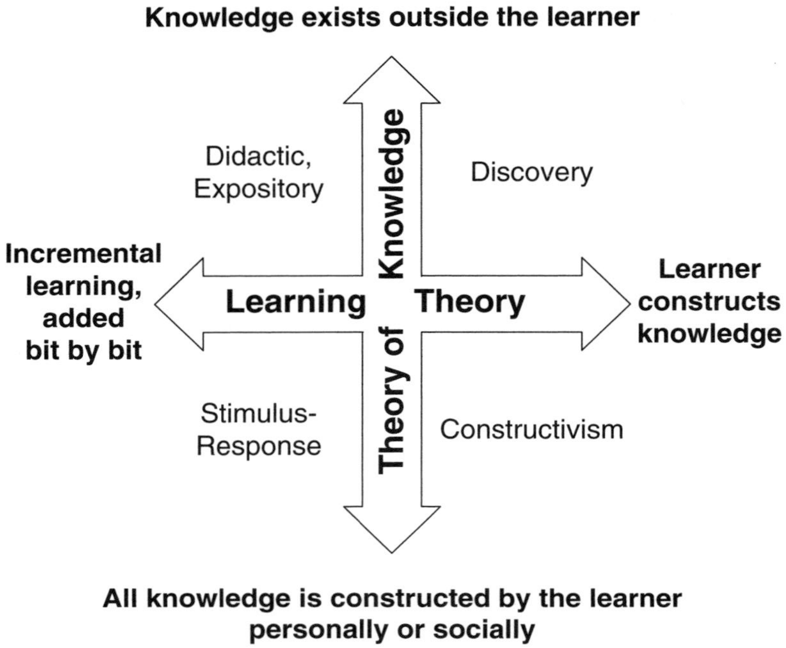

**Knowledge exists outside the learner**

Didactic,
Expository

Discovery

**Knowledge**

**Incremental
learning,
added
bit by bit**

**Learning    Theory**

**Learner
constructs
knowledge**

**Theory of**

Stimulus-
Response

Constructivism

**All knowledge is constructed by the learner
personally or socially**

*FIGURE 2-2   Educational theories.*
*SOURCE: Hein (1998).*

of a museum or exhibit into one of four quadrants, on the basis of the kinds of learning environments they offer visitors. For example, quadrant 1 experiences are "didactic expository." They assume scientific knowledge should be conveyed as factual and confirmed and that learners should be driven through this body of knowledge (rather than invited to think about and apply knowledge). In contrast, quadrant 2 is exemplified by experiences in a "discovery museum," in which understanding emerges through self-directed interactions with the world and representations of the world. Quadrant 3, exemplified by the "constructivist museum," portrays an environment in which individuals construct their knowledge of the world through integration of existing and new conceptions, making personal sense of what they learn. Quadrant 4, which Hein refers to as behaviorist, defines environments in which learners build knowledge of an external world by mastering "pieces" of knowledge incrementally.

### Place-Centered Lens

In important ways, learning can be thought of as happening within and across particular places. Sociocultural perspectives argue that the physical features, the available materials, and the typical activities associated with specific places centrally influence learning processes and outcomes. The expert use of artifacts (e.g., an apparatus in a museum exhibit, a scientific representation of data) for responding to problems or accomplishing projects that people engage in can be viewed as a desired form of intelligent human performance in its own right (Hutchins, 1995). For example, researchers have studied how science-related interests are pursued across different physical settings, social groups, and hobbyist endeavors associated with amateur astronomy (Azevedo, 2004). There are specific tools (e.g., telescopes, astronomical databases), locations (e.g., hillsides, hobbyist group meetings), and activities (e.g., conducting observations, building computer models) associated with learning science in these informal environments and experiences. Analysis centered on the use of artifacts that mediate learning and desired performance in specific contexts and places is regarded as a "practice turn" in theoretical and empirical accounts of human learning, development, and performance (Jessor, 1996; Shweder, 1996; Lave and Wenger, 1991; Rogoff, 2003). This turning away from studying internalized mental learning outcomes to analyzing social practice is evident in the science studies literature as well—in accounts of sophisticated scientific activities that emerge or develop as a result of particular arrangements of resources in specific places like labs or field sites (Latour and Woolgar, 1986; Rouse, 1999). In this view, the material and technological objects, including visual representations of data and technological tools, constitute the foundational resources through which people individually and collectively engage in learning activities.

Hutchins summarizes the role of artifacts in distributed cognition as follows: "The properties of *groups of minds* in interaction with each other, or the properties of the interaction between *individual minds and artifacts in the world*, are frequently at the heart of intelligent human performance" (Hutchins, 1995, p. 62). The ubiquity of human interaction with artifacts in the shaping of learning and thought has been documented for many decades in ecological research focused on understanding human activity in physical settings (Gibson, 1986; Shaw, Turvey, and Mace, 1982). A group of visitors on a museum floor makes sense of exhibits through forms of talk and physical activities that are fundamentally shaped by the nature of the material and technological objects they encounter in those places (Heath, Luff, von Lehn, Hindmarsh, and Cleverly, 2002). Scientists and other professionals conduct their measurements and engage in practical "intelligent routines" in concert with the features of the specific material objects and representations with which they work (Latour, 1995; Pea, 1993; Scribner, 1984; Traweek, 1988). Archaeologists conducting fieldwork, for example, go through sophisticated sequences of interaction and gesturing around the physical objects of their inquiry to develop their theoretical inferences about past cultures and local settings (Goodwin, 1994). Children working as street vendors know how to perform sophisticated arithmetic operations in conjunction with the specific currencies and retail products (Saxe, 1988; Nunes, Schlieman, and Carraher,

1993). The instrumental use of artifacts in the course of mediating everyday cognition and learning is pervasive.

In the context of everyday learning, people frequently develop unique arrangements of artifacts and associated practices in order to respond to the pressing problems or opportunities at hand. This assemblage and use of artifacts can take on both happenstance and patterned qualities in terms of how people come to respond to a situation over time, given the locally available and culturally recognized resources. In this view, learning is seen as "adaptive organization in a complex system" (Hutchins, 1995). For example, designers of informal education exhibits frequently build in ways for museum-goers to alter and customize their experience with an exhibit—and sometimes museum-goers develop their own innovative changes in order to support their own preferred way to engage.

Learning artifacts and associated activities often turn up in some spaces more than others. For example, science centers often try to cultivate use of unique physical and electronic objects that are focused on exploration, sense-making, and social interaction. Those same objects and activities are not as easily made available in other locations (e.g., in a neighborhood park or in a home). In this way, specific forms of science learning are often associated with particular spaces.

Media also represent a rich layer of learning artifacts. The various forms of media available in society—interactive, multiplayer video games, television, print—provide a specific infrastructure for learning that is historically unique. Arrays of related information and perspectives have become broadly available through online resources and communities. Electronic gadgets have become a pervasive fixture of the toolkit of personal activity and learning. Many people routinely develop and share media objects that involve sophisticated learning and social interaction.

At a different scale, in many social niches in society, the natural environment itself becomes an infrastructure and focus for learning (e.g., as groups immerse themselves in ecosystems). Science is learned in relation to these broader physical contexts (e.g., the interdependencies of natural systems, the influence of human society on the environment). The material world, with its rich place-specific features and processes, becomes the focus of inquiry and learning. For example, children reared in rural agricultural communities are often brought into an understanding of the living world through intense, sustained engagement with agricultural practices and the flora and fauna of specific ecosystems.

### Culture-Centered Lens

One of the most important theoretical shifts in education research in the past few decades has been the recognition that all learning is a cultural process. Cultural theories regarding the nature of the mind, of intelligence, and of knowing and learning shape educational practices in a process through which they are more or less designed to conform with those theories. The theories, in turn, explain the practices. As Bruner summarized the situation: "How a people believe the mind works will, we now know, have a profound effect on how it is *compelled* to work if anybody is to get on in a culture. And that fact, ironically, may indeed turn out to be a robust cultural universal" (Bruner, 1996, p. xvii). To truly examine learning from a cultural perspective, these underlying—and often tacit—theories themselves "must be explained, accounted for, and confronted" (McDermott and Varenne, 1996).

Foundational work by Vygotsky, a contemporary of Piaget, offers insight into the cultural origin of human development. The current prominence of sociocultural perspectives grew out of long-standing concerns with a nearly exclusive focus on individual thinking and learning. Instead, sociocultural theory explores how individuals develop through their involvement in cultural practices (e.g., Cole, 1996; Heath, 1983; Rogoff, 2003). Culture is an admittedly contested terrain, which is notoriously difficult to define. However, most scholars agree that it is constituted by the strong social affiliations of learners through which they access and voice their own ideas, values, and practices. They develop specific skills, commitments, knowledge, and their identity as they become proficient in practices that are valued in specific communities. As we discuss more thoroughly in Chapter 7, two aspects of the current view of culture are critical to understanding its relevance to learning. First, culture is bidirectional and dynamic. In addition to acquiring culturally valued skills, knowledge, and identities, individuals also influence the cultural systems they participate in. They bring their own prior experiences and knowledge to cultural groups. They have agency in carrying out their own agendas. Through these mechanisms they influence the values, practices, and knowledge of cultural groups. Rogoff (2003, pp. 3-4) captures this succinctly: "People develop as participants in cultural communities. Their development can be understood only in light of the cultural practices and circumstances of their communities—which also change." Second, culture is also distributed variably among group members, and individuals frequently participate in many cultural communities. Culture is not equivalent to ethnicity, occupation, or social class. In any cultural group, some individuals affiliate more strongly with a particular cultural identity than their peers.

This view encompasses both individual and collective activity. As noted, individual development unfolds in cultural contexts (although culture itself is neither uniform nor static). Simultaneously, through individuals' actions, culture itself is modified and transformed. Hence, from a strictly cognitive perspective, science is a series of processes that generate and validate knowledge. From the perspective of mediated activity, science is a collective practice of generating worthy questions about the natural world and pursuing answers through empirical analysis using specific cognitive tools. Participating in even simple practices can afford learners the development of fluency with particular cultural practices some of which are closely connected to science. For example, in many homes, dinnertime conversations encourage children to weave narratives, hold and defend positions, and otherwise articulate points of view (Ochs, Taylor, Rudolph, and Smith, 1992). How parents encourage and shape children's language and question-asking about the world can be foundational for helping them view science as a form of communication and collaborative sense-making.

Conceiving of culture as shared repertoires of practices sometimes leads researchers to refer to membership in almost any type of group as membership in a culture. In particular, in this volume, scientists are frequently treated as a cultural group, in which people share common commitments to questions, research perspectives, notions of what is a viable scientific stance, and how one makes arguments. They also use specialized tools (or artifacts) to carry out their work and spend significant effort coordinating and refining their practices.

This conceptualization of culture is highly relevant to the ecology of science learning contexts. Educators often hold stereotyped notions of what counts as scientific reasoning and privilege a subset of sense-making practices at the expense of others (Ballenger, 1997). Yet research on scientific discussions and in active research groups reveals that many practices in which scientists engage are not recognized as useful or as a part of science in the classroom. For example, scientists regularly use visual and discursive resources whereby they imagine themselves inside physical events and processes to explore the ways in which they may behave (Ochs, Gonzales, and Jacoby, 1996; Wolpert and Richards, 1997). These and other findings undermine the view that typical scientific practices are largely abstract logical derivations that are disassociated from everyday experience of the natural world. The observation that science and science learning are richly social also underlines the opportunity of educators working with designed environments to take better advantage of the cultural practices that a diverse set of learners might bring to the environment (Cobb, Confrey, diSessa, Lehrer, and Schauble, 2003; Bricker and Bell, 2008; Warren et al., 2001).

Many children who fail in school, including those who are from nondominant cultural or lower socioeconomic groups, may show competence on the same subject matter in out-of-school contexts (McLaughlin, Irby, and Langman, 2001). These asymmetries raise questions about the design of school-based instruction, and they invite analyses of factors that facilitate success in less formal settings. Freedom from a timetable that dictates a schedule for learning, for example, may allow children to explore scientific phenomena in ways that are personally more comfortable and intellectually more engaging than they would be in school (Bell, Zimmerman, Bricker, and Lee, no date). A central issue is how to integrate experiences across settings to develop synergies in learning—in other words, how to maximize the ecological connections among learning experiences toward outcomes and competencies of interest or of consequence (Bell et al., 2006).

A cultural lens makes salient a broad set of aspects of learning experiences that can be harnessed (e.g., by educators, facilitators, parents) to interpret, extend, and support learning. These include attending to the resources for learning that learners bring to a learning environment (e.g., specialized forms of talking and argumentation), the ways in which learners relate to and identify with the natural world, the models of disciplinary and everyday science they encounter in their communities, the material resources and activities that are familiar and available to them, and the community goals and needs related to science learning. For example, in a classic study by Heath (1983), fifth-grade children were supported in conducting a science investigation related to food production that engaged them in different aspects of community life. The children acted as ethnographers of local agricultural activities and engaged with a range of community members about food production. In the process they learned how to scientifically obtain, verify, and communicate information, and their oral and written language demonstrated that their understanding of relevant scientific concepts developed over the course of their inquiry.

### *Critical Issues*

An ecological approach underlines two critical issues for understanding the context of learning. One is that the intellectual, knowledge-focused domain cannot be isolated from the domain of social identity. Identity development and elaboration are linked to affective and motivational issues that catalyze learning (Resnick, 1987; Schauble, Leinhardt, and Martin, 1998; Hull and Greeno, 2006). The second, as discussed above, is that there is a shift in focus from the individual learner in isolation to culturally variable participation structures, such as apprenticeship learning and legitimate peripheral participation, the process through which individuals move from simpler tasks at the periphery of group activity to higher level and more central positions of responsibility and expertise as they learn new capabilities (Rogoff, 1990, 2003; Lave and Wenger, 1991).

This study explores the broad range of learning settings and outcomes found in the literatures on learning science in informal environments. We examine the role of personal psychology, places, and cultural practices on science learning. In the next section we define the kinds of outcomes that are especially relevant to informal environments for science learning.

## GOALS OF SCIENCE LEARNING

Learning science in informal environments is a diverse enterprise and serves a broad range of intended outcomes. These include inspiring emotional reactions, reframing ideas, introducing new concepts, communicating the social and personal value of science, promoting deep experiences of natural phenomena, and showcasing cutting-edge scientific developments.

This book recognizes several principles:

- Knowledge, practice, and science learning commence early in life, continue throughout the life span, and are inherently cultural.
- Science is a system of acquiring knowledge through systematic observation and experimentation.
- The body of scientific knowledge that has been established is continually being extended, refined, and revised by the community of scientists.
- Science and scientific practice weave together content and process features.
- Effective science education reflects the ways in which scientists actually work.

Science learning involves much more than the acquisition of disciplinary content knowledge and process skills. Like the scientific proficiencies enumerated in *Taking Science to School* (National Research Council, 2007), science learning can be envisioned as strands of a rope intertwined to produce experiences, environments, and social interactions that provide strong connections to pull people of all ages and backgrounds toward greater scientific understanding, fluency, and expertise. Informal science learning experiences often occur in situations that immediately serve peoples' interests and prepare them for their future learning in unanticipated ways. Learning experiences in informal settings also grab learners' attention, provoke emotional responses, and support direct experience with phenomena. In this sense, informal settings occupy an important and unique space in the overarching infrastructure of science learning. At a broad level, informal environments have strengths that are unique and complementary to the strengths of schools.

There are also differences and junctures between informal environments and other venues for science learning, such as K-12 schools, universities, and workplaces. Identifying their respective goals and specific ways in which they do (and do not) intersect can promote thoughtful analysis and coordination of the overarching infrastructure. For example, it is common for schools and science centers to partner with respect to school group visits, teacher education, and summer programs. Despite this overlap, informal environments also have their own distinct mission and mandate. Unlike K-12 schools, they typically do not compel participation. Nor do they have the historical mandate to improve the learning of academic forms of science—especially as measured in terms of standardized achievement indicators—as is increasingly common for formal education. Thus, while informal science learning can be integrated with K-12 science curriculum, the fit is not seamless.

That is why the model of science learning we present here places special emphasis on providing entrée to, and sustained engagement with, science—reflecting the purview of informal learning—while keeping an eye on its potential to support a broad range of science-specific learning outcomes and intersect with related institutional players and broader societal interests. Here we introduce, and in Chapter 3 we expand upon, six interweaving strands that describe goals and practices of science learning (see Box 2-2). It is important to note that while these strands reflect conceptualizations developed in research, as a set they have not been systematically applied and analyzed. The strands are interdependent—advances in one are closely associated with advances in the others. Taken together they represent the ideal that all institutions that create and provide informal environments for people to learn science can strive for in their programs and facilities.

## Strand 1: Developing Interest in Science

Strand 1 addresses motivation to learn science, emotional engagement with it, curiosity, and willingness to persevere over time despite encountering challenging scientific ideas and procedures over time. Research suggests that personal interest and enthusiasm are important for supporting children's

---

### BOX 2-2  Strands of Informal Science Learning

Learners who engage with science in informal environments . . .

Strand 1: Experience excitement, interest, and motivation to learn about phenomena in the natural and physical world.

Strand 2: Come to generate, understand, remember, and use concepts, explanations, arguments, models, and facts related to science.

Strand 3: Manipulate, test, explore, predict, question, observe, and make sense of the natural and physical world.

Strand 4: Reflect on science as a way of knowing; on processes, concepts, and institutions of science; and on their own process of learning about phenomena.

Strand 5: Participate in scientific activities and learning practices with others, using scientific language and tools.

Strand 6: Think about themselves as science learners and develop an identity as someone who knows about, uses, and sometimes contributes to science.

participation in learning science (Jolly, Campbell, and Perlman, 2004). Tai and colleagues' nationally representative study of factors associated with science career choices, in fact, suggests that an expressed interest in science during early adolescence is a strong predictor of science degree attainment (Tai, Liu, Maltese, and Fan, 2006). Even though early interest does not guarantee extended learning, early engagement can trigger the motivation to explore the broader educational landscape to pursue additional experiences that may persist throughout life. Youth-focused hobby or interest groups, designed exhibits, and after-school programs are commonly organized and planned to support this strand of science learning. They allow for the extended pursuit of learning agendas, the refinement of interests, the sharing of relevant learning resources and feedback, access to future learning experiences, and opportunities to be identified as having science-related interests.

Adults, including older adults, choose to learn science in informal environments often because of a personal interest, a specific need for science-related information, or to introduce children in their care to aspects of the scientific enterprise.

## Strand 2: Understanding Science Knowledge

Strand 2 addresses learning about the main scientific theories and models that frame Western civilization's understanding of the natural world. Associated educational activities address how people construct or understand the models and theories that scientists construct by generating, interpreting, and refining evidence. Concepts, explanations, arguments, models, and facts are the knowledge products of scientific inquiry that collectively aid in the description and explanation of natural systems when they are integrated and articulated into highly developed and well-established theories.

## Strand 3: Engaging in Scientific Reasoning

Asking and answering questions and evaluating evidence are central to doing science and to successfully navigating through life (e.g., looking at nutrition labels to decide which food items to purchase, understanding the impact of individual and collective decisions related to the environment, diagnosing and addressing personal health issues, testing different possible causes of malfunction in technological systems). The generation and explanation of evidence is at the core of scientific practice; scientists are constantly refining theories and constructing new models based on observations and experimental data. Understanding the connections, similarities, and differences between evidence evaluation in daily living and the practice of science is an important contribution that is easily introduced and delivered in informal everyday settings.

We also note that this strand is related to engineering design process, which is parallel to, but distinct from, scientific inquiry. Although engineers apply scientific concepts and mathematics in their work, they also apply engineering design principles, such as the idea of trade-offs, the recognition that most problems have several possible solutions, and the idea that new technologies may have unanticipated effects. These ideas can also be communicated through experiences in informal settings. As described in Chapter 4, connecting natural types of evaluation to everyday experience—such as posing and answering commonsense questions and making predictions based on observational data concerning interesting phenomena—can support learners in developing an understanding of science. Deepening these experiences to include mathematical and conceptual tools to analyze data and further refine the questions, observations, and experimental design may also result in participants' developing strong understanding of the practice of science.

## Strand 4: Reflecting on Science

The practice of science revolves around the dynamic refinement of scientific understanding of the natural world. New evidence can always emerge, existing theories are continuously questioned, explanatory models are constantly refined or enlarged, and scientists argue about how the evidence can be interpreted. The appreciation of how profoundly exciting this is has attracted some of the best and brightest minds to the practice of science. Strand 4 focuses on learners' understanding of science as a way of knowing—as a social enterprise that advances scientific understanding over time. It includes an appreciation of how the thinking of scientists and scientific communities changes over time as well as the learners' sense of how his or her own thinking changes.

Informal learning environments and programming seem to be particularly well suited to providing opportunities for children, youth, and adults to experience some of the excitement of participation in a process that is constantly open to revision. Understanding of how scientific knowledge develops can be imparted in museums and media by creative reconstruction of the history of scientific ideas or the depiction of contemporary advances. Because the stakes can be high and scientists are human, there are many compelling personal stories in science (e.g., Galileo Galilei, Benjamin Franklin, Charles Darwin, Marie Curie, James Watson, Francis Crick, and Barbara McClintock).

Creating and delivering opportunities for participants to assume the role of a scientist can be a powerful way for them to come to understand science as a way of knowing, though learners require significant support (e.g., to stimulate reflection and facilitate knowledge integration) to do so. Engaging in scientific practice can create the recognition that diverse methods and tools are used, there are multiple interpretations of the same evidence, multiple theories are usually advanced, and a passionate defense of data often occurs in searching for core explanations of an event or phenomena. With guidance, this process can lead participants to reflect on their own state of knowledge and how it was acquired. Rich media representations (e.g., large screen documentaries) and digital technologies, such as simulations and immersive environments (e.g., visualizations, interactive virtual reality, games), can expand more traditional hands-on approaches to engage the public in authentic science activities.

## Strand 5: Engaging in Scientific Practice

Because scientific practice is a complex endeavor and depends on openness to revision, it is done by groups of people operating in a social system with specific language apparatus, procedures, social practices, and data representations. Participation in the community of science requires knowledge of the language, tools, and core values. Changing the inaccurate stereotype of the lone scientist working in isolation in his laboratory to the accurate perception of groups of people interacting with each other to achieve greater understanding of a problem or phenomenon is critical to creating a positive attitude toward science learning. Strand 5 focuses on how learners in informal environments come to appreciate how scientists communicate in the context of their work as well as building learners' own mastery of the language, tools, and norms of science as they participate in science-related inquiry.

## Strand 6: Identifying with the Scientific Enterprise

Not only can educational activities develop the knowledge and practices of individuals and groups, they can also help people develop identities as science learners and, in some cases, as scientists—by helping them to identify and solidify their interests, commitments, and social networks, thereby providing access to scientific communities and careers. This strand pertains to how learners view themselves with respect to science. Strand 6 is relevant to the small number of people who, over the course of a lifetime, come to view themselves as scientists as well as the great majority of people who do not become scientists. For the latter group, it is an important goal that all members of society identify themselves as being comfortable with, knowledgeable about, or interested in science.

We note that in the strand framework in *Taking Science to School* (National Research Council, 2007), the development of identity was not a separate strand but was construed as a component of participation in science (Strand 5 here, Strand 4 in the previous volume). While we do not disagree that participation and identity development are closely related, we see identity as worthy of its own focus here with particular importance to informal settings, which engage learners of all ages. Identity is developed over the life span and so incorporates the dimension of time. We urge the community of informal science education to support identity development over time by creating opportunities for sustained participation and engagement over the life span.

# VENUES FOR SCIENCE LEARNING

We are interested in a broad array of settings that can capture lifelong, life-wide, and life-deep learning. We organize our discussion of environments across three venues or configurations for learning: everyday informal environments, designed environments, and out-of-school and adult programs.

All learning environments, including school and nonschool settings, can be said to fall on a continuum of educational design or structure (see Figure 2-3). Although what makes a learning environment informal is the subject of much debate, informal environments are generally defined as including learner choice, low consequence assessment, and structures that build on the learners' motivations, culture, and competence. Furthermore, it is generally accepted that informal environments provide a safe, nonthreatening, open-ended environment for engaging with science. In this report we limit our analysis to nonschool informal environments out of a felt need to promote careful analysis and research in this area, which has often taken a back seat to research in school settings.

## Everyday and Family Learning

Everyday learning is pervasive in people's lives and includes a range of experiences that may extend over a lifetime, such as family or peer discussions and activities, personal hobbies, and mass media engagement and technology use. The agenda and manner of interaction in the environment are largely selected, organized, and coordinated by the learners and thus vary across and within cultures. Assessment is most often structured as immediate feedback through situated responses. Doing, learning, knowing, and demonstrating knowledge are typically intertwined and not easily distinguished

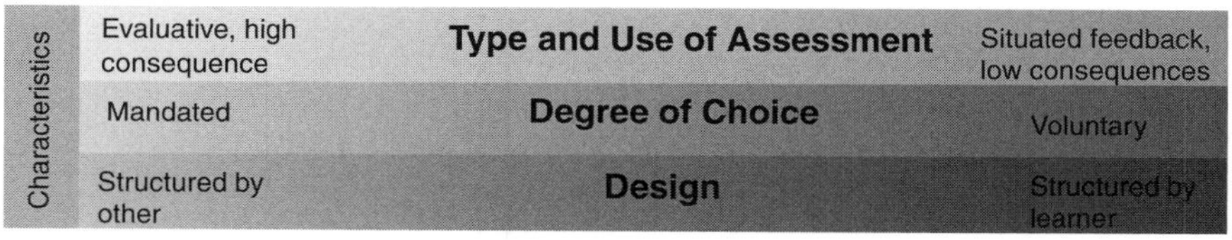

| Characteristics | Evaluative, high consequence | **Type and Use of Assessment** | Situated feedback, low consequences |
| | Mandated | **Degree of Choice** | Voluntary |
| | Structured by other | **Design** | Structured by learner |

FIGURE 2-3 Continuum of learning environments.

from each other. In these environments, demonstrating competence often results in a more central role in the learning configuration. For example, as children who grow up in an agricultural society develop greater knowledge and skill, their duties may shift. Feeding animals and cleaning stalls may give way to tending animal wounds and monitoring well-being.

## Designed Environments

Examples of designed environments include museums, science centers, botanical gardens, zoos, aquariums, and libraries. Artifacts, media, and signage are primarily used to guide the learner's experience. While these environments are structured by institutions, the nature of the learner's interaction with the environment is often determined by the individual. Learners enter these environments primarily by choice, either their own personal choice or the choice of an adult (e.g., parent or teacher). Learners also have significant choice in setting their own learning agenda by choosing to attend to only exhibits or aspects of exhibits that align with their interests. Typically, learners' engagement is short-term and sporadic in the setting, and learning takes place in peer, family, or mentor interactions. However, there is increasing interest in extending the impact of these experiences over time through post-visit web experiences, traveling exhibits, and follow-up mail or e-mail contact.

## After-School and Adult Programs

Examples of after-school and adult programs include summer programs, clubs, science center programs, Elderhostel programs, volunteer groups, and learning vacations. Often program content includes a formal curriculum that is organized and designed to address the concerns of the sponsoring institutions. The curriculum and activities are focused primarily on content knowledge or skills, but they also may focus on attitudes and values and using science to solve applied problems. Activities are often designed to serve those seen to be in need of support, such as economically disadvantaged children and adults. Like designed spaces, individuals most often participate in these activities either by their own choice or the choice of a parent or teacher. They attend programs that align with their interests or needs. Experiences in these environments are typically guided and monitored by a trained facilitator and include opportunities for hands-on, collaborative experiences. The time scale of these learning experiences ranges from being sustained, long-term programs with in-depth engagement to brief, targeted, short-term programs. Assessments are often used, and may affect the participants' reputation or status in the program, however they are not typically meant to judge individual attainment or progress in comparison to institutional expectations.

## Heterogeneity Within Each Venue

Assessment, choice, and design characteristics define each type of informal learning venue. Yet it is important to note that there is great variability within each of the types of venue we have described. Consider everyday learning environments—which also frequently include use of materials and activities designed (or repurposed) to support science learning (e.g., commercially available science kits, locally fashioned and commercially available products associated with hobbies, collections of science-related media). Everyday learning environments are the most learner-driven and least externally structured of the three. Yet everyday learning can also be heavily structured by someone other than the learner, such as a parent or sibling. Others play a critical role in facilitating learning—asking questions, providing resources. It is also important to note that what may begin as one learner's incidental inquiry, say about insects, can turn into something fundamentally different. For example, it is easy to imagine a parent or older sibling turning a child's curious musing about the insects she has seen into a mini-assessment of the child's technical knowledge of insect names or body parts. In this case, with the purpose and structure of the activity defined externally, the event can easily shift the learning focus and shut down the original inquiry and the child's learning.

## CONCLUSION

In this chapter we have argued science learning should be viewed as a lifelong, life-wide, and life-deep endeavor that occurs across a range of venues focused on multiple outcome strands of interest. We have observed that there are a range of perspectives in research on learning science in informal environments which, despite clear similarities and areas of overlap, have not been well integrated into a common body of knowledge. We see this as a critical goal for the advancement of learning science in informal environments as an area of educational practice and inquiry. We described an ecological framework that might hold some potential for researchers, designers, and educators to collectively view the informal learning of science as relating to the details of learning processes, mechanisms, and outcomes associated with people, places, and cultures. We have also introduced the organizational scheme of this report, which reflects the theoretical commitments we have introduced. Our analysis spans diverse venues and configurations, and a broad array of science learning outcomes and processes as indicated in the strands. The strands also reflect an effort to integrate the range of learning practices and outcomes used in prominent sociocultural and cognitive studies of learning and to focus these in science-specific ways. We hope that these perspectives may serve as the kernel of a shared framework to guide the accumulation of research findings on science learning and the design knowledge related to powerful educational practice in service of diverse communities of learners.

## REFERENCES

Ash, D. (2003). Dialogic inquiry in life science conversations of family groups in museums. *Journal of Research in Science Teaching, 40*, 138-162.

Astor-Jack, T., Whaley, K.K., Dierking, L.D., Perry, D., and Garibay, C. (2007). Understanding the complexities of socially mediated learning. In J.H. Falk, L.D. Dierking, and S. Foutz (Eds.), *In principle, in practice: Museums as learning institutions*. Walnut Creek, CA: AltaMira Press.

Azevedo, F.S. (2004). *Serious play: A comparative study of learning and engagement in hobby practices*. Berkeley: University of California Press.

Ballenger, C. (1997). Social identities, moral narratives, scientific argumentation: Science talk in a bilingual classroom. *Language and Education, 19*(1), 1-14.

Banks, J.A., Au, K.H., Ball, A.F., Bell, P., Gordon, E.W., Gutiérrez, K., Heath, S.B., Lee, C.D., Lee, Y., Mahiri, J., Nasir, N.S., Valdes, G., and Zhou, M. (2007). *Learning in and out of school in diverse environments: Lifelong, life-wide, life-deep*. Seattle: Center for Multicultural Education, University of Washington.

Barron, B. (2006). Interest and self-sustained learning as catalysts of development: A learning ecology perspective. *Human Development, 49*(4), 153-224.

Bell, P., Bricker, L.A., Lee, T.R., Reeve, S., and Zimmerman, H.T. (2006). Understanding the cultural foundations of children's biological knowledge: Insights from everyday cognition research. In S.A. Barab, K.E. Hay, and D. Hickey (Eds.), *Proceedings of the seventh international conference of the learning sciences (ICLS)* (pp. 1029-1035). Mahwah, NJ: Lawrence Erlbaum Associates.

Bell, P., Zimmerman, H.T., Bricker, L.A., and Lee, T.R. (no date). *The everyday cultural foundations of children's biological understanding in an urban, high-poverty community*. Everyday Science and Technology Group, University of Washington.

Borun, M., Dritsas, J., Johnson, J.I., Peter, N.E., Wagner, K.F., Fadigan, K., Jangaard, A., Stroup, E., and Wenger, A. (1998). *Family learning in museums: The PISEC perspective*. Philadelphia: Franklin Institute.

Bricker, L.A., and Bell, P. (2008). Conceptualizations of argumentation from science studies and the learning sciences and their implications for the practices of science education. *Science Education, 92*(3), 473-498.

Bronfenbrenner, U. (1977). Toward an experimental ecology of human development. *American Psychologist, 35*, 513-531.

Bruner, J. (1996). Foreword. In B. Shore (Ed.), *Culture in mind: Cognition, culture, and the problem of meaning*. New York: Oxford University Press.

Cobb, P., Confrey, J., diSessa, A., Lehrer, R., and Schauble, L. (2003). Design experiments in education research. *Education Researcher, 32*(1), 9-13.

Cole, M. (1996). *Cultural psychology: A once and future discipline*. Cambridge, MA: Belknap Press.

Crowley, K., and Galco, J. (2001). Family conversations and the emergence of scientific literacy. In K. Crowley, C. Schunn, and T. Okada. (Eds.), *Designing for science: Implications from everyday, classroom, and professional science* (pp. 393-413). Mahwah, NJ: Lawrence Erlbaum Associates.

Crowley, K., and Jacobs, M. (2002). Islands of expertise and the development of family scientific literacy. In G. Leinhardt, K. Crowley, and K. Knutson (Eds.), *Learning conversations in museums*. Mahwah, NJ: Lawrence Erlbaum Associates.

Eisenhart, M., and Edwards, L. (2004). Red-eared sliders and neighborhood dogs: Creating third spaces to support ethnic girls' interests in technological and scientific expertise. *Children, Youth and Environments, 14*(2), 156-177. Available: http://www.colorado.edu/journals/cye/ [accessed October 2008].

Ellenbogen, K.M. (2002). Museums in family life: An ethnographic case study. In G. Leinhardt, K. Crowley, and K. Knutson (Eds.), *Learning conversations in museums*. Mahwah, NJ: Lawrence Erlbaum Associates.

Ellenbogen, K.M. (2003). From dioramas to the dinner table: An ethnographic case study of the role of science museums in family life. *Dissertation Abstracts International, 64*(03), 846A. (University Microfilms No. AAT30-85758.)

Ellenbogen, K.M., Luke, J.J., and Dierking, L.D. (2004). Family learning research in museums: An emerging disciplinary matrix? Available: http://www3.interscience.wiley.com/cgi-bin/fulltext/109062559/PDFSTART [accessed March 2009].

Falk, J.H. (2006). An identity-centered approach to understanding museum learning. *Curator, 49*(2), 151-166.

Falk, J.H., and Dierking, L.D. (2000). *Learning from museums: The visitor experience and the making of meaning*. Walnut Creek, CA: AltaMira Press.

Falk, J.H., and Storksdieck, M. (2005). Using the contextual model of learning to understand visitor learning from a science center exhibition. *Science Education, 89*, 744-778.

Gibson, J.J. (1986). *An ecological approach to visual perception*. Hillsdale, NJ: Lawrence Erlbaum Associates.

Goodwin, C. (1994). Professional vision. *American Anthropologist, 96*(3), 606-633.

Greeno, J.G., Collins, A., and Resnick, L.B. (in press). Cognition and learning. In D. Berliner and R. Calfee (Eds.), *Handbook of educational psychology*. Macmillan Library Reference. New York: Simon and Schuster Macmillan.

Gutiérrez, K.D. (2008). Developing a sociocritical literacy in the third space. *Reading Research Quarterly, 43*(2), 148-164.

Heath, C., Luff, P., vom Lehn, D., Hindmarsh, J., and Cleverly, J. (2002). Crafting participation: Designing ecologies, configuring experience. *Visual Communication, 1*(1), 9-34.

Heath, S.B. (1983). *Ways with words: Language life and work in communities and classrooms*. Cambridge, England: Cambridge University Press.

Hein, G.E. (1998). *Learning in the museum*. New York: Routledge.

Hull, G.A., and Greeno, J.G. (2006). Identity and agency in nonschool and school worlds. In Z. Bekerman, N. Burbules, and D.S. Keller (Eds.), *Learning in places: The informal education reader* (pp. 77-97). New York: Peter Lang.

Hutchins, E. (1995). *Cognition in the wild*. Cambridge, MA: MIT Press.

Jessor, R. (1996). Ethnographic methods in contemporary perspective. In R. Jessor, A. Colby, and R.A. Shweder (Eds.), *Ethnography and human development*. Chicago: University of Chicago Press.

Jolly, E., Campbell, P., and Perlman, L. (2004). *Engagement, capacity, continuity: A trilogy for student success*. St. Paul: GE Foundation and Science Museum of Minnesota.

Koke, J., and Dierking, L.D. (2007). *Engaging America's youth: The long-term impact of Institute for Museum and Library Services' youth-focused programs*. Unpublished technical report, Annapolis, MD, Institute for Learning Innovation.

Latour, B. (1995). The "Pédofil" of Boa Vista: A photo-philosophical montage. *Common Knowledge, 4*, 144-187.

Latour, B., and Woolgar, S. (1986). *Laboratory life: The social construction of scientific facts*. Princeton, NJ: Princeton University Press.

Lave, J., and Wenger, E. (1991). *Situated learning: Legitimate peripheral participation*. New York: Cambridge University Press.

Lee, C. (2008). The centrality of culture to the scientific study of learning and development: How an ecological framework in education research facilitates civic responsibility. *Educational Researcher, 37*(5), 267-279.

Luke, J.J., Stein, J., Kessler, C. and Dierking, L.D. (2007). Making a difference in the lives of youth: Connecting the impacts of museum programs to the "six Cs" of positive youth development. *Curator, 50*(4).

McDermott, R., and Varenne, H. (1996). Culture, development, disability. In R. Jessor, A. Colby, and R.A. Shweder (Eds.), *Ethnography and human development* (pp. 101-126). Chicago: University of Chicago Press.

McLaughlin, M., Irby, M.A., and Langman, J. (2001). *Urban sanctuaries: Neighborhood organizations in the lives and futures of inner-city youth*. San Francisco: Jossey Bass.

National Research Council. (1999). *How people learn: Brain, mind, experience, and school*. Committee on Developments in the Science of Learning. J.D. Bransford, A.L. Brown, and R.R. Cocking (Eds.). Washington, DC: National Academy Press.

National Research Council. (2007). *Taking science to school: Learning and teaching science in grades K-8*. Committee on Science Learning, Kindergarten Through Eighth Grade. R.A. Duschl, H.A. Schweingruber, and A.W. Shouse (Eds.). Washington, DC: The National Academies Press.

Nunes, T.N., Schliemann, A.D., and Carraher, D.W. (1993). *Street mathematics and school mathematics*. New York: Cambridge University Press.

Ochs, E., Gonzales, P., and Jacoby, S. (1996). "When I come down I'm in the domain state": Grammar and graphic representation of the interpretive activity of physicists. In E. Ochs, E.A. Schegloff, and S.A. Thompson (Eds.), *Interaction and grammar* (pp. 328-369). New York: Cambridge University Press.

Ochs, E., Taylor, C., Rudolph, D., and Smith, R. (1992). Storytelling as a theory-building activity. *Discourse Processes, 15*(1), 37-72.

Packard, B.W. (2003). Student training promotes mentoring awareness and action. *Career Development Quarterly, 51*, 335-345.

Pea, R.D. (1993). Practices of distributed intelligence and designs for education. In G. Salomon (Ed.), *Distributed cognitions* (pp. 47-87). New York: Cambridge University Press.

Reeve, S., and Bell, P. (in press). Children's self-documentation and understanding of the concepts "healthy" and "unhealthy." Submitted to *International Journal of Science Education*.

Resnick, L.B. (1987). Learning in school and out. *Educational Researcher, 16*(9), 13-20.

Rogoff, B. (1990). *Apprenticeship in thinking: Cognitive development in social context.* New York: Oxford University Press.

Rogoff, B. (2003). *The cultural nature of human development.* New York: Oxford University Press.

Rounds, J. (2006). Doing identity work in museums. *Curator, 49*(2), 133-150.

Rouse, J. (1999). Understanding scientific practices: Cultural studies of science as a philosophical program. In M. Biagioli (Ed.), *The science studies reader* (pp. 442-457). New York: Routledge.

Saxe, G.B. (1988). Candy selling and math learning. *Educational Researcher, 17*(6), 14-21.

Schauble, L., Leinhardt, G., and Martin, L. (1998). Organizing a cumulative research agenda in informal learning contexts. *Journal of Museum Education, 22*(2 and 3), 3-7.

Scribner, S. (1984). Studying working intelligence. In B. Rogoff and J. Lave (Eds.), *Everyday cognition: Its development in social context* (pp. 9-40). Cambridge, MA: Harvard University Press.

Shaw, R., Turvey, M.T., and Mace, W.M. (1982). Ecological psychology: The consequence of a commitment to realism. In W. Weimer and D. Palermo (Eds.), *Cognition and the symbolic processes II* (pp. 159-226). Hillsdale, NJ: Lawrence Erlbaum Associates.

Shweder, R.A. (1996). True ethnography: The lore, the law, and the lure. In R. Jessor, A. Colby, and R.A. Shweder (Eds.), *Ethnography and human development.* Chicago: University of Chicago Press.

Stake, J.E., and Mares, K.R. (2005). Evaluating the impact of science-enrichment programs on adolescents' science motivation and confidence: The splashdown effect. *Journal of Research in Science Teaching, 42*(4), 359-375.

Stevens, R. (no date). *The learning in informal and formal environments (LIFE) center.* Available: http://life-slc.org/?page_id=124 [accessed February 2009].

Tai, R.H., Liu, C.Q., Maltese, A.V., and Fan, X. (2006). Planning early for careers in science. *Science, 312,* 1143-1144.

Tate, E., and Linn, M.C. (2005). How does identity shape the experiences of women of color engineering students? *Journal of Science Education and Technology, 14*(5-6), 483-493.

Thorndike, E.L. (1931). *Human learning.* New York: Century.

Traweek, S. (1988). *Beamtimes and lifetimes: The world of high energy physicists.* Cambridge, MA: Harvard University Press.

Warren, B., Ballenger, C., Ogonowski, M., Rosebery, A., and Hudicourt-Barnes, J. (2001). Rethinking diversity in learning science: The logic of everyday sense-making. *Journal of Research in Science Teaching, 38,* 1-24.

Wiggins, D.W. (1989). "Great speed but little stamina": The historical debate over black athletic superiority. *Journal of Sport History, 16,* 158-185.

Wolpert, L., and Richards, A. (1997). *Passionate minds: The inner world of scientists.* New York: Oxford University Press.

# 3

# Assessment

Assessment is commonly thought of as the means to find out whether individuals have learned something—that is, whether they can demonstrate that they have learned the information, concepts, skills, procedures, etc., targeted by an educational effort. In school, examinations or tests are a standard feature of students' experience, intended to measure the degree to which someone has, for example, mastered a subtraction algorithm, developed a mental model of photosynthesis, or appropriately applied economic theory to a set of problems. Other products of student work, such as reports and essays, also serve as the basis for systematic judgments about the nature and degree of individual learning.

Informal settings for science learning typically do not use tests, grades, class rankings, and other practices commonly used in schools and workplace settings to document achievement. Nevertheless, the informal science community has embraced the cause of assessing the impact of out-of-school learning experiences, seeking to understand how everyday, after-school, museum, and other types of settings contribute to the development of scientific knowledge and capabilities.[1] This chapter discusses the evidence for outcomes from engagement in informal environments for science learning,

---

[1]The educational research community generally makes a distinction between *assessment*—the set of approaches and techniques used to determine what individuals learn from a given instructional program—and *evaluation*—the set of approaches and techniques used to make judgments about a given instructional program, approach, or treatment, improve its effectiveness, and inform decisions about its development. Assessment targets what learners have or have not learned, whereas evaluation targets the quality of the intervention.

focusing on the six strands of scientific learning introduced earlier and addressing the complexities associated with what people know based on their informal learning experiences.

In both informal and formal learning environments, assessment requires plausible evidence of outcomes and, ideally, is used to support further learning. The following definition reflects current theoretical and design standards among many researchers and practitioners (Huba and Freed, 2000, p. 8):

> Assessment is the process of gathering and discussing information from multiple and diverse sources in order to develop a deep understanding of what students know, understand, and can do with their knowledge as a result of their educational experiences; the process culminates when assessment results are used to improve subsequent learning.

Whether assessments have a local and immediate effect on learning activities or are used to justify institutional funding or reform, most experts in assessment agree that the improvement of outcomes should lie at the heart of assessment efforts. Yet assessing learning in ways that are true to this intent often proves difficult, particularly in informal settings. After reviewing some of the practical challenges associated with assessing informal learning, this chapter offers an overview of the types of outcomes that research in informal environments has focused on to date, how these are observed in research, and grouping these outcomes according to the strands of science learning. Appendix B includes discussion of some technical issues related to assessment in informal environments.

## DIFFICULTIES IN ASSESSING SCIENCE LEARNING IN INFORMAL ENVIRONMENTS

Despite general agreement on the importance of collecting more and better data on learning outcomes, the field struggles with theoretical, technical, and practical aspects of measuring learning. For the most part, these difficulties are the same ones confronting the education community more broadly (Shepard, 2000; Delandshere, 2002; Moss, Giard, and Haniford, 2006; Moss, Pullin, Haertel, Gee, and Young, in press; Wilson, 2004; National Research Council, 2001). Many have argued that the diversity of informal learning environments for science learning further contributes to the difficulties of assessment in these settings; they share the view that one of the main challenges is the development of practical, evidence-centered means for assessing learning outcomes of participants across the range of science learning experiences (Allen et al., 2007; Falk and Dierking, 2000; COSMOS Corporation, 1998; Martin, 2004).

For many practitioners and researchers, concerns about the appropriateness of assessment tasks in the context of the setting are a major constraint on assessing science learning outcomes (Allen et al., 2007; Martin, 2004).[2] It stands roughly as a consensus that the standardized, multiple-choice test—what Wilson (2004) regrets has become a "monoculture" species for demonstrating outcomes in the K-12 education system—is at odds with the types of activities, learning, and reasons for participation that characterize informal experiences. Testing can easily be viewed as antithetical to common characteristics of the informal learning experience. Controlling participants' experiences to isolate particular influences, to arrange for pre- and post-tests, or to attempt other traditional measures of learning can be impractical, disruptive, and, at times, impossible given the features, norms, and typical practices in informal environments.

To elaborate: Visits to museums and other designed informal settings are typically short and isolated, making it problematic to separate the effects of a single visit from the confluence of factors contributing to positive science learning outcomes. The very premise of engaging learners in activities largely for the purposes of promoting future learning experiences beyond the immediate environment runs counter to the prevalent model of assessing learning on the basis of a well-defined educational treatment (e.g., the lesson, the unit, the year's math curriculum). In addition, many informal learning spaces, by definition, provide participants with a leisure experience, making it essential that the experience conforms to expectations and that events in the setting do not threaten self-esteem or feel unduly critical or controlling—factors that can thwart both participation and learning (Shute, 2008; Steele, 1997).

Other important features of informal environments for science learning include the high degree to which contingency typically plays a role in the unfolding of events—that is, much of what happens in these environments emerges during the course of activities and is not prescribed or predetermined. To a large extent, informal environments are learner-centered specifically because the agenda is mutually set across participants—including peers, family members, and any facilitators who are present—making it difficult to consistently control the exposure of participants in the setting to particular treatments, interventions, or activities (Allen et al., 2007). It may well be that contingency, insofar as it allows for spontaneous alignment of personal goals and motivations to situational resources, lies at the heart of some of the most powerful learning effects in the informal domain. Put somewhat differently, the freedom and flexibility that participants have in working with people and materials in the environment often make informal learning settings particularly attractive.

Another feature that makes many informal learning environments attractive is the consensual, collaborative aspect of deciding what counts as success: for example, what children at a marine science camp agree is a

---

[2]This is also an issue of great importance among educators and education researchers concerned with classroom settings.

good design for a submersible or an adequate method for measuring salinity. In some instances, determining a workable standard for measuring success ahead of time—that is, before the learning activities among participants take place—can be nearly impossible. The agenda that arises, say, in a family visit to a museum may include unanticipated episodes of identity reinforcement, the telling of stories, reminders of personal histories, rehearsals of new forms of expression, and other nuanced processes—all of which support learning yet evade translation into many existing models of assessment.

The type of shared agency that allows for collaborative establishment of goals and standards for success can extend to multiple aspects of informal learning activities. Participants in summer camps, science centers, family activities, hobby groups, and such are generally encouraged to take full advantage of the social resources available in the setting to achieve their learning goals. The team designing a submersible in camp or a playgroup engineering a backyard fort can be thought of as having implicit permission to draw on the skills, knowledge, and strengths of those present as well as any additional resources available to get their goals accomplished. "Doing well" in informal settings often means acting in concert with others. Such norms are generally at odds with the sequestered nature of the isolated performances characteristic of school. Research indicates that these sequestered assessments lead to systematic undermeasurement of learning precisely because they fail to allow participants to draw on material and human resources in their environment, even though making use of such resources is a hallmark of competent, adaptive behavior (Schwartz, Bransford, and Sears, 2005).

Despite the difficulties of assessing outcomes, researchers have managed to do important and valuable work. In notable ways, this work parallels the "authentic assessment" approaches taken by some school-based researchers, employing various types of performances, portfolios, and embedded assessments (National Research Council, 2000, 2001). Many of these approaches rely on qualitative interpretations of evidence, in part because researchers are still in the stages of exploring features of the phenomena rather than quantitatively testing hypotheses (National Research Council, 2002). Yet, as a body of work, assessment of learning in informal settings draws on the full breadth of educational and social scientific methods, using questionnaires, structured and semistructured interviews, focus groups, participant observation, journaling, think-aloud techniques, visual documentation, and video and audio recordings to gather data.

Taken as a whole, existing studies provide a significant body of evidence for science learning in informal environments as defined by the six strands of science learning described in this report.

## TYPES OF OUTCOMES

A range of outcomes are used to characterize what participants learn about science in informal environments. These outcomes—usually described as particular types of knowledge, skills, attitudes, feelings, and behaviors—can be clustered in a variety of ways, and many of them logically straddle two or more categories. For example, the degree to which someone shows persistence in scientific activity could be categorized in various ways, because this outcome depends on the interplay between multiple contextual and personal factors, including the skills, disposition, and knowledge the person brings to the environment. Similarly, studies focusing on motivation might emphasize affect or identity-related aspects of participation. In Chapter 2, we described the goals of science learning in terms of six interweaving conceptual strands. Here our formulation of the strands focuses on the science-related behaviors that people are able to engage in because of their participation in science learning activities and the ways in which researchers and evaluators have studied them.

## Strand 1: Developing Interest in Science

### Nature of the Outcome

Informal environments are often characterized by people's excitement, interest, and motivation to engage in activities that promote learning about the natural and physical world. A common characteristic is that participants have a choice or a role in determining what is learned, when it is learned, and even how it is learned (Falk and Storksdieck, 2005). These environments are also designed to be safe and to allow exploration, supporting interactions with people and materials that arise from curiosity and are free of the performance demands that are characteristic of schools (Nasir, Rosebery, Warren, and Lee, 2006). Engagement in these environments creates the opportunity for learners to experience a range of positive feelings and to attend to and find meaning in relation to what they are learning (National Research Council, 2007).

Participation is often discussed in terms of interest, conceptualized as both the state of heightened affect for science and the predisposition to reengage with science (see Hidi and Renninger, 2006).[3] Interest includes the excitement, wonder, and surprise that learners may experience and the knowledge and values that make the experience relevant and meaningful. Recent research on the relationship between affect and learning shows that the emotions associated with interest are a major factor in thinking and learning, helping people learn as well as helping with what is retained and how long it is remembered (National Research Council, 2000). Interest may even have a neurological basis (termed "seeking behavior," Panksepp,

---

[3]Whereas motivation is used to describe the will to succeed across multiple contexts (see Eccles, Wigfield, and Schiefele, 1998), interest is not necessarily focused on achievement and is always linked to a particular class of objects, events, or ideas, such as science (Renninger, Hidi, and Krapp, 1992; Renninger and Wozniak, 1985).

1998), suggesting that all individuals can be expected to have and to be able to develop interest.[4] In addition, interest is an important filter for selecting and focusing on relevant information in a complex environment (Falk and Dierking, 2000). In this sense, the psychological state of mind referred to as interest can be viewed as an evolutionary adaptation to select what is perceived as important or relevant from the environment. People pay attention to the things that interest them, and hence interest becomes a strong filter for what is learned.

When people have a more developed interest for science—sometimes described in terms of hobbies or personal excursions (Azevedo, 2006), islands of expertise (Crowley and Jacobs, 2002), passions (Neumann, 2006), or identity-related motivations (Ellenbogen, Luke, and Dierking, 2004; Falk and Storksdieck, 2005; Falk, 2006)—they are inclined to draw more heavily on available resources for learning and use systematic approaches to seek answers (Engle and Conant, 2002; Renninger, 2000). This line of research suggests that the availability or existence of stimulating, attractive learning environments can generate the interest that leads to participation (Falk et al., 2007). People with an interest in science are also likely to be motivated learners in science; they are more likely to seek out challenge and difficulty, use effective learning strategies, and make use of feedback (Csikszentmihalyi, Rathunde, and Whalen, 1993; Lipstein and Renninger, 2006; Renninger and Hidi, 2002). These outcomes help learners continue to develop interest, further engaging in activity that promotes enjoyment and learning. People who come to informal environments with developed interests are likely to set goals, self-regulate, and exert effort easily in the domains of their interests, and these behaviors often come to be habits, supporting their ongoing engagement (Lipstein and Renninger, 2006; Renninger and Hidi, 2002; Renninger, Sansone, and Smith, 2004).

### Methods of Researching Strand 1 Outcomes

Although self-report data are susceptible to various forms of bias on the part of the research participant, they are nonetheless frequently used in studying outcomes with affective and attitudinal components because of the subjective nature of these outcomes. Self-report studies are typically based on questionnaires or structured interviews developed to target attitudes, beliefs, and interests regarding science among respondents in particular age groups, with an emphasis on how these factors relate to school processes and outcomes (e.g., Renninger, 2003; Moore and Hill Foy, 1997; Weinburgh and Steele, 2000). Methods linking prior levels of interest and motivation to outcomes have been used in research as well.

---

[4] It should be noted that all normatively functioning individuals might be expected to have interest; Travers (1978) points out that lack of interest accompanies pathology.

Researchers have also used self-report techniques to investigate whether prior levels of interest were related to learning about conservation (Falk and Adelman, 2003; Taylor, 1994). Falk and Adelman (2003), for example, showed significant differences in knowledge, understanding, and attitudes for subgroups of participants based on their prior levels of knowledge and attitudes. Researchers replicated this approach with successful results in a subsequent study at Disney's Animal Kingdom (Dierking et al., 2004).

Studies of public understanding of science have used questionnaires to assess levels of interest on particular topics. For example, they have documented variation in people's reported levels of interest in science topics: The general adult population in both the United States and Europe is mildly interested in space exploration and nuclear energy; somewhat more than mildly interested in new scientific discoveries, new technologies, and environmental issues; and fairly interested in medical discoveries (European Commission, 2001; National Science Board, 2002).

An important component of interest, as noted, is positive affect (Hidi and Renninger, 2006). Whereas positive affect toward science is often regarded as a primary outcome of informal learning, this outcome is notoriously difficult to assess. Positive affect can be transient and can develop even when conscious attention is focused elsewhere making it difficult for an observer to assess. Various theoretical models have attempted to map out a space of emotional responses, either in terms of a small number of basic emotions or emotional dimensions, such as pleasure, arousal, and dominance, and to apply these in empirical research (Plutchik, 1961; Russell and Mehrabian, 1977; Isen, 2004).

Analysis of facial expressions has been a key tool in studying affect, with mixed results. Ekman's seven facial expressions have been used to assess fleeting emotional states (Ekman and Rosenberg, 2005). Dancu (2006) used this method in a pilot study to assess emotional states of children as they engaged with exhibits and compared these observations to reports by children and their caregivers, finding low agreement among all measures. Kort, Reilly, and Picard (2001) have created a system of analyzing facial expressions suited to capturing emotions relevant to learning (such as flow, frustration, confusion, eureka), but her methods require special circumstances (e.g., the subject must sit in a chair) and do not allow for naturalistic study in large spaces, thus complicating application of this approach many informal settings. Ma (2006) used a combination of open-ended and semantic-differential questions, in conjunction with a self-assessment mannequin. Physiological measures (skin conductance, posture, eye movements, EEG, EKG) relevant to learning are being developed (Mota and Picard, 2003; Lu and Graesser, in press; Jung, Makeig, Stensmo, and Seinowski, 1997).

Discourse analysis has been another important method for naturalistic study of emotion during museum visits. Allen (2002), for example, coded visitors' spontaneous articulations of their emotions using three categories of affect: positive, negative, and neutral. Both spontaneous comments and comments elicited by researchers have similarly been coded to show differences in emotional response during museum visits. Clipman (2005), for example, used the Positive and Negative Affect Schedule to show that visitors leaving a Chihuly exhibit of art glass reported being more happy and inspired than visitors to a quilting exhibit in the same museum (Clipman, 2005). Myers, Saunders, and Birjulin (2004) used Likert and semantic-differential measures to show that zoo visitors had stronger emotional responses to gorillas than other animals on display. Raphling and Serrell (1993) asked visitors to complete the sentence "It reminded me that . . ." as a part of an exit questionnaire for exhibitions on a range of topics, and they reported that this prompt tends to elicit affective responses from visitors, including wonderment, imagining, reminiscences, convictions, and even spiritual connection (such as references to the power of God or nature).

In studies of informal learning, interest and related positive affect are also often inferred on the basis of behavior displayed. That is, participants who seem engaged in informal learning activities are presumed to be interested. In this sense, interest and positive affect are often not treated as outcomes, but rather as preconditions for engagement. Studies that document children spontaneously asking "why" questions, for example, take as a given that children are curious about, interested in, and positively predisposed to engaging in activity that entails learning about the natural world (e.g., Heath, 1999). Studies that focus on adult behavior, such as engaging in hobbies, are predicated on a similar assumption—that interest can be assumed for the people and the context being studied (e.g., Azevedo, 2006). A meta-analysis of the types of naturally occurring behavior thought to provide evidence of individuals' interest in informal learning activities could be useful for developing systematic approaches to studying interest. Such an analysis also could be useful in showing how interest is displayed and valued among participants in informal learning environments, providing an understanding of interest as it emerges and is made meaningful in social interaction.

## Strand 2: Understanding Science Knowledge

### Nature of the Outcome

As progressively more research shows, learning about natural phenomena involves ordinary, everyday experiences for human beings from the earliest ages (National Research Council, 2007). The types of experiences common across the spectrum of informal environments, including everyday settings, do more than provide enjoyment and engagement: they provide substance on which more systematic and coherent conceptual understanding and content structures can be built. Multiple models exist of the ways in which scientific understanding is built over time. Some (e.g., Vosniadou and Brewer, 1992) argue that learners build coherent theories, much like scientists, by integrating their experiences, and others (e.g., diSessa, 1988) argue that scientific knowledge is often constructed of many small fragments that are brought to mind in relevant situations. Either way, small pieces of insight, inferences, or understanding are accepted as vital components of scientific knowledge-building.

Most traditionally valued aspects of science learning fall into this strand: models, fact, factual recall, and application of memorized principles. These aspects of science learning can be abstract and highly curriculum-driven; they are often not the primary focus of informal environments. Assessments that focus on Strand 2 frequently show little or no positive change of Strand 2 outcomes for learners. However, there are several studies that have shown positive learning outcomes, suggesting that even a single visit to an informal learning setting (e.g., an exhibition) may support development or revision of knowledge (Borun, Massey, and Lutter, 1993; Fender and Crowley, 2007; Guichard, 1995; Korn, 2003; McNamara, 2005).

At the same time, studies of informal environments for science learning have explored cognitive outcomes that are more compatible with experiential and social activities: perceiving, noticing, and articulating new aspects of the natural world, understanding concepts embedded in interactive experiences, making connections between scientific ideas or experiences and everyday life, reinforcing prior knowledge, making inferences, and building an experiential basis for future abstractions to refer to. Informal experiences have also been shown to be quite memorable over time (see, e.g., Anderson and Piscitelli, 2002; Anderson and Shimizu, 2007).

While the knowledge of most learners is often focused on topics of personal interest, it is important to note that most people do not learn a great deal of science in the context of a single, brief "treatment." However, this ought not to be considered an entirely negative finding. Consider that learning in school is rarely assessed on the basis of a one- or two-hour class, yet science learning in informal environments is often assessed after exposures that do not exceed one to two hours. Falk and Storksdieck (2005) found that a single visit to an exhibition did increase the scientific content knowledge of at least one-third of the adult visitors, particularly those with low prior knowledge. However, even participants whose learning is not evident in a pre-post design may take away something important: The potential to learn later—what *How People Learn* refers to as preparation for future learning (National Research Council, 2000). For example, visitors whose interest is sparked (Strand 1) presumably are disposed to build on this experience in the months that follow a science center visit by engaging in other informal learning experiences.

## Methods of Researching Strand 2 Outcomes

Outcomes in this category can be the most "loaded" for learners. If not carefully designed, assessments of content knowledge can make learners feel inadequate, and this throws into question the validity of the assessment, going against the expectations of learners in relation to norms of the setting and the social situation. The traditional method for measuring learning (or science literacy) has been to ask textbook-like questions and to judge the nearness of an individual's answer to the expert's version of the scientific story. In terms of what researchers know about the nature of learning, this is a limited approach to documenting what people understand about the world around them. This outcome category is also vulnerable to false negatives, because cognitive change is highly individual and difficult to assess in a standardized way. An essential element of informal environments is that learners have some choice in what they attend to, what they take away from an experience, what connections they make to their own lives. Consequently, testing students only on recall of knowledge can cause researchers to miss key learning outcomes for any particular learner, since these outcomes are based on the learner's own experience and prior knowledge.

To avert the ethical, practical, and educational pitfalls related to assessing content knowledge, many researchers and evaluators working in informal environments put effort into generating assessments that have nonthreatening content, a breadth of possible responses, comfortable delivery mechanisms, a conversational tone, and appropriateness to the specific audience being targeted. Also, these assessments leave room for unexpected and emergent outcomes. Questions asked with an understanding of the ways in which people are likely to have incorporated salient aspects of a scientific idea into their own lives appropriately measure their general level of science knowledge and understanding. Yet we also acknowledge that while such measures are well aligned with the goals of informal environments, they lack objectivity of standardized measures.

An important method for assessing scientific knowledge and understanding in informal environments is the analysis of participants' conversations. Researchers interested in everyday and after-school settings study science-related discourse and behavior as it occurs in the course of ordinary, ongoing activity (Bell, Bricker, Lee, Reeve, and Zimmerman, 2006; Callanan, Shrager, and Moore, 1995; Sandoval, 2005). Researchers focused on museums and other designed environments have used a variety of schemes to classify these conversations into categories that show that people are doing cognitive work and engaging in sense-making. The categories used in these classification schemes have included: identify, describe, interpret/apply (Borun, Chambers, and Cleghorn, 1996); list, personal synthesis, analysis, synthesis, explanation (Leinhardt and Knutson, 2004); perceptual, conceptual, affective, connecting, strategic (Allen, 2002); and levels of metacognition (Anderson and Nashon,

2007). Most of these categorizations have some theoretical basis, but they are also partly emergent from the data.

A great deal of research has been conducted on the new information, ideas, concepts, and even skills acquired in museums and other designed settings. Some museum researchers have measured content knowledge using think-aloud protocols. In these protocols, a participant goes through a learning experience and talks into a microphone while doing it. O'Neil and Dufresne-Tasse (1997) used a talk-aloud method to show that visitors were very cognitively active when looking at objects, even objects passively displayed. The principal limitation of this method is that it is likely to disrupt the learning processes to some degree, not least of which is the elimination of conversation in a visiting group. Beaumont (2005) used a variation of this technique with whole groups by inviting families to think aloud "when appropriate" during their visit to an exhibition. When studying children, clinical interviews may be helpful for eliciting the ways in which they think about concepts embedded in exhibits, as well as the ways in which their understanding may be advanced or hindered. For example, Feher and Rice (1987, 1988) interviewed children using a series of museum exhibits about light and color, to identify common conceptions and suggest modifications to the exhibit.

Several methods are used to elicit the concepts, explanations, arguments, models and facts related to science that participants generate, understand, and remember after engaging in science learning experiences. These include structured self-reports, in the form of questionnaires, interviews, and focus groups (see Appendix B for a discussion of individual and group interviews). Self-reports can be used to assess understanding and recall of an individual's experiences, syntheses of big ideas, and information that the respondent says he or she "never knew." For example, a summative evaluation of Search for Life (Korn, 2006) showed that visitors had understood a challenging big idea (that the search for life on other planets begins by looking at extreme environments on Earth that may be similar) and also showed they had not thought deeply about issues regarding space exploration or life on other planets. Researchers also sometimes engage visitors to museums, science centers, and other designed environments in conversations; they ask them to talk about their experience in relation to particular issues of interest to the institution to better understand the overlap between the agendas of the institution's staff and the visitors. For example, for each of an exhibition's five primary themes, Leinhardt and Knutson (2004) gave visitors a picture and a statement and coded the ensuing discussion as part of their assessment of learning in the exhibition. Rubrics have been used to code the quality of visitors' descriptions of a particular topic or concept of interest. Perry called these "knowledge hierarchies" (1993) and used them to characterize both baseline understandings and learning from an exhibition. One important underlying assumption in this research is the relationship between thought and language. However, mapping the relationship between language and thought is complex and not fully developed.

Several types of learning outcomes assessments used in museums and other designed spaces engage participants in activities that require them to demonstrate what they learned by producing a representation or artifact. Concept maps are often used to characterize an individual's knowledge structure before and after a learning experience. They are particularly well suited to informal environments in that they allow for personalization of both prior knowledge and knowledge-building during the activity and are less threatening than other cognitive assessments. However, they require a longer time commitment than a traditional exit interview, are time-consuming to code, are difficult to administer and standardize, and may show a bias unless a control group has been used (see Appendix B). While a variety of concept mapping strategies have been used in these settings (Anderson, Lucas, Ginns, and Dierking, 2000; Gallenstein, 2005; Van Luven and Miller, 1993), perhaps the most commonly used in museum exhibitions is Personal Meaning Mapping (Falk, Moussouri, and Coulson, 1998), in which the dimensions of knowledge assessed are extent, breadth, depth, and mastery. Personal Meaning Mapping is typically presented to learners in paper format, although Thompson and Bonney (2007) created an online version to assess the impact of a citizen science project.

Drawing tasks can be an important way to broaden research participants' modes of communication and may enable some to articulate ideas and observations that they could not in spoken or written language. Drawings can capture visitors' memories of their experience (e.g., map study), or show their understanding of a science concept (Guichard, 1995). Typically, a drawing is annotated or discussed so that the meaning of the various parts is clear to the researcher. Moussouri (1997) has shown how drawings can be used to capture different stages of children's reasoning. Jackson and Leahy (2005) have similarly used drawing and creative writing tasks to study how a museum theater experience may influence children's learning.

Sorting tasks, which typically involve cards, photos, or other objects, are yet another means through which participants can demonstrate their conceptual learning after visiting a museum, zoo, or other designed setting. To be compelling proof of learning, this method requires some kind of control group and preferably also a pretest. Sorting tasks have the advantage that they do not publicly reveal that a given answer is scientifically incorrect and can usually be done with the same participants more than once.

E-mail or phone interviews, often done weeks, months, or even years after a visit or program, are particularly important in informal learning environments because they are often the only way to test two key assumptions: (1) that the experiences are highly memorable and (2) that learners integrate the experiences into the rest of their lives and build on them over time. Typical follow-up questions probe these two aspects of the learning by asking what the participants remember about their experience and what they have done in relation to the content since. For example, Falk, Scott, Dierking, Rennie, and Cohen-Jones (2004) used follow-up interviews to explore how the cognitive outcomes of a visit to a museum varied over time. Anderson and Shimizu (2007) showed that many people remembered details of what they had done at a world's fair or exposition decades previously, and Allen (2004) found that it was not unusual for visitors to say that a single exhibit experience changed the way they think about something in their lives. Spock (2000) lists some of the trade-offs of doing follow-up interviews soon versus long after the event and points to the connection between more profound potential outcomes and a longer time frame.

When learners are participating in an extended program (e.g., docents or watchers of a TV series), it may be feasible to conduct pre- and posttests of conceptual learning, similar to those used in schools, to test their learning of formal concepts. For example, Rockman Et Al (1996) used a series of multiple-choice questions to show that children who watched *Bill Nye the Science Guy* made significant gains in understanding that Bernoulli's principle explains how airplanes fly. Another means by which researchers have assessed learning over extended time frames is by asking participants to write reflections in a journal, possibly to discuss with others and to share with researchers. Leinhardt, Tittle, and Knuston (2002) used this method to showcase the deep connections and knowledge-building done by frequent museum-goers.

## Strand 3: Engaging in Scientific Reasoning

### *Nature of the Outcome*

This strand focuses on the activities and skills of science—including inquiry and reasoning skills, which are intimately related and often explored in research simultaneously with conceptual knowledge. However, we focus here on the ways in which researchers go after activities and skills of science specifically. Informal environments often provide opportunities for learners to engage in authentic inquiry using a range of resources, without pressure to cover particular content, yet with access to engaging phenomena and staff ready to support them in their own explorations and discoveries. The outcomes in this strand include scientific inquiry skills, such as asking questions, exploring, experimenting, applying ideas, predicting, drawing conclusions from evidence, reasoning, and articulating one's thinking in conversation with others. Other outcomes are skills related to learning in the particular informal environment: how to use an interactive exhibit, how to navigate a website, how to draw relevant information from a large body of text, how to learn effectively with others of different skill levels—sharing resources, teaching, scaffolding, negotiating activity.

## Methods of Researching Strand 3 Outcomes

Developmental studies based on observations of children's spontaneous behavior show that their approach to natural phenomena shows similarities to science: exploratory, inquiry-oriented, evidence-seeking (Beals, 1993; Callanan and Oakes, 1992). Controlled studies result in similar findings, indicating that everyday thinking entails reasoning about causality and complex relations among variables as discussed in Chapter 4.

This strand of outcomes is almost always assessed by examining the participant's learning process rather than a pre-post measure of outcome. This is because the only way to do a pre-post measurement requires that learners demonstrate what they are able to do in the "pre" condition. Pretesting requires that learners be put on the spot in a manner that is inconsistent with the leisure-oriented and learner-centered nature of most informal environments. Instead, skills are usually assessed as they are practiced, and the assumption is made that practicing a skill leads to greater expertise over time.

Research focused on assessing practical and discursive inquiry skills in informal environments often rely on video and audio recordings made during activities that are later analyzed for evidence of such skills as questioning, interpreting, inferring, explaining, arguing, and applying ideas, methods, or conjectures to new situations (see Appendix B for a discussion of video- and audiotaping). For example, Humphrey and Gutwill (2005), analyzing the kinds of questions visitors asked each other and the ways they answered them, found that visitors using "active prolonged engagement" exhibits asked more questions that focused on using or understanding the exhibits than visitors using the more traditional planned discovery exhibits. Randol (2005), assessing visitors' use of scientific inquiry skills at a range of interactive exhibits, found that the inquiry could be characterized equally well by holistic measures or small-scale behavioral indicators (such as "draws a conclusion") as long as the sophistication of the behaviors was measured rather than their number. Meisner et al. (2007) and vom Lehn, Heath, and Hindmarsh (2001, 2002) studied short fragments of video to reveal the ways in which exhibits enable particular forms of coparticipation, modeling, and interactions with strangers. Researchers have used video analysis to investigate a large range of behaviors related to how learners make sense of the natural and physical world, including interacting appropriately with materials and showing others how to do something. Stevens and colleagues (Stevens, 2007; Stevens and Hall, 1997; Stevens and Toro-Martell, 2003) used a video annotation system on the museum floor to prompt visitors to reflect on how they and others interacted with an interactive science display, leaving a durable video trace of their activity and reflections for others to explore and discuss as they come to the display. The traces then serve as data for subsequent interactional analysis of learning.

Researchers have also asked learners after participation in science learning activities to provide self-reports of their own (or each other's) skill levels. Sometimes museum visitors will spontaneously report that they or a member of their group (typically a child) learned a new skill while participating in the activity. While this approach wants for direct evidence to back up such claims, it may be the only kind of evidence of a change in skill level that can be collected given the social norms of many environments and, in certain cases, without risking discomfort for participants. Although it may be possible to pretest the skill levels of learners in certain settings, in general such testing is a high-risk assessment practice for informal environments. Campbell (2008) points out the dangers of doing this in youth programs, in which learners may experience themselves as failing and consequently never return.

## Strand 4: Reflecting on Science

### *Nature of the Outcome*

A fundamental goal of science education is to improve learners' understanding of what science is—that is, to increase understanding of the nature of the scientific enterprise. The outcomes targeted in this strand address issues related to how scientific knowledge is constructed, and how people, including the learner herself, come to know about natural phenomena and how the learner's ideas change. Direct experience with the process of knowledge construction through the types of inquiry-based activities characteristic of informal environments can serve as an important point of departure for the outcomes in this strand: recognizing that people are involved in the interpretive aspects of evaluating theories, evidence, and the relationship between the two; that scientific knowledge is uncertain and changeable; and that a diversity of strategies and methods are employed in scientific research.

Whether or not a person becomes a professional scientist, the forms of scientific understanding associated with Strand 4 outcomes are considered by many to be crucial for having an informed citizenry given public debates about political issues related to science (American Association for the Advancement of Science, 1993). Although lay people will always rely on the work of professional scientists, a view of scientific knowledge as fundamentally constructed from evidence rather than merely factual or received from authoritative sources can provide a critical stance from which the public can evaluate claims in relation to evidence (Brossard and Shanahan, 2003; Miller, 2004). Presumably, such a public can thereby make better judgments about public policy related to such issues as global warming or the teaching of intelligent design. The body of research on the topic indicates that young children, youth, and even adults do not have a strong understanding of the nature of science per se and what is entailed by disciplinary methods of knowing and learning (Osborne et al., 2003).

There is evidence that such limits in understanding derive from a lack of exposure to appropriate opportunities to learn in these areas (American Association for the Advancement of Science, 1993). When people are provided with opportunities to learn about the problematic nature of scientific knowledge construction (Smith, Maclin, Houghton, and Hennessey, 2000), to understand the processes of modeling and testing (Penner, Giles, Lehrer, and Schauble, 1997), or to reflect on or explicitly investigate epistemological issues (Bell and Linn, 2002), their understanding of the nature of scientific practice, process, and knowledge improves. Research into practical or everyday epistemologies provides some preliminary evidence suggesting that informal environments provide appropriate opportunities for learning about the nature of science (Sandoval, 2005). The degree to which they promote these outcomes has not been heavily researched, but the inquiry-oriented experiences afforded by most informal environments may provide cultural and educational resources for promoting better understanding of the nature of science.

### *Methods of Researching Strand 4 Outcomes*

Studies regarding conceptions of the nature of science, typically using either questionnaires or structured interview protocols, have been conducted in schools, often with the aim of drawing relationships between children's conceptions of what real scientists do and their own classroom activities (Abd-El-Khalick and Lederman, 2000; Bartholomew, Osborne, and Ratcliffe, 2004; Schwartz and Lederman, 2002). These studies generate information about children's epistemological reasoning that ostensibly reflects how they individually think about the nature of knowledge and warrants for claims, regardless of the activity setting. In their study of 9-, 12-, and 16-year-olds, Driver, Leach, Millar, and Scott (1996), for example, used interview data based on specific probes to identify three levels of reasoning about the nature of science. According to their analysis, at the lowest level, students' reasoning is grounded in phenomena; at the mid-level, students reason about the relationships between quantities or variables; and at the highest level, students reason with and about imagined models. Interestingly, the researchers were able to engage only the 16-year-olds in discussions of science as a social enterprise. Similar studies show the difficulty with which younger adolescents and children conceive of science as a social process (Abd-El-Khalick and Lederman, 2000; Bartholomew et al., 2004; Schwartz and Lederman, 2002).

Some researchers have specifically tried to link conceptions of science and scientific practice to the learning setting (Bell and Linn, 2002; Carey and Smith, 1993; Hammer and Elby, 2003; Rosenberg, Hammer, and Phelan, 2006; Sandoval, 2005; Songer and Linn, 1991). Sandoval specifies four types of difficulties students have understanding the constructed and changeable nature of disciplinary science in the school setting, positing "practical epistemologies" that inhere in the organizational structures of institutions and activities rather than trait-like or stage-like personal epistemologies that belong to individuals. Sandoval argues, in effect, that the practical epistemologies at play in everyday settings allow students to take a more self-reflective and nuanced view of scientific process.

## Strand 5: Engaging in Scientific Practices

### Nature of the Outcome

This strand builds on and expands the notion of participation discussed in *Taking Science to School* (National Research Council, 2007). In that report participation meant learners participating in normative scientific practices akin to those that take place in and govern scientific work. For example, whereas young learners may understand argumentation in a range of contexts outside of science (e.g., resolving conflicts at home or on the playground), they typically must learn how to argue in scientific ways (e.g, using evidence to support claims). Participating in science meant, among other things, appropriating scientific ways of arguing. As that report established, there is a substantial body of evidence that illustrates how even young learners can develop the knowledge, skills, and commitments necessary to participate in a classroom scientific culture. That literature also indicates that learning to participate in science requires that learners have copious opportunities to do science plus substantial instructional support over long periods of time. An important difference in the construal of participation in this report is that we are focusing on nonschool settings where the development of shared norms and practices is typically not afforded by the goals and constraints of the educational experience. Thus, we take a broader and admittedly somewhat less clearly defined view of participation in order to capture important ways in which informal environments can contribute to this goal.

Participation in informal learning environments is generally voluntary at many scales (coming to an event, staying for its duration, using an exhibit thoroughly or repeatedly, returning to more events, etc.). By analogy with measuring time spent, attendance can be used as a measure of learning, either as a necessary minimal condition or as an indicator (assuming learning increases with number of returns or as a direct assessment of learning as participation in a community). For this reason, environments for science learning pay particular attention to keeping track of the demographics, motivations, and expectations of the people who arrive and return to use their educational offerings. St. John and Perry (1993) take this argument to a much broader scale, arguing that the entire infrastructure of environments for science learning should be assessed, at least in part, on the basis of its voluntary usage by the public as a learning resource.

A common goal across informal contexts is for participants to experience pleasure while working with tasks that allow exploration and do not overwhelm (e.g., Allen, 2004; Martin, 2004). The objective is for participants to have conversations, explore, and have fun in and around science. The expectation is that participation in informal contexts involves learning science and that science learning will follow. In other words, if there is participation, then learning is assumed to be occurring (see Lave, 1996); if there is enjoyment, then return to science and possible identification with science is anticipated. Recent work by Falk et al. (2007) suggests that visitors to zoos and aquariums who already identify themselves as participants in science learning anticipate that their visits will enhance and strengthen this identity—which appears to be the case.

While short-term participation in well-defined programs is relatively easy to assess, long-term and cumulative progressions are much more challenging to document, due primarily to the difficulties of tracking learners across time, space, and range of activity. Nevertheless, researchers must accept this challenge, because a key assumption in the field (e.g., Crowley and Jacobs, 2002) is that effective lifelong learning is a cumulative process that incorporates a huge variety of media and settings (everyday life in the home, television, Internet, libraries, museum programs, school courses, after-school programs, etc.). Thus, longitudinal studies are particularly useful.

In assessing Strand 5 outcomes, culturally responsive evaluation techniques help to maximize validity, since members of a community may identify their levels of participation in quite different ways from researchers who may be outside it. For example, in a study by Garibay (2006) researchers had to broaden their definitions of "parent involvement" to fit the norms of a community they were unfamiliar with.

### Methods of Researching Strand 5 Outcomes

Because learner choice is such a key element in most informal learning environments and the extent to which learners engage in science over time is a key element of learning to participate in science, data on who enrolls in a program, attends an event or offering, joins science clubs and related affinity groups, or uses websites or other forms of media or tools for science learning is important to track. Often, researchers collect demographic data (e.g., Diamond, 1999) in conjunction with attendance data. Collecting accurate data on participation, especially degrees of participation, is notoriously difficult in many informal settings, such as after-school programs and community-based organizations (Chaput, Little, and Weiss, 2004).

To study participation at a finer scale, researchers interested in designed settings—museums, science centers, community gardens, and other community-based organizations—record the detailed movements of visitors through a public space or exhibit, showing their degree of engagement throughout the area as well as the relative attracting and holding powers of the individual designed elements (see Appendix B for a discussion of holding time). Although tracking studies have been done for nearly a century (Robinson, 1928; Melton, 1935), Serrell's (1998) meta-analysis served to standardize some of the methods and definitions, including a "stop" (planting the feet and attending to an exhibit for at least 2-3 seconds), a "sweep rate" (the speed with which visitors move through a region of exhibits), and a "percentage of diligent visitors" (the percentage of visitors who stop at more than half of the elements). It also suggests benchmarks of success for various types of exhibit format (dioramas, interactives, etc.).

Some researchers have modified the traditional "timing and tracking" approach, creating an unobtrusive structured observation based on holistic measures. These measures recognize that although the amount of time spent in an exhibition is a good quantitative indicator of visitors' use of a gallery space or exhibit element, it often poorly reflects the quality of their experience with an exhibition. Therefore, to complement quantitative measures, researchers have developed a ranking scale with which they can assess the quality of interactions that visitors have in various sections of an exhibition or at specific exhibit components (Leinhardt and Knutson, 2004). The scale involves time to some degree but not solely.

Participants' submissions to websites, through comment cards, and even via visitor guest books provide evidence that learners are willing and able to participate in a dialogue with the institution or people who generated the learning resource. Feedback mechanisms have become well established in museums and have been increasingly displayed openly rather than collected through a comment box or other means for staff to review privately. These methods have been assisted by the development of technological systems for automatically caching and displaying a select number of visitor responses, as well as wiki models of distributed editing. For example, the Association of Science-Technology Centers hosts ExhibitFiles, a community site for designers and developers to share their work; the Liberty Science Center has created Exhibit Commons, a website that invites people to submit contributions for display in the museum; and the Tech Museum of Innovation is using Second Life as an open source platform for exhibit design, with plans to replicate some of the best exhibits in its real-world museum.

These means of collecting data may be useful for research as well as for institutional and practical reasons, so it is important to be clear when they are appropriately construed in a science learning framework. Showing up is important and the scale of research of informal learning institutions speaks to their capacity, but making claims about participation in science is not the same as making claims about how many people passed through a particular setting.

Issues of accessibility are important when assessing participation rates in informal environments. Participation may be reduced because activities or environments are inaccessible to some learners, physically or intellectually. Reich, Chin, and Kunz (2006) and Moussouri (2007) suggest ways to build relationships with museum visitors with disabilities who can serve as testers or codevelopers, as well as techniques for conducting interviews with these audiences in particular, to determine participatory outcomes. Similarly, Garibay (2005) suggests ways to design assessment techniques to be culturally responsive to a target audience, even for a single activity.

Ways of assessing participation in media-based activity vary. Web resource usage can be assessed by number of users, duration of use, pages viewed, path of exploration, and entry points from other sites (e.g., Rockman Et Al, 2007). Surveys are used to assess broadcast audiences for TV and radio. Ways to assess depth of participation or integration of experiences are especially important, and these methods are varied. One aspect of progression in an activity is personal ownership and creativity—that is, not just going through the motions of a predefined activity but creating something original in it. For example, Gration and Jones (2008) developed a coding scheme for innovation. Others have focused on evidence of creativity or self-initiated activity. To document participation across settings, events, media, and programs, Ellenbogen (2002) conducted case studies showing examples of families who use many resources in a highly integrated fashion.

Some researchers have investigated extended engagement in science practices by studying home discussions or activities related to science. For example, Ellenbogen showed that frequent users of a science museum continued their discussions and activities in the home and other settings, engaging in integrated, multisetting learning. Other researchers have taken a prospective approach to studying anticipated actions. Clipman (2005) has designed and tested a Visit Inspiration Checklist that asks visitors to anticipate what actions they might take following their visit, including further resources they might use, connections they might make, and activities they might undertake to extend their experience.

Taking a longitudinal approach to data collection allows researchers to get a more complete picture of the role of these learning experiences in peoples' lives. Researchers have repeatedly shown that many of the conversations that begin in the museum continue once families are back at home (see Astor-Jack et al., 2007).

Ethnographic case studies that involved a long-term relationship between the researcher and a set of families who visited museums frequently, allowing for repeated observations and interviews before, during, and after museum visits (Ellenbogen, 2002, 2003), have suggested that conversational connections between museum experiences and real-world contexts are frequent yet must be examined carefully, since the connections are not always obvious to those outside the family. Perhaps the most important and interesting work on participatory structures in informal environments is ethnographic, allowing for an analysis of particular discourse practices in relation to cultural norms and meanings that are enacted in the setting (Rogoff, 2003; McDermott and Varenne, 1995).

## Strand 6: Identifying with the Scientific Enterprise

Scientific identity typically refers to a person's concept of herself as a potential scientist (Brickhouse, Lowery, and Schultz, 2000, 2001; Calabrese Barton, 2003). Research in this strand also pertains to the ways in which people experience and recognize their own agency in relation to activities associated with learning or doing science (Holland, Lachicotte, Skinner, and Cain, 1998; Hull and Greeno, 2006). Identity is often equated with a subjective sense of belonging—to a community, in a setting, or in an activity related to science. The changes in community affiliation and related behaviors that can signal changes in identity usually require extended time frames of involvement with a program or community (e.g., Beane and Pope, 2002; McCreedy, 2005). Brossard, Lewenstein, and Bonney (2005) showed that citizen scientists not only increased their knowledge, but also were able to suggest revisions to scientists' protocols when they did not work. Identity changes often are reflected in the behaviors of others in the learners' lives, such as parents, caregivers, and the institutional staff involved.

A sense of agency or belonging can be experienced retrospectively when reflecting on past events, it can be experienced in relation to current activities, and it can be projected into the future through imaginative acts regarding what one might become. To a greater or lesser degree, identity can be more a matter of embodied experience than of explicit labels for what someone can do or who one is. A child, for example, may engage fluently and comfortably with her family's gardening practices, yet not think of herself or be referred to by others as a gardener, a budding botanist, etc. Another might gain qualitative understandings of Newtonian mechanics based on observations of everyday phenomena, and, as a consequence, engage in activities that build on this understanding, but not make explicit associations to various possible labels relating to her capabilities.

Although researchers in the field generally agree that identity affects science participation and learning (National Research Council, 2007; Leinhardt and Knutson, 2004; Falk, 2006; Anderson, 2003), there are varied and disparate theoretical frameworks that address issues of identity. Some conceptions of identity emphasize personal beliefs and attitudes, for example, measured by the degree to which participants endorse such statements as "I have a good feeling toward science" or "I could be a good scientist" (Roth and Li, 2005; Weinburgh and Steele, 2000). Other conceptions of identity focus on the way that identity is created through talk and other features of moment-to-moment interactions that position people among the roles and statuses available in particular situations (Jacoby and Gonzales, 1991; Brown, Reveles, and Kelly, 2004; Hull and Greeno, 2006; Holland, Lachicotte, Skinner, and Cain, 1998; Holland and Lave, 2001; Rounds, 2006). This latter conception emphasizes that the type of person one can be in a setting—e.g., competent, skilled, creative, or lacking in these qualities—both depends on the way these types are defined in social context and determines the possible identities someone can have. The ways that people interact with material resources (e.g., instruments, tools, notebooks, media) and other participants (e.g., through speaking, gesture, reading, writing) combine to assign individuals to the available identities (Hull and Greeno, 2006; Jacoby and Gonzales, 1991; Brown, Reveles, and Kelly, 2004).

There seems to be a strong relationship between science-related identity and the kinds of activities people engage in, usually with others. Gutiérrez and Rogoff (2003), for example, emphasize the repertoires of practice (ways of participating in activities) that people come to know through participation in diverse communities, each with its own goals, needs, routines, and norms. These repertoires of practice serve as resources and help define who a person is, in terms of their social identity, in any given situation. Brown's research (2004) demonstrates the links between communication practices and the building of scientific identity, charting the complexities of negotiating between in-school and out-of-school practices and identities. Hull and Greeno (2006) describe identity changes for workers in a circuit board factory that co-occurred with the introduction of a new system of participation, symbolic representation, communication, management, and personal recognition at the site. This body of work illustrates the importance of considering the practical, experiential, and embodied aspects of scientific identity. Generally, the research on scientific identity emphasizes the opportunities that learners have to encounter and make use of the ideas, images, communities, resources, and pathways that can lead to progressively greater involvement in the practices of science.

### Methods of Researching Strand 6 Outcomes

In many cases, research on scientific identity has relied on questionnaires and structured interviews regarding beliefs about oneself, one's experiences, and the supports for science learning that exist in one's school and community (Barron, 2006; Beane and Pope, 2002; Moore and Hill Foy, 1997; Schreiner and Sjoberg, 2004; Weinburgh and Steele, 2000). Longer term studies focusing on changes in behavior or community affiliation have also been conducted using self-report measures based on questionnaires and structured interviews (Fadigan and Hammrich, 2004; Falk, 2008; Gupta and Siegel, 2008). In settings where long-term participation has led to evidence of changes in learners' identity, parents, caregivers, and the institutional staff have provided self-reports on how these changes were related to their own perceptions and behaviors (Barron, 2006; McCreedy, 2005; Falk, 2008). Studies of increasing levels of involvement and interest have included questionnaires, interviews, ethnographic methods, and analysis of learner artifacts (e.g., Barron, 2006; Bell et al., 2006; Brown, 2004; Nasir, 2002; Warren, Ballenger, Ogonowski, Rosebery, and Hudicourt-Barnes, 2001). Zoos and aquariums, which are particularly interested in documenting behavior change related to conservation and the environment, typically question visitors about their intended behaviors, following up with phone calls or Internet-based interviews.

The effect of science experience on career choice for children is a major Strand 6 outcome, but it is also very difficult to assess because the time frame involved is so long. Logistical difficulties include tracking individuals, securing long-term funding, and the many intervening factors that can alter the research plan (Allen et al., 2007). In most circumstances, it may be more feasible to look at the immediate choices that lead toward a potential science career, such as choice of school courses, after-school activities, reading material, games and hobbies, and the like. Some researchers have capitalized on extant datasets to conduct longitudinal analyses. In looking at career paths of youth first questioned in middle school and then followed into their adult lives, Tai, Liu, Maltese, and Fan (2006) document the importance of career expectations for young adolescents and suggest that early elementary experiences (before eighth grade) may be of importance. This research also supports the idea that the labels or plans people appropriate for themselves may be an important motivator for participation in activities associated with the label. Sachatello-Sawyer et al. (2002) suggest that being labeled a "museum lover" motivates attendance for adult program participants.

## PERSPECTIVES, DIRECTIONS, AND CONCLUSIONS

The outcomes discussed in this chapter represent a broad view of the ways in which practitioners and researchers characterize and measure the effects of science learning experiences. The six strands cover a wide range of approaches to studying and understanding individual learning, from those most focused on cognitive and conceptual change to those most focused on shifts in participation and identity. Although there is a diversity of thought in the informal science learning community about what outcomes are most important and what means of measurement are most appropriate, a rough and emerging consensus exists around some core assumptions about the nature of informal science learning outcomes.

**Outcomes can include a broad range of behaviors.** We have noted many of the key types of individual outcomes investigated. This kind of research could be designed to allow for varied personal learning trajectories and outcomes that are complex and holistic, rather than only those that are narrowly defined.

**Outcomes can be unanticipated.** Outcomes can be based on the goals and objectives of a program (and therefore closely tied to its design), or they can be unplanned and unanticipated, developing contingently on the basis of what is most valuable to the participant. In informal settings, outcomes are often guided by the learners themselves. Research can target outcomes that emerge in these experiences, not only those that are defined a priori.

**Outcomes can become evident at different points in time.** Short-term outcome measures have long been used to assess the impact of informal learning experiences, but these experiences can also have enduring, long-term impacts that differ from the short-term ones.

**Outcomes can occur at different scales.** Outcomes defined on the level of individual participants answer the question: How is the individual influenced by the experience? Most of the outcomes discussed in this chapter and in the literature generally focus at this level. But it is also useful to ask: How is the entire social group in the environment influenced? For example, did group members learn about one another, reinforce group identity and history, or develop new strategies for collaborating together? We can also define outcomes on the community scale: How does the activity, exhibition, or program influence the local community?

These assumptions regarding outcomes align with three high-level criteria that the evidence suggests are essential in the development of assessments appropriate for science learning in informal environments. First, the assessments must address not only cognitive outcomes, but also the range of intellectual, attitudinal, behavioral, social, and participatory capabilities that informal environments effectively promote (Jolly, Campbell, and Perlman, 2004; Hein, 1998; Schauble et al., 1995; Csikszentmihalyi and Hermanson, 1995). Second, assessments should fit with the kind of participant experiences that make these environments attractive and engaging; that is, any assessment activities undertaken in informal settings should not undermine the features that make for effective learning there (Allen, 2002; Martin, 2004). Third, the assessments used must be valid, measuring what they purport to be measuring—that is, outcomes from those science learning experiences (National Research Council, 2001).

Assessment must also be valid in terms of construct validity—that it measures what it purports to measure—and in terms of the ecological validity—that it aligns with the opportunities for learning that are present in the learning environment (Moss et al., in press). In light of the tendency to use conventional academic outcomes to study learning in informal settings, it is important for researchers and practitioners to carefully consider ecological validity of such measures for informal settings. Measures must ensure that the same kinds of material, social, cognitive, and other features of the activities designed to promote learning in an informal setting should be part of the assessment, serving as cues for activating the capabilities and dispositions that participants have or might have learned. Before drawing conclusions about whether the informal experiences have led to particular outcomes, researchers and practitioners should ask themselves: Are the assessment activities similar in relevant ways to the learning activities in the environment? Are the assessments based on the same social norms as those that promote engagement in the learning activities? Overall, is it clear that learners in a setting have had ample opportunity to both learn and demonstrate desired outcomes? Without such clarity, it is difficult to make fair inferences about what has been learned or the effectiveness of the environment for promoting learning.

To a significant extent, the ability to answer these questions depends on how well the research community is able to describe the nature of participants' experience in particular types of informal learning environments, with an eye to eventually understanding what is consistent and systematic across these environments. An in-depth understanding of key features of the environments (e.g., what are the physical and social resources? What are the norms of behavior?), ways in which learning is framed or organized (e.g., what activities are presumed to lead to learning? How is learning supported? What does it mean to be knowledgeable in this setting?), and the capacities being built (e.g., what skills, knowledge, or concepts are learners engaging with?) can lead to critical insights regarding the particular contributions of informal experiences to science learning, therefore highlighting the outcomes one would most expect and want to see.

As important as it is to document the unique and valuable contributions of informal opportunities for learning, there is a tension in the field regarding the degree to which one can or should try to direct outcomes. On one hand, the field has an overarching commitment to valuing the great diversity of ways in which informal learning experiences can positively affect participants. Researchers and practitioners are receptive to acknowledging the many types of outcomes, anticipated or not, that emerge from the interplay of people and resources as they engage in science learning activities. This receptivity to contingencies, George Hein explains, is "a matter of ideology" (1995, p. 199).

> By framing the questions as we do, we leave ourselves open for the broader responses, for noting unexpected behaviors, and we do not shut out the possibility of documenting learning that is distinct from the teaching intended. By leaving our list of issues deliberately vague and general, we do not exclude the possibility of learning something about the . . . experience that may be outside the framework of . . . expectations.

Hein's formulation suggests that informal environments are oriented toward providing learning experiences that are relevant to the interests and needs of the people they serve. One can argue then, that, as institutions, informal environments for science learning are characterized by a flexibility and openness to changes in the communities, societies, and cultures of which they are a part. In order to do justice to both informal environments and those served by them, efforts to identify, measure, and document learning should be expansive enough to accommodate the full range of what and how they may help people learn.

At the same time, researchers and practitioners recognize the importance of building consensus in the field regarding standards for research methods and learning outcomes (Bitgood, Serrell, and Thompson, 1994; Loomis, 1989). Without a common framework specifying outcomes and approaches, it is difficult to show gains in learning that occur across localities or across time frames, and attempts to portray the contributions of infrastructure for science learning that exists across varied institutions and activities will continue to be hindered. Efforts to create more rigorous, meaningful, and equitable opportunities for science learning depend on understanding what opportunities for science learning exist across the educational landscape, what the nature of this learning is in the variety of environments, how outcomes currently complement and build on one another, and how designs, processes, and practices for supporting learning can be improved in the future. Developing new ways to document learning outcomes that are both appropriate in informal environments and useful across the range of them would create greater opportunity to leverage their potency to improve science learning for all.

## REFERENCES

Abd-El-Khalick, F., and Lederman, N.G. (2000). The influence of history of science courses on students' views of nature of science. *Journal of Research in Science Teaching, 37*(10), 1057-1095.

Allen, S. (2002). Looking for learning in visitor talk: A methodological exploration. In G. Leinhardt, K. Crowley, and K. Knutson (Eds.), *Learning conversations in museums* (pp. 259-303). Mahwah, NJ: Lawrence Erlbaum Associates.

Allen, S. (2004). Designs for learning: Studying science museum exhibits that do more than entertain. *Science Education, 88*(Suppl. 1), S17-S33.

Allen, S., Gutwill, J., Perry, D.L., Garibay, C., Ellenbogen, K.M., Heimlich, J.E., Reich, C.A., and Klein, C. (2007). Research in museums: Coping with complexity. In J.H. Falk, L.D. Dierking, and S. Foutz (Eds.), *In principle, in practice: Museums as learning institutions* (pp. 229-245). Walnut Creek, CA: AltaMira Press.

American Association for the Advancement of Science. (1993). *Benchmarks for science literacy.* New York: Oxford University Press.

Anderson, D. (2003). Visitors' long-term memories of world expositions. *Curator, 46*(4), 401-420.

Anderson, D., and Nashon, S. (2007). Predators of knowledge construction: Interpreting students' metacognition in an amusement park physics program. *Science Education, 91*(2), 298-320.

Anderson, D., and Piscitelli, B., (2002). Parental recollections of childhood museum visits. *Museum National, 10*(4), 26-27.

Anderson, D., and Shimizu, H. (2007). Factors shaping vividness of memory episodes: Visitors' long-term memories of the 1970 Japan world exposition. *Memory, 15*(2), 177-191.

Anderson, D., Lucas, K.B., Ginns, I.S., and Dierking, L.D. (2000). Development of knowledge about electricity and magnetism during a visit to a science museum and related post-visit activities. *Science Education, 84*(5), 658-679.

Astor-Jack, T., Whaley, K.K., Dierking, L.D., Perry, D., and Garibay, C. (2007). Understanding the complexities of socially mediated learning. In J.H. Falk, L.D. Dierking, and S. Foutz (Eds.), *In principle, in practice: Museums as learning institutions.* Walnut Creek, CA: AltaMira Press.

Azevedo, F.S. (2006). Personal excursions: Investigating the dynamics of student engagement. *International Journal of Computers for Mathematical Learning, 11*(1), 57-98.

Barron, B. (2006). Interest and self-sustained learning as catalysts of development: A learning ecology perspective. *Human Development, 49*(4), 193-224.

Bartholomew, H., Osborne, J., and Ratcliffe, M. (2004). Teaching students ideas about science: Five dimensions of effective practice. *Science Education, 88*(5), 655-682.

Beals, D.E. (1993). Explanatory talk in low-income families' mealtime. Preschoolers' questions and parents' explanations: Causal thinking in everyday parent-child activity. *Hispanic Journal of Behavioral Sciences, 19*(1), 3-33.

Beane, D.B., and Pope, M.S. (2002). Leveling the playing field through object-based service learning. In S. Paris (Ed.), *Perspectives on object-centered learning in museums* (pp. 325-349). Mahwah, NJ: Lawrence Erlbaum Associates.

Beaumont, L. (2005). *Summative evaluation of wild reef-sharks at Shedd.* Report for the John G. Shedd Aquarium. Available: http://www.informalscience.com/download/case_studies/report_133.doc [accessed October 2008].

Bell, P., and Linn, M.C. (2002). Beliefs about science: How does science instruction contribute? In B.K. Hofer and P.R. Pintrich (Eds.), *Personal epistemology: The psychology of beliefs about knowledge and knowing.* Mahwah, NJ: Lawrence Erlbaum Associates.

Bell, P., Bricker, L.A., Lee, T.F., Reeve, S., and Zimmerman, H.H. (2006). Understanding the cultural foundations of children's biological knowledge: Insights from everyday cognition research. In A. Barab, K.E. Hay, and D. Hickey (Eds.), *7th international conference of the learning sciences, ICLS 2006* (vol. 2, pp. 1029-1035). Mahwah, NJ: Lawrence Erlbaum Associates.

Bitgood, S., Serrell, B., and Thompson, D. (1994). The impact of informal education on visitors to museums. In V. Crane, H. Nicholson, M. Chen, and S. Bitgood (Eds.), *Informal science learning: What the research says about television, science museums, and community-based projects* (pp. 61-106). Deadham, MA: Research Communication.

Borun, M., Chambers, M., and Cleghorn, A. (1996). Families are learning in science museums. *Curator, 39*(2), 123-138.

Borun, M., Massey, C., and Lutter, T. (1993). Naive knowledge and the design of science museum exhibits. *Curator, 36*(3), 201-219.

Brickhouse, N.W., Lowery, P., and Schultz, K. (2000). What kind of a girl does science? The construction of school science identities. *Journal of Research in Science Teaching, 37*(5), 441-458.

Brickhouse, N.W., Lowery, P., and Schultz, K. (2001). Embodying science: A feminist perspective on learning. *Journal of Research in Science Teaching, 18*(3), 282-295.

Brossard, D., and Shanahan, J. (2003). Do they want to have their say? Media, agricultural biotechnology, and authoritarian views of democratic processes in science. *Mass Communication and Society, 6*(3), 291-312.

Brossard, D., Lewenstein, B., and Bonney, R. (2005). Scientific knowledge and attitude change: The impact of a citizen science program. *International Journal of Science Education, 27*(9), 1099-1121.

Brown, B. (2004). Discursive identity: Assimilation into the culture of science and its implications for minority students. *Journal of Research in Science Teaching, 41*(8), 810-834.

Brown, B., Reveles, J., and Kelly, G. (2004). Scientific literacy and discursive identity: A theoretical framework for understanding science learning. *Science Education, 89*(5), 779-802.

Calabrese Barton, A. (2003). *Teaching science for social justice.* New York: Teachers College Press.

Callanan, M.A., and Oakes, L. (1992). Preschoolers' questions and parents' explanations: Causal thinking in everyday activity. *Cognitive Development, 7*, 213-233.

Callanan, M.A., Shrager, J., and Moore, J. (1995). Parent-child collaborative explanations: Methods of identification and analysis. *Journal of the Learning Sciences, 4*(1), 105-129.

Campbell, P. (2008, March). Evaluating youth and community programs: In the new ISE framework. In A. Friedman (Ed.), *Framework for evaluating impacts of informal science education projects* (pp. 69-75). Available: http://insci.org/docs/ Eval_Framework.pdf [accessed October 2008].

Carey, S., and Smith, C. (1993). On understanding the nature of scientific knowledge. *Educational Psychologist, 28*(3), 235-251.

Chaput, S.S., Little, P.M.D., and Weiss, H. (2004). Understanding and measuring attendance in out-of-school time programs. *Issues and Opportunities in Out-of-School Time Evaluation, 7*, 1-6.

Clipman, J.M. (2005). *Development of the museum affect scale and visit inspiration checklist.* Paper presented at the 2005 Annual Meeting of the Visitor Studies Association, Philadelphia. Available: http://www.visitorstudiesarchives.org [accessed October 2008].

COSMOS Corporation. (1998). *A report on the evaluation of the National Science Foundation's informal science education program.* Washington, DC: National Science Foundation. Available: http://www.nsf.gov/pubs/1998/nsf9865/nsf9865. htm [accessed October 2008].

Crowley, K., and Jacobs, M. (2002). Islands of expertise and the development of family scientific literacy. In G. Leinhardt, K. Crowley, and K. Knutson (Eds.), *Learning conversations in museums* (pp. 333-356). Mahwah, NJ: Lawrence Erlbaum Associates.

Csikszentmihalyi, M., and Hermanson, K. (1995). Intrinsic motivation in museums: Why does one want to learn? In J.H. Falk and L.D. Dierking (Eds.), *Public institutions for personal learning: Establishing a research agenda.* Washington, DC: American Association of Museums.

Csikszentmihalyi, M., Rathunde, K., and Whalen, S. (1993). *Talented teenagers: The roots of success and failure.* New York: Cambridge University Press.

Dancu, T. (2006). *Comparing three methods for measuring children's engagement with exhibits: Observations, caregiver interviews, and child interviews.* Poster presented at 2006 Annual Meeting of the Visitor Studies Association, Grand Rapids, MI.

Delandshere, G. (2002). Assessment as inquiry. *Teachers College Record, 104*(7), 1461-1484.

Diamond, J. (1999). *Practical evaluation guide: Tools for museums and other educational settings.* Walnut Creek, CA: AltaMira Press.

Dierking, L.D., Adelman, L.M., Ogden, J., Lehnhardt, K., Miller, L., and Mellen, J.D. (2004). Using a behavior change model to document the impact of visits to Disney's Animal Kingdom: A study investigating intended conservation action. *Curator, 47*(3), 322-343.

diSessa, A. (1988). Knowledge in pieces. In G. Forman and P. Pufall (Eds.), *Constructivism in the computer age* (pp. 49-70). Mahwah, NJ: Lawrence Erlbaum Associates.

Driver, R., Leach, J., Millar, R., and Scott, P. (1996). *Young people's images of science.* Buckingham, England: Open University Press.

Eccles, J.S., Wigfield, A., and Schiefele, U. (1998). Motivation to succeed. In N. Eisenberg (Ed.), *Handbook of child psychology: Social, emotional, and personality development* (5th ed., pp. 1017-1095). New York: Wiley.

Ekman, P., and Rosenberg, E. (Eds.). (2005). *What the face reveals: Basic and applied studies of spontaneous expression using the facial action coding system.* New York: Oxford University Press.

Ellenbogen, K.M. (2002). Museums in family life: An ethnographic case study. In G. Leinhardt, K. Crowley, and K. Knutson (Eds.), *Learning conversations in museums.* Mahwah, NJ: Lawrence Erlbaum Associates.

Ellenbogen, K.M. (2003). From dioramas to the dinner table: An ethnographic case study of the role of science museums in family life. *Dissertation Abstracts International, 64*(3), 846A. (University Microfilms No. AAT30-85758.)

Ellenbogen, K.M., Luke, J.J., and Dierking, L.D. (2004). Family learning research in museums: An emerging disciplinary matrix? *Science Education, 88*(Suppl. 1), S48-S58.

Engle, R.A., and Conant, F.R. (2002). Guiding principles of fostering productive disciplinary engagement: Explaining an emergent argument. *Cognition and Instruction, 20*(4), 399-483.

European Commission. (2001). *Eurobarometer 55.2: Europeans, science and technology.* Brussels, Belgium: Author.

Fadigan, K.A., and Hammrich, P.L. (2004). A longitudinal study of the educational and career trajectories of female participants of an urban informal science education program. *Journal of Research on Science Teaching, 41*(8), 835-860.

Falk, J.H. (2006). The impact of visit motivation on learning: Using identity as a construct to understand the visitor experience. *Curator, 49*(2), 151-166.

Falk, J.H. (2008). Calling all spiritual pilgrims: Identity in the museum experience. *Museum* (Jan/Feb.). Available: http://www.aam-us.org/pubs/mn/spiritual.cfm [accessed March 2009].

Falk, J.H., and Adelman, L.M. (2003). Investigating the impact of prior knowledge and interest on aquarium visitor learning. *Journal of Research in Science Teaching, 40*(2), 163-176.

Falk, J.H., and Dierking, L.D. (2000). *Learning from museums*. Walnut Creek, CA: AltaMira Press.

Falk, J.H., and Storksdieck, M. (2005). Using the "contextual model of learning" to understand visitor learning from a science center exhibition. *Science Education, 89*(5), 744-778.

Falk, J.H., Moussouri, T., and Coulson, D. (1998). The effect of visitors' agendas on museum learning, *Curator, 41*(2), 107-120.

Falk, J.H., Reinhard, E.M., Vernon, C.L., Bronnenkant, K., Deans, N.L., and Heimlich, J.E. (2007). *Why zoos and aquariums matter: Assessing the impact of a visit*. Silver Spring, MD: Association of Zoos and Aquariums.

Falk, J.H., Scott, C., Dierking, L.D., Rennie, L.J., and Cohen-Jones, M.S. (2004). Interactives and visitor learning. *Curator, 47*(2), 171-198.

Feher, E., and Rice, K. (1987). Pinholes and images: Children's conceptions of light and vision. I. *Science Education, 71*(4), 629-639.

Feher, E., and Rice, K. (1988). Shadows and anti-images: Children's conceptions of light and vision. II. *Science Education, 72*(5), 637-649.

Fender, J.G, and Crowley, K. (2007). How parent explanation changes what children learn from everyday scientific thinking. *Journal of Applied Developmental Psychology, 28*(3), 189-210.

Gallenstein, N. (2005). Never too young for a concept map. *Science and Children, 43*(1), 45-47.

Garibay, C. (2005, July). *Visitor studies and underrepresented audiences*. Paper presented at the 2005 Visitor Studies Conference, Philadelphia.

Garibay, C. (2006, January). *Primero la Ciencia remedial evaluation*. Unpublished manuscript, Chicago Botanic Garden.

Gration, M., and Jones, J. (2008, May/June). Learning from the process: Developmental evaluation within "agents of change." *ASTC Dimensions, From Intent to Impact: Building a Culture of Evaluation*. Available: http://www.astc.org/blog/2008/05/16/from-intent-to-impact-building-a-culture-of-evaluation/ [accessed April 2009].

Guichard, H. (1995). Designing tools to develop the conception of learners. *International Journal of Science Education, 17*(2), 243-253.

Gupta, P., and Siegel, E. (2008). Science career ladder at the New York Hall of Science: Youth facilitators as agents of inquiry. In R.E. Yaeger and J.H. Falk (Eds.), *Exemplary science in informal education settings: Standards-based success stories*. Arlington, VA: National Science Teachers Association.

Gutiérrez, K., and Rogoff, B. (2003). Cultural ways of learning: Individual traits or repertoires of practice. *Educational Researcher, 32*(5), 19-25.

Hammer, D., and Elby, A. (2003). Tapping epistemological resources for learning physics. *Journal of the Learning Sciences, 12*(1), 53-90.

Heath, S.B. (1999). Dimensions of language development: Lessons from older children. In A.S. Masten (Ed.), *Cultural processes in child development: The Minnesota symposium on child psychology* (Vol. 29, pp. 59-75). Mahwah, NJ: Lawrence Erlbaum Associates.

Hein, G.E. (1995). Evaluating teaching and learning in museums. In E. Hooper-Greenhill (Ed.), *Museums, media, message* (pp. 189-203). New York: Routledge.

Hein, G.E. (1998). *Learning in the museum.* New York: Routledge.

Hidi, S., and Renninger, K.A. (2006). The four-phase model of interest development. *Educational Psychologist, 41*(2), 111-127.

Holland, D., and Lave, J. (Eds.) (2001). *History in person: Enduring struggles, contentious practice, intimate identities.* Albuquerque, NM: School of American Research Press.

Holland, D., Lachicotte, W., Skinner, D., and Cain, C. (1998). *Identity and agency in cultural worlds.* Cambridge, MA: Harvard University Press.

Huba, M.E., and Freed, J. (2000). *Learner-centered assessment on college campuses: Shifting the focus from teaching to learning.* Needham Heights, MA: Allyn and Bacon.

Hull, G.A., and Greeno, J.G. (2006). Identity and agency in nonschool and school worlds. In Z. Bekerman, N. Burbules, and D.S. Keller (Eds.), *Learning in places: The informal education reader* (pp. 77-97). New York: Peter Lang.

Humphrey, T., and Gutwill, J.P. (2005). *Fostering active prolonged engagement: The art of creating APE exhibits.* San Francisco: The Exploratorium.

Isen, A.M. (2004). Some perspectives on positive feelings and emotions: Positive affect facilitates thinking and problem solving. In A.S.R. Manstead, N. Frijda, and A. Fischer (Eds.), *Feelings and emotions: The Amsterdam symposium* (pp. 263-281). New York: Cambridge University Press.

Jackson, A., and Leahy, H.R. (2005). "Seeing it for real?" Authenticity, theater and learning in museums. *Research in Drama Education, 10*(3), 303-325.

Jacoby, S., and Gonzales, P. (1991). The constitution of expert-novice in scientific discourse. *Issues in Applied Linguistics, 2*(2), 149-181.

Jolly, E.J., Campbell, P.B., and Perlman, L. (2004). *Engagement, capacity, and continuity: A trilogy for student success.* St. Paul: GE Foundation and Science Museum of Minnesota.

Jung, T., Makeig, S., Stensmo, M., and Sejnowski, T.J. (1997). Estimating alertness from the EEG power spectrum. *Biomedical Engineering, 44*(1), 60-69.

Korn, R. (2003). *Summative evaluation of "vanishing wildlife."* Monterey, CA: Monterey Bay Aquarium. Available: http://www.informalscience.org/evaluations/report_45.pdf [accessed October 2008].

Korn, R. (2006). *Summative evaluation for "search for life."* Queens: New York Hall of Science. Available: http://www.informalscience.org/evaluation/show/66 [accessed October 2008].

Kort, B., Reilly, R., and Picard, R.W. (2001). An affective model of interplay between emotions and learning: Reengineering educational pedagogy—Building a learning companion. In *Proceedings of IEEE International Conference on Advanced Learning Technologies,* Madison, WI.

Lave, J. (1996). Teaching, as learning, in practice. *Mind, Culture, and Activity, 3*(3), 149-164.

Leinhardt, G., and Knutson, K. (2004). *Listening in on museum conversations.* Walnut Creek, CA: AltaMira Press.

Leinhardt, G., Tittle, C., and Knutson, K. (2002). Talking to oneself: Diaries of museum visits. In G. Leinhardt, K. Crowley, and K. Knutson (Eds.), *Learning conversations in museums* (pp. 103-133). Mahwah, NJ: Lawrence Erlbaum Associates.

Lipstein, R., and Renninger, K.A. (2006). "Putting things into words": The development of 12-15-year-old students' interest for writing. In P. Boscolo and S. Hidi (Eds.), *Motivation and writing: Research and school practice* (pp. 113- 140). New York: Kluwer Academic/Plenum.

Loomis, R.J. (1989). The countenance of visitor studies in the 1980's. *Visitor Studies, 1*(1), 12-24.

Lu, S., and Graesser, A.C. (in press). An eye tracking study on the roles of texts, pictures, labels, and arrows during the comprehension of illustrated texts on device mechanisms. Submitted to *Cognitive Science.*

Ma, J. (2006). *Philosopher's corner.* Unpublished report. Available: http://www.exploratorium.edu/partner/pdf/philCorner_rp_02.pdf [accessed October 2008].

Martin, L.M. (2004). An emerging research framework for studying informal learning and schools. *Science Education, 88*(Suppl. 1), S71-S82.

McCreedy, D. (2005). Youth and science: Engaging adults as advocates. *Curator, 48*(2), 158-176.

McDermott, R., and Varenne, H. (1995). Culture as disability. *Anthropology and Education Quarterly, 26*(3), 324-248.

McNamara, P. (2005). *Amazing feats of aging: A summative evaluation report.* Portland: Oregon Museum of Science and Industry. Available: http://www.informalscience.org/evaluation/show/82 [accessed October 2008].

Meisner, R., vom Lehn, D., Heath, C., Burch, A., Gammon, B., and Reisman, M. (2007). Exhibiting performance: Co-participation in science centres and museums. *International Journal of Science Education, 29*(12), 1531-1555.

Melton, A.W. (1935). *Problems of installation in museums of art.* Washington, DC: American Association of Museums.

Miller, J.D. (2004). Public understanding of and attitudes toward scientific research: What we know and what we need to know. *Public Understanding of Science, 13*(3), 273-294.

Moore, R.W., and Hill Foy, R.L. (1997). The scientific attitude inventory: A revision (SAIII). *Journal of Research in Science Teaching, 34*(4), 327-336.

Moss, P.A., Girard, B., and Haniford, L. (2006). Validity in educational assessment. *Review of Research in Education, 30,* 109-162.

Moss, P.A., Pullin, D., Haertel, E.H., Gee, J.P., and Young, L. (Eds.). (in press). *Assessment, equity, and opportunity to learn.* New York: Cambridge University Press.

Mota, S., and Picard, R.W. (2003). *Automated posture analysis for detecting learner's interest level.* Paper prepared for the Workshop on Computer Vision and Pattern Recognition for Human-Computer Interaction, June, Madison, WI. Available: http://affect.media.mit.edu/pdfs/03.mota-picard.pdf [accessed March 2009].

Moussouri, T. (1997). The use of children's drawings as an evaluation tool in the museum. *Museological Review, 4,* 40-50.

Moussouri, T. (2007). Implications of the social model of disability for visitor research. *Visitor Studies, 10*(1), 90-106.

Myers, O.E., Saunders, C.D., and Birjulin, A.A. (2004). Emotional dimensions of watching zoo animals: An experience sampling study building on insights from psychology. *Curator, 47*(3), 299-321.

Nasir, N.S. (2002). Identity, goals, and learning: Mathematics in cultural practices. *Mathematical Thinking and Learning, 4*(2 & 3), 213-247.

Nasir, N.S., Rosebery, A.S., Warren B., and Lee, C.D. (2006). Learning as a cultural process: Achieving equity through diversity. In R.K. Sawyer (Ed.), *The Cambridge handbook of the learning sciences* (pp. 489-504). New York: Cambridge University Press.

National Research Council. (1996). *National science education standards.* National Committee on Science Education Standards and Assessment. Washington, DC: National Academy Press.

National Research Council. (2000). *How people learn: Brain, mind, experience, and school* (expanded ed.). Committee on Developments in the Science of Learning. J.D. Bransford, A.L. Brown, and R.R. Cocking (Eds.). Washington, DC: National Academy Press.

National Research Council. (2001). *Knowing what students know: The science and design of educational assessment.* Committee on the Foundations of Assessment. J.W. Pellegrino, N. Chudowsky, and R. Glaser (Eds.). Washington, DC: National Academy Press.

National Research Council. (2002). *Scientific research in education.* Committee on Scientific Principles for Education Research. R.J. Shavelson and L. Towne (Eds.). Washington, DC: National Academy Press.

National Research Council. (2007). *Taking science to school: Learning and teaching science in grades K-8.* Committee on Science Learning, Kindergarten Through Eighth Grade. R.A. Duschl, H.A. Schweingruber, and A.W. Shouse (Eds.). Washington, DC: The National Academies Press.

National Science Board. (2002). *Science and engineering indicators—2002* (NSB-02-1). Arlington, VA: National Science Foundation. Available: http://www.nsf.gov/statistics/seind02/pdfstart.htm [accessed October 2008].

Neumann, A. (2006). Professing passion: Emotion in the scholarship of professors in research universities. *American Educational Research Journal, 43*(3), 381-424.

O'Neill, M.C., and Dufresne-Tasse, C. (1997). Looking in everyday life/Gazing in museums. *Museum Management and Curatorship, 16*(2), 131-142.

Osborne, J., Collins, S., Ratcliffe, M., Millar, R., and Duschl, R. (2003). What "ideas-about-science" should be taught in school science? A Delphi study of the expert community. *Journal of Research in Science Teaching, 40*(7), 692-720.

Panksepp, J. (1998). *Affective neuroscience: The foundations of human and animal emotions.* New York: Oxford University Press.

Penner, D., Giles, N.D., Lehrer, R., and Schauble, L. (1997). Building functional models: Designing an elbow. *Journal of Research in Science Teaching, 34*(2), 125-143.

Perry, D. L. (1993). Measuring learning with the knowledge hierarchy. *Visitor Studies: Theory, Research and Practice: Collected Papers from the 1993 Visitor Studies Conference, 6,* 73-77.

Plutchik, R. (1961). Studies of emotion in the light of a new theory. *Psychological reports, 8,* 170.

Randol, S.M. (2005). *The nature of inquiry in science centers: Describing and assessing inquiry at exhibits.* Unpublished doctoral dissertation, University of California, Berkeley.

Raphling, B., and Serrell, B. (1993). Capturing affective learning. *Current Trends in Audience Research and Evaluation, 7,* 57-62.

Reich, C., Chin, E., and Kunz, E. (2006). Museums as forum: Engaging science center visitors in dialogue with scientists and one another. *Informal Learning Review, 79,* 1-8.

Renninger, K.A. (2000). Individual interest and its implications for understanding intrinsic motivation. In C. Sansone and J.M. Harackiewicz (Eds.), *Intrinsic motivation: Controversies and new directions* (pp. 373-404). San Diego: Academic Press.

Renninger, K.A. (2003). Effort and interest. In J. Gutherie (Ed.), *The encyclopedia of education* (2nd ed., pp. 704-707). New York: Macmillan.

Renninger, K.A., and Hidi, S. (2002). Interest and achievement: Developmental issues raised by a case study. In A. Wigfield and J. Eccles (Eds.), *Development of achievement motivation* (pp. 173-195). New York: Academic Press.

Renninger, K.A., and Wozniak, R.H. (1985). Effect of interests on attentional shift, recognition, and recall in young children. *Developmental Psychology, 21*(4), 624-631.

Renninger, K., Hidi, S., and Krapp, A. (1992). *The role of interest in learning and development.* Mahwah, NJ: Lawrence Erlbaum Associates.

Renninger, K.A., Sansone, C., and Smith, J.L. (2004). Love of learning. In C. Peterson and M.E.P. Seligman (Eds.), *Character strengths and virtues: A handbook and classification* (pp. 161-179). New York: Oxford University Press.

Robinson, E.S. (1928). *The behavior of the museum visitor.* New Series, No. 5. Washington, DC: American Association of Museums.

Rockman Et Al. (1996). *Evaluation of Bill Nye the Science Guy: Television series and outreach.* San Francisco: Author. Available: http://www.rockman.com/projects/124.kcts.billNye/BN96.pdf [accessed October 2008].

Rockman Et Al. (2007). *Media-based learning science in informal environments.* Background paper for the Learning Science in Informal Environments Committee of the National Research Council. Available: http://www7.nationalacademies.org/bose/Rockman_et%20al_Commissioned_Paper.pdf [accessed October 2008].

Rogoff, B. (2003). *The cultural nature of human development.* New York: Oxford University Press.

Rosenberg, S., Hammer, D., and Phelan, J. (2006). Multiple epistemological coherences in an eighth-grade discussion of the rock cycle. *Journal of the Learning Sciences, 15*(2), 261-292.

Roth, E.J., and Li, E. (2005, April). *Mapping the boundaries of science identity in ISME's first year.* A paper presented at the annual meeting of the American Educational Research Association, Montreal.

Rounds, J. (2006). Doing identity work in museums. *Curator, 49*(2), 133-150.

Russell, J.A., and Mehrabian, A. (1977). Evidence for a three-factor theory of emotions. *Journal of Research in Personality, 11,* 273-294.

Sachatello-Sawyer, B., Fellenz, R.A., Burton, H., Gittings-Carlson, L., Lewis-Mahony, J., and Woolbaugh, W. (2002). *Adult museum programs: Designing meaningful experiences.* American Association for State and Local History Book Series. Blue Ridge Summit, PA: AltaMira Press.

Sandoval, W.A. (2005). Understanding students' practical epistemologies and their influence on learning through inquiry. *Science Education, 89*(4), 634-656.

Schauble, L., Glaser, R., Duschl, R., Schulze, S., and John, J. (1995). Students' understanding of the objectives and procedures of experimentation in the science classroom. *Journal of the Learning Sciences, 4*(2), 131-166.

Schreiner, C., and Sjoberg, S. (2004). *Sowing the seeds of ROSE. Background, rationale, questionnaire development and data collection for ROSE (relevance of science education)—A comparative study of students' views of science and science education.* Department of Teacher Education and School Development, University of Oslo.

Schwartz, D.L., Bransford, J.D., and Sears, D. (2005). Efficiency and innovation in transfer. In J.P. Mestre (Ed.), *Transfer of learning from a modern multidisciplinary perspective* (pp. 1-51). Greenwich, CT: Information Age.

Schwartz, R.S., and Lederman, N.G. (2002). "It's the nature of the beast": The influence of knowledge and intentions on learning and teaching nature of science. *Journal of Research in Science Teaching, 39*(3), 205-236.

Serrell, B. (1998). *Paying attention: Visitors and museum exhibitions.* Washington, DC: American Association of Museums.

Shepard, L. (2000). The role of assessment in a learning culture. *Educational Researcher, 29*(7), 4-14.

Shute, V.J. (2008). Focus on formative feedback. *Review of Educational Research, 78*(1), 153-189.

Smith, C.L., Maclin, D., Houghton, C., and Hennessey, M.G. (2000). Sixth-grade students' epistemologies of science: The impact of school science experiences on epistemological development. *Cognition and Instruction, 18*, 349-422.

Songer, N.B., and Linn, M.C. (1991). How do students' views of science influence knowledge integration? *Journal of Research in Science Teaching, 28*(9), 761-784.

Spock, M. (2000). *On beyond now: Strategies for assessing the long term impact of museum experiences.* Panel discussion at the American Association of Museums Conference, Baltimore.

St. John, M., and Perry, D.L. (1993). Rethink role, science museums urged. *ASTC Newsletter, 21*(5), 1, 6-7.

Steele, C.M. (1997). A threat in the air: How stereotypes shape the intellectual identities and performance of women and African Americans. *American Psychologist, 52*(6), 613-629.

Stevens, R. (2007). Capturing ideas in digital things: the traces digital annotatin medium. In R. Goldman, B. Barron, and R. Pea (Eds.), *Video research in the learning sciences.* Cambridge: Cambridge University Press.

Stevens, R., and Hall, R.L. (1997). Seeing the "tornado": How "video traces" mediate visitor understandings of natural spectacles in a science museum. *Science Education, 81*(6), 735-748.

Stevens, R., and Toro-Martell, S. (2003). Leaving a trace: Supporting museum visitor interpretation and interaction with digital media annotation systems. *Journal of Museum Education, 28*(2), 25-31.

Tai, R.H., Liu, C.Q., Maltese, A.V., and Fan, X. (2006). Planning early for careers in science. *Science, 312*(5777), 1143-1144.

Taylor, R. (1994). The influence of a visit on attitude and behavior toward nature conservation. *Visitor Studies, 6*(1), 163-171.

Thompson, S., and Bonney, R. (2007, March). Evaluating the impact of participation in an on-line citizen science project: A mixed-methods approach. In J. Trant and D. Bearman (Eds.), *Museums and the web 2007: Proceedings.* Toronto: Archives and Museum Informatics. Available: http://www.archimuse.com/mw2007/papers/thompson/thompson.html [accessed October 2008].

Travers, R.M.W. (1978). *Children's interests.* Kalamazoo: Michigan State University, College of Education.

Van Luven, P., and Miller, C. (1993). Concepts in context: Conceptual frameworks, evaluation and exhibition development. *Visitor Studies, 5*(1), 116-124.

vom Lehn, D., Heath, C., and Hindmarsh, J. (2001). Exhibiting interaction: Conduct and collaboration in museums and galleries. *Symbolic Interaction, 24*(2), 189-216.

vom Lehn, D., Heath, C., and Hindmarsh, J. (2002). Video-based field studies in museums and galleries. *Visitor Studies, 5*(3), 15-23.

Vosniadou, S., and Brewer, W.F. (1992). Mental models of the earth: A study of conceptual change in childhood. *Cognitive Psychology, 24*(4), 535-585.

Warren, B., Ballenger, C., Ogonowski, M., Rosebery, A., and Hudicourt-Barnes, J. (2001). Rethinking diversity in learning science: The logic of everyday sense-making. *Journal of Research in Science Teaching, 38,* 529-552.

Weinburgh, M.H., and Steele, D. (2000). The modified attitude toward science inventory: Developing an instrument to be used with fifth grade urban students. *Journal of Women and Minorities in Science and Engineering, 6*(1), 87-94.

Wilson, M. (Ed.). (2004). *Towards coherence between classroom assessment and accountability.* Chicago: University of Chicago Press.

# Part II

# Venues and Configurations

4

# Everyday Settings and Family Activities

Everyday science learning is not really a single setting at all—it is the constellation of everyday activities and routines through which people often learn things related to science. What distinguishes everyday and family learning from the other venues represented in this volume is that a significant portion of it occurs in settings in which there is not necessarily any explicit goal of teaching or learning science—at least not part of an institutional agenda to engage in science education. In many situations, scientific content, ways of thinking, and practices are opportunistically encountered and identified, without any particular prior intention to learn about science. In this way, science learning is simply woven into the fabric of the everyday activities or problems.

An individual could be asked to make a health-related decision, contingent on a set of scientific concepts and complex underlying models, while keeping a routine doctor's appointment. A family might stumble across a science-related event—like a robotics or science fair put on by avid hobbyists—while on a weekend outing. An individual may have to learn about some detailed aspect of computer technology in order to resolve a problem with a computer or network. A group of children might decide to construct an elaborate treehouse one summer, necessitating that they develop a deeper understanding of materials and structural mechanics. Or community members may decide to canvass their neighborhood to educate and involve others responding to an environmental hazard that has been uncovered. As each of these examples illustrates, moments for science learning and teaching surface in people's everyday lives in unpredictable and opportunistic ways. The research reviewed in this chapter raises intriguing questions about how such everyday moments can figure importantly into a longer developmental pathway that leads to an increasingly sophisticated understanding of science.

A typical scenario for everyday science learning might be a child learning from a parent, or children and adults learning from the media, siblings, peers, and coworkers. Everyday science learning can even appear in the structure of schools and the workplace. For example, some have argued that many child-oriented preschools and apprentice-like graduate programs have in common a kind of situated learning embedded in meaningful activities characteristic of everyday learning (Tharp and Gallimore, 1989). In some school classrooms, as well, children engage with science concepts and activities in informal ways (Brown and Campione, 1996). Many adults learn a great deal about science in the workplace. The science learning we focus on in this chapter, however, occurs in less structured settings.

An important distinction can be made between two categories of everyday science learning. First, there are spontaneous, opportune moments of learning that come up unexpectedly. Second, there are more deliberate and focused pursuits that involve science learning and may grow into more stable interests and activity choices. These types establish two ends of a continuum, with a range of activities falling in between.

Virtually all people participate in spontaneous everyday science learning. A classic example is when a preschool-age child asks a parent a question during everyday activities. For example in one study, while fishing with his dad, a four-year-old boy asked, "Why do fish die outside the water?" While watching a movie about dinosaurs, another four-year-old boy asked, "Why do dinosaurs grow horns?" A five-year-old girl eating dinner with her family asked, "When you die what is your body like?" (Callanan, Perez-Granados, Barajas, and Goldberg, no date). Such questions often emerge in conversations that become potential learning situations for children. Although the children themselves are not likely to be thinking about the domain of science, their questions engage other people in the exploration of ideas, creating an important context for early thinking about science.

Of course, young children are not the only ones to engage with science ideas in these spontaneous ways. Every adult has had experiences in which they pick up some new idea or new way of understanding something scientific through a casual conversation, or through a newspaper article or television show. Conversational topics one might casually encounter range from what causes earthquakes, to how new television screen technology works, to the best way to determine what food may be causing allergic reactions in a child. What these examples have in common is that science learning may be occurring without any particular goal of learning.

Not everyone participates in the second, more deliberate type of everyday science activity. But many do: children become "experts" in particular domains (dinosaurs, birds, stars), adults pursue science hobbies (computers, ham radio, gardening), and other focused pursuits emerge because of life circumstances (caring for a family member with a particular condition, dealing with a local environmental hazard). In these more deliberate pursuits, there is a learning goal, although it might be quite different from the goals held by science teachers for their students. For example, an adult with a hobby of flying model planes learns a great deal about aerodynamics, and a child who develops a keen interest in dinosaurs gains expertise in understanding biological adaptation. The focused pursuits that are based on life circumstances also involve learning and teaching—for example, a young woman who searches the Internet to better understand her mother's cancer diagnosis, as well as the community member who learns about water contamination because of a local hazard. Agricultural communities and families engage in sophisticated science learning related to environmental conditions and botany in specific ecosystems. Hobbyists and volunteers can spend hundreds of hours each year engaging in science-related elective pursuits, from astronomy and robotics to animal husbandry and environmental stewardship (Sachatello-Sawyer et al., 2002). A parent might decide to structure significant portions of weekend family time around a science-related practice like systematic mixing to make perfumes or cross-pollination experiments with house plants (Bell et al., 2006).

In contrast to the more opportunistic experiences described first, these deliberate educational opportunities are more systematic, more sustained, more likely to involve the development of social groups to support the activities (e.g., hobby groups), and more likely to link with institutions that make the pursuits possible (e.g., equipment manufacturers, government agencies). Furthermore, sustained learning is more of a central goal in these activities than in the spontaneous ones. But notice that the learning and teaching that occurs in these examples is not defined by the goal of becoming expert in a domain of science or in science as a global concept. The learning is much more specific, more focused, and more connected to the deeply motivated interests and goals of the learner. These everyday pursuits, while they involve sustained individual inquiry, are also often intensive social practices in which individuals share expertise and combine their distributed expertise to reach goals that include solving problems, increasing expertise, and enjoyment.

## SETTINGS FOR EVERYDAY LEARNING

The settings in which everyday and family science learning occur vary a great deal in terms of physical setting, the degree to which a particular location is obviously marked as science-oriented, and the relationship to science learning institutions and programs.

Some settings for everyday and family learning are clearly tied to science content—activities like fishing, berry picking, agricultural practices, and gardening, for example. Although participants in these settings may not view their activities as relevant to science, it is not difficult to make the case that they are potentially interesting places for science learning as they are linked to scientific domains (e.g., berry picking can overlap with questions of botany). Other everyday activities are even more explicitly focused on learning science content; these include reading books about science topics, or watching videos and television shows about such topics (e.g., the Discovery Channel). When children are a bit older, homework activities with parents (e.g., science fair projects) are possible venues for science conversations, as well as conversations related to literacy and other school topics (McDermott, Goldman, and Varenne, 1984; Valle and Callanan, 2006).

Some settings for everyday and family science learning may occur in or build on settings designed for science learning—science or natural history museums, zoos, science centers, environmental centers, school experiences, and the like. Although we discuss experiences in designed settings at length in Chapter 5, it is important to note that the distinction between everyday learning and learning in designed settings is blurry and imperfect. After all, family groups are among the most common social configurations of participants in these settings. Conversations about these events and activities occur as the experiences are unfolding in both unstructured family settings and institutionally organized, designed settings. For example, Crowley and Galco (2001) report on the ways that parents, through conversations with their children in museums, seem to extend children's exploration and provide brief explanations of the phenomena they are observing. Reflection on those experiences often extends after these experiences and is observed in future family activities in a variety of home and other settings (Bell et al., 2006; Bricker and Bell, no date).

A third type of setting—the unanticipated incidental experiences of family life—are in some sense not obviously linked to a scientific setting. Dinner table conversation is one activity that has been studied by a number of researchers (Ochs, Smith, and Taylor, 1996). Other activities, such as driving in the car, can also provide opportunities for reflection on the events of the day or on issues that come to mind (Callanan and Oakes, 1992). Goodwin (2007) discusses "occasioned knowledge exploration," in which, for example, a family on an evening walk might encounter events that lead to explanation. She discusses one family walk on which each family member pretended to be a different animal, and this engendered open-ended discussion of a number of topics, such as camouflage, how fireflies' lights work, and the behavior of snakes.

A crucial point to make here is that the features of the settings for everyday science learning are likely to vary a great deal depending on the cultural community, as well as the particular family in question. Some individuals, families, and communities live in ways that give them regular exposure to living animals, while others are limited to encountering only pictures of animals, along with pets and occasional zoo visits. People, especially children, also vary a great deal in their exposure to different types of technology (such as computers, automobile mechanics, and construction equipment). In addition, there is diversity in the patterns of interaction of children and adults in families. Some communities value storytelling, others focus more on explanation, others focus more on intent observation of ongoing activity without as much verbal commentary (Heath, 1983; Rogoff et al., 2003). All of these issues have importance for the ways in which groups of people tend to engage with the natural and technological world and the ways in which young children master, as well as learn to identify as normal, habitual modes of interacting with one another and with science and the natural world. We return to this in greater detail in Chapter 7.

# WHO LEARNS IN EVERYDAY SETTINGS

Virtually all people develop skills, interests, and knowledge relevant to science in everyday and family settings. The nature of learning varies over time as development, maturation, and the life course unfold. Particular interests and abilities arise through development that shape pursuits of learning, as well as the intellectual and social resources individuals draw on to learn science. People develop new interests and manage new tasks that arise through the life course. Being a sibling, entering the workforce, caring for one's self, one's children, and one's aging parents, for example, often demand that one navigate and explore new scientific terrain. Here we briefly sketch out a life-course developmental view of science learning as it unfolds in everyday and family settings.

At birth, children begin to build the basis for science learning. By the end of the first two years of life, individuals have acquired a remarkable amount of knowledge about the physical aspects of their world (Baillargeon, 2004; Cohen and Cashon, 2006). This "knowledge" is not formal science knowledge, but rather a developing intuitive grasp of regularity in the natural world. It is derived from the child's own experimentation with objects, rather than through planned learning by adults. In accidentally dropping something from a high chair or crib, for example, the child begins to recognize the effects of gravity. These early experiences do not always lead to accurate interpretations or understandings of the physical world (Krist, Fieberg, and Wilkening, 1993). As children acquire new or deeper knowledge about physical objects and events, some of their learning will correct false or incomplete inferences that they have made earlier.

As a child masters language and becomes more mobile, opportunities for science learning expand. Informal and unplanned discoveries of scientific phenomena (e.g., scrutinizing bugs in the backyard) are supplemented by more programmatic learning (e.g., bedtime reading by parents, family visits to museums or science centers, science-related activities in child care or preschool settings). These lead to the development of scientific concepts (Gelman and Kalish, 2006), which are enhanced by the child's expanding reasoning skills (Halford and Andrews, 2006). Even in these initial years of life, children display preferences for some phenomena more than others. Such preferences can evolve into specific science interests (e.g., dinosaurs, insects, flight, mechanics) that can be nurtured when parents or others provide experiences or resources related to the interests (Chi and Koeske, 1983; Crowley and Jacobs, 2002).

By the time they enter formal school environments, most children have developed an impressive array of cognitive skills, along with an extensive body of knowledge related to the natural world (National Research Council, 2007). It is also likely that they have become familiar with numerous modalities for acquiring scientific information other than formal classroom instruction: reading, surfing the Internet, watching science-related programs on television, speaking with peers or adults who have some expertise on a topic, or exploring the environment on their own (Korpan, Bisanz, Bisanz, and Lynch, 1998). These activities continue throughout the years in which young people and young adults are engaged in formal schooling, as well as later in life (Farenga and Joyce, 1997).

It is also common for elementary schoolchildren to bring the classroom home, to regale parents with stories of what happened in school that day and involve them in homework assignments. These events help to alert parents to a child's specific intellectual interests and may inspire family activities that feature these interests. A child's comments about a science lesson at school may encourage parents to work with the child on the Internet or take him or her to a zoo or museum or concoct scientific experiments with household items in order to gather more information. In these ways, informal experiences can supplement and complement school-based science education.

As young people move into adolescence, they tend to express a desire to pursue activities independently of adults (Falk and Dierking, 2002). This does not necessarily mean that relationships with parents grow more distant (Zimmer-Gembeck and Collins, 2003), but young people do spend less time with parents or other adult relatives and more time with peers or alone (Csikszentmihalyi and Larson, 1984). Attachment to teachers also wanes across adolescence (Eccles, Lord, and Buchanan, 1996). Despite such alterations in relationships with adults who have organized or supervised their learning experiences in previous years, many young people continue to engage in many activities outside school that can involve science learning. Individuals' interests in and motivations to pursue scientific learning change during adolescence. Yet especially for those with strong personal interests in scientific areas, learning experiences in informal settings potentially continue to supplement classroom science instruction.

As individuals move into adult roles, they usually reserve a reasonable amount of time for leisure pursuits. Those with hobbies related to science, technology, engineering, or mathematics are especially likely to continue with intentional, self-directed learning activities in that area (Barron, 2006). Science learning may also continue in more unintentional ways, such as watching television shows or movies with scientific content or falling into conversation with friends or associates about science-related issues. Some adults may focus especially on scientific issues related to their occupation or career, and in many cases their pursuit of scientific topics will be influenced by personal interests or (in later years) the school-related needs of their children.

Beginning in middle age and continuing through later adulthood, individuals are often motivated by events in their own lives or the lives of significant others to obtain health-related information (Flynn, Smith, and Freese, 2006). Health-related concerns draw many adults into a new domain of science learning. At the same time, with retirement, older adults have more time to devote to personal interests. Their science learning addresses long-standing scientific interests as well as new areas of interest (Kelly, Savage, Landman, and Tonkin, 2002).

In sum, although the nature and extent of science-related learning may vary considerably from one life stage to another, most people develop relevant capabilities and intuitive knowledge from the days immediately after birth and expand on these in later stages of their life. In this sense, science learning in informal environments is definitely a lifelong enterprise (Falk and Dierking, 2002). To date, no one has compiled reliable information on the amount of information about the natural world acquired by infants and toddlers through everyday interactions in the world or through more programmed learning contexts (e.g., preschool activities, television shows). Information is equally scant on the amount of scientific knowledge that young people acquire in school classrooms in comparison to other venues. It is safe to say, however, that the sheer number of hours in which individuals encounter scientific information outside school over the life span is far greater than the number of hours of science education in formal classroom environments.

## WHAT IS LEARNED

This section focuses on the science knowledge, skills, and interests that children and adults develop in everyday learning. We organize this discussion according to the strands of our framework, focusing specifically on the evidence of learning in everyday and family settings. The strands serve as a means of pulling apart the evidence in ways that make the stronger claims more evident. We devote varied amounts of space to the strands. In most cases, this variability reflects the quantity of work that has examined the strand in a particular venue. Here and in subsequent chapters, we often discuss the strands individually for analytic purposes. Yet we hope to keep sight of how the strands are interrelated and mutually supportive in practice. Tizard and Hughes (1984), for example, offer an illustrative example of an almost-4-year-old's conversation with her mother (see Box 4-1). In this short thread, we see the child using her parent as source of information (Strand 5)

---

**BOX 4-1  Example of a Parent-Child Incidental Science Conversation**

Child:   Is our roof a sloping roof?

Mother:  Mmm. We've got two sloping roofs, and they sort of meet in the middle.

Child:   Why have we?

Mother:  Oh, it's just the way our house is built. Most people have sloping roofs, so that the rain can run off them. Otherwise, if you have a flat roof, the rain would sit in the middle of the roof and make a big puddle, and then it would start coming through.

Child:   Our school has a flat roof, you know.

Mother:  Yes it does actually, doesn't it?

Child:   And the rain sits there and goes through?

Mother:  Well, it doesn't go through. It's probably built with drains so that the water runs away. You have big blocks of flats with rather flat sort of roofs. But houses that were built at the time this house was built usually had sloping roofs.

Child:   Does Lara have a sloping roof? [Lara is her friend]

Mother:  Mmm. Lara's house is very like ours. In countries where they have a lot of snow, they have even more sloping roofs. So that when they've got a lot of snow, the snow can just fall off.

Child:   If you have a flat roof, what would it do? Would it just have a drain?

Mother:  No, then it would sit on the roof, and when it melted it would make a big puddle.

SOURCE: Tizard and Hughes (1984).

---

as she explores a "why" question (Strand 1) and tries to explain the role of pitched roofs in drainage (Strand 2).

## Strand 1: Developing Interest in Science

What sets everyday learning apart from other learning is the sense of excitement and pure intrinsic interest that often underlies it (Hidi and Renninger,

2006). One potential advantage of everyday informal settings is that they may be more likely to support learners' interest-driven and personally relevant exploration than are more structured settings, such as classrooms and other designed educational settings.

Children's cause-seeking "why" questions have been argued to be one sign of their intense curiosity about the world (see Heath, 1999; Gopnik, Meltzoff, and Kuhl, 1999; Tizard and Hughes, 1984). Simon (2001) compares these questions to the creative thought and exploratory thinking of scientists. Similarly, Gopnik (1998) suggests that explanation seeking is a basic human process. Some children become so interested in one domain that they are described as experts—for example a great deal of research has characterized the activities of preschool-age dinosaur experts, as well as experts in other domains relevant to science or technology (Chi, Hutchinson, and Robin, 1989; Johnson et al., 2004). Such children may also develop social reputations as experts in a particular science domain (Palmquist and Crowley, 2007). These social reputation systems can serve to further the child's learning, in that adults, peers, and siblings may call on the child to perform as an expert (e.g., to produce and refine an explanation of a natural phenomenon) or provide them with specialized topic-related learning resources to further their learning (Barron, 2006; Bell et al., 2006). Similarly, adult experts often develop their knowledge through informal channels.

Adult science learning in everyday settings is also usually self-motivated and tightly connected to individual interest and problem solving. For example, adult learners often learn about science in the context of hobbies, such as bird watching or model airplane building (Azevedo, 2006). A sociocultural perspective on adult learning highlights how learning is often initiated in direct response to a current life problem or issue (Spradley, 1980). Environmental science learning often occurs in the context of local conflicts that threaten neighborhoods, such as pesticide use, industrial waste, effects of severe weather, or introduction of new industries in an area (Ballantyne and Bain, 1995). Also, a great deal of adult learning about human physiology and medicine tends to occur because of immediate and strong motivation to learn about illnesses experienced by the learner or someone close to them (Flynn, Smith, and Freese, 2006). Indeed, one conclusion from the literature is that adult learners tend not to be generalists in their learning of science; rather, they tend to become experts in one particular domain of interest (Sachatello-Sawyer, 2006).

Even when science learning is of the momentary type (rather than sustained or expert-like), keen interest is likely to be behind it. The research on adults' medical knowledge is one strong example; that knowledge often comes from deep questioning of health care providers and intense searches of literature (and, more recently, the Internet) when one is facing a medical crisis (for either oneself or a loved one). The motivation to understand in the context of such a crisis is strong and persistent (Dickerson et al., 2004; Flynn et al., 2006; Pereira et al., 2000).

Some have argued that schools and science centers should learn from the authentic moments of curiosity and exploration seen in everyday learning—and try to recreate them in their settings (Falk and Storksdieck, 2005; Hall and Schaverien, 2001; National Research Council, 2000). While pursuit of scientific questions for the sake of pure interest is often a goal in planning curriculum or museum exhibits, visitors may not have that goal. Yet the personal histories of scientists suggest that sustained everyday experiences are often seen as a crucial influence on their expertise development (Csikszentmihalyi, 1996; Simon, 2001). If learning experiences in informal settings are to be linked more productively with formal education, a fundamental challenge is to systematically explore the effectiveness of ways of offering resources and supports that allow learners to pursue their own deeply held interests.

## Strand 2: Understanding Scientific Knowledge

As noted, throughout the life span, people learn a myriad of facts, ideas, and explanations that are relevant to a variety of scientific domains. Studies of early cognitive development suggest that young children, prior to the age at which they enter school, make great strides in understanding regularities in the natural world, which can be developed into more robust understanding of science (National Research Council, 2007). Their earliest experiences of learning about the natural world begin in infancy. Even in the first days of life, infants' physical encounters with objects and people begin to give them information about the nature of their new world. Newborns' contacts with surfaces and objects give them an intuitive understanding of motion which later may be drawn on in the study of physics (Baillargeon, 2004; Spelke, 2002; von Hofsten, 2004). For example, when presented with a person holding an object, 4-month-old babies look longer when the person lets go and the object stays stationary than when the object drops, suggesting that they are surprised when the typical effects of gravity are violated (Baillargeon, 2004). Throughout the first year of life, babies' simple behaviors, such as looking in anticipation for the movement of a rolling ball, show that they have begun to develop expectations about the behaviors of physical objects, as well as the actions of other people (Luo and Baillargeon, 2005; Saxe, Tzelnic, and Carey, 2007).

Much of young children's early understanding of the natural world grows out of experiences in everyday settings. Consider, for example, research on children's learning about two scientific questions: (1) What kinds of things are alive? (2) What is the shape of the earth? These are two areas in which extensive research has uncovered patterns in children's early understanding, as well as developmental changes in their concepts over time.

The developing understanding of distinctions between living and nonliving things has been explored in infancy and early childhood using a number of methodologies (Bullock, Gelman, and Baillargeon, 1982; Gelman and Gottfried, 1996; National Research Council, 2007; Springer and Keil, 1991). It is evident from this work that many of children's earliest ideas about the natural world seem to focus on a distinction between social, intentional creatures as distinct from nonintentional, inanimate things (Carey, 1985). Indeed, it takes many years for children to accept plants as living things (Waxman, 2005).

Laboratory studies of children's inferences about living things first suggested that they think about animals in terms of their relation to people (Carey, 1985). When told that people have a particular organ (e.g., a spleen) and asked whether a series of animals have that organ, children as old as 7 years often seemed to make decisions based on how similar the animal was to humans; a monkey would be judged as more likely to have the organ than would a butterfly, for example. Such findings were taken to suggest that children did not have a "naïve theory" of biology, but rather thought in terms of a "naïve psychology" with humans as the prototype. Later studies, however, have shown that Carey's sample of mostly urban majority children reason differently on this task than do children from communities with more firsthand experience with nature. Both rural American Indian children from the Menominee community and rural majority children made inferences that indicate reasoning about biological kinds without anthropomorphism (Ross, Medin, Coley, and Atran, 2003). Furthermore, Tarlowski (2006) found that children whose parents are expert biologists were more likely to reason about animals in terms of biological categories, and Inagaki and Hatano (1996) found that children who had experience raising goldfish were more likely to reason in terms of biology than those who had not.

Research on children's understanding of evolution has also revealed some interesting influences of learning about biology in families. Evans (2001, 2005) found some ways that developmental phases in understanding the origin of species are similar for children from different family backgrounds. She finds that many young children give "creationist" explanations, and then, as they get older, their families' beliefs seem to influence children from fundamentalist and nonfundamentalist households to differentiate their beliefs about evolution.

These findings demonstrate that while there are trends related to age, children's particular experiences, including cultural experiences outside school, are likely to have impact on their thinking about the domain of living things. Less is known about precisely how specific experiences actually affect their thinking. What does seems clear, however, is that much of this learning occurs in informal settings, and that it is likely to involve conversations with peers (Howe, McWilliam, and Cross, 2005; Howe, Tolmie, and Rodgers, 1992; Lumpe, 1995), parents, and other important people in children's lives (Jipson and Gelman, 2007; Waxman and Medin, 2007).

Children's understanding of the shape of the earth is another area in which research has uncovered developmental patterns that suggest the importance of everyday learning and cultural context (Agan and Sneider, 2004; Nussbaum and Novak, 1976; Vosniadou and Brewer, 1992). While perceptual experience tells children that the earth is a flat surface, even 4- and 5-year-olds show evidence of knowing that the earth is in fact round. Vosniadou and Brewer (1992) demonstrated that children's lived experience of the earth conflicts with what they are told—that the earth is round—and that children attempt to reconcile this conflict by creating hybrid mental models of the earth that bridge what they learn through observation versus through conversation. Using interview questions designed to uncover children's solutions to this conflicting information, Vosniadou and Brewer identified a number of different models in children's answers. For example, some children answered questions in ways that suggested a dual-earth model, in which they distinguished the flat earth on which they walk from the round earth up in the sky. Another model was the hollow-earth model, in which children seemed to think that the earth is round, but that people live on a surface on the inside of the globe (with the top of the globe sometimes seen as the sky). Other studies have found cultural variation in the kinds of models children describe (Samarapungavan, Vosniadou, and Brewer, 1996), showing that experiences, cultural values, and interactions with other people are likely to influence children as they make sense of their world and revise their understanding over time. For example, Samarapungavan and colleagues analyzed the cosmological beliefs of Indian children ages 5-9. They found that, in generating explanations for cosmological phenomena, children commonly conflated the physical characteristics of heavenly bodies (e.g., shape, angle, location) with local folkloric explanations.

Just as children learn science in everyday settings, so do adults. The clearest examples are health- and environment-related information. In seeking information about these issues, adults often turn to various sources besides such traditional experts as health practitioners. Additional modes of health information-seeking now commonly include the mass media and the use of local experts. The use of mass media for health information is well documented. A review of three national surveys conducted before the Internet's rapid growth showed that mass media, including magazines, newspapers, other printed publications, television, radio, street signs, and billboards were cited as the predominant source of health news for the majority of the respondents (Brodie et al., 2003). More recent studies confirmed those findings, with the Internet (whether defined as a resource or as a mass medium) growing dramatically in importance (Fox, 2006; Madden and Fox, 2006). Mass media play a substantial role in defining health and illness, detailing products and services designed to assist individuals in negotiating their health and well-being, and providing models of others with particular health concerns for consumers.

Local experts (i.e., individuals who have tangible experience in the health care profession or who themselves once experienced a particular medical condition) are also a major source of information for adults (Tardy and Hale, 1998). Tardy and Hale (1998) found that lay individuals are often sought out because they appear approachable and amicable and are integrated into their local communities. People often feel more comfortable seeking health information from them than from their health care providers. Epstein (1996) documented the process by which AIDS activists, initially relatively naïve about technical aspects of AIDS research, became sufficiently expert in the science of AIDS to contribute meaningfully to research policy, research funding, and research design. Epstein's work is part of a qualitative, case study-oriented sociological tradition that highlights ways in which nonexperts can learn technical information when relevant to their needs and indeed may contribute to the production of knowledge in ways that are unavailable to traditional scientific experts. This focus on lay knowledge comes largely from British explorations of the public understanding of science in the 1980s and 1990s (Irwin and Wynne, 1996; Layton, 1993).

Although these studies document health information-seeking through mass media and local experts, neither they nor other well-developed literatures have provided evidence for conclusions about the specific impact such behaviors are having on adults' understanding of health, illness, and medicine. Nonetheless, in the presence of so much information gathering and with demonstrable behaviors, such as health care actions, as a result of the information gathering, we believe it evident that learning takes place in these everyday settings.

These examples of developing understanding of scientific domains in both adults and children help support our contention about the importance of everyday learning. It is worth noting, however, that there is some disagreement about exactly *what* is learned. Much of the developmental psychology research approaches conceptual development with the assumption that particular symbolic concepts and causal theories are acquired at particular ages (Gopnik and Wellman, 1992; Gelman, 2003). A related approach focuses on misconceptions or alternative frameworks that children and adults have about science topics, which need to be corrected through intervention (e.g., Treagust, 1988). Finally, perhaps because research on learning in informal environments often focuses on naturalistic data (rather than laboratory tasks or intervention studies), sociocultural-historical approaches have been an important approach in this field (Cole, 2005; Rogoff, 2003).

The emergence of the theory approach in developmental psychology has had the positive effect of acknowledging the coherence and internal consistency of children's thinking, even when their reasoning is different from adults. As argued in *Taking Science to School* (National Research Council, 2007), Piaget's assumptions about children's early illogical thinking were not supported when their logic was examined on its own terms (Carey, 1985; Gelman and Baillargeon, 1983). Gelman (2003) argues further that both children and adults are essentialist thinkers who develop understanding of biological categories, for example, guided by an assumption that these important categories have inherent essences. Thus, learning something about a particular animal (e.g., that it eats bamboo leaves) leads them to make the inference that it is not just of that individual, but of all animals of the same type.

In parallel with the changes in developmental psychology, for the past 25 years science education researchers have focused substantial attention on the details of children's conceptual understanding of disciplinary science topics. The range of people's ideas that differ from the understanding in the discipline are often framed as misconceptions, preconceptions, or alternative conceptions that need to be replaced with more normative understandings (e.g., Treagust, 1988; Snyder and Ohadi, 1998).

An emerging cognitive perspective, which complicates this model, involves focusing on the pieces of knowledge present in a complex knowledge system of an individual that need to be brought into a coherent understanding (diSessa, 1988). It acknowledges that refinement may take place over a significant period of time. As we noted earlier, everyday experiences with natural phenomena are important for developing these pieces of knowledge.

From this knowledge construction and refinement perspective, it might be more educationally useful to think about children's ideas as productive resources that they can reorganize and apply to specific contexts and problems in more scientific ways (Smith, diSessa, and Roschelle, 1993). In this view, arriving quickly at correct subject matter responses is less important than following a scientific knowledge-building process in one's conceptual change. Educational experiences might benefit from focusing on the individual variation in children's thinking—the knowledge fragments that are brought in to make sense of a particular context—in that they can serve as leverage points for further knowledge refinement, as opposed to looking only at central tendencies in thinking (e.g., coherent accounts generated systematically across many individuals).

The sociocultural-historical approach has become very influential in the field, as discussed in Chapter 2. Especially when considering the everyday contexts in which science concepts are encountered, some argue that it may be more productive to characterize what is learned in terms of situated thinking as it arises in meaningful action and interaction, rather than in terms of stable cognitive abilities that are absent at one point and present at a later point (Cole, 1996; Rogoff, 2003). In the interdisciplinary literature that touches on everyday science learning, the disagreement seems to focus on the role of everyday experiences in children's developing scientific thinking. Some research suggests that children may develop misconceptions about science from everyday experiences (Ioannides and Vosniadou, 2002; National Research Council, 2007; Snir, Smith, and Raz, 2003). Other research suggests that children may deploy more sophisticated reasoning about science and the natural world in everyday settings than they do in school settings (Bell et al., 2006; Sandoval, 2005). Further research is needed in order to reveal the subtleties of the interaction between thinking about science in everyday and in school settings.

## Strand 3: Engaging in Scientific Reasoning

Another important focus of research on science learning in informal settings has been on how people employ the types of reasoning involved in science in their everyday activities. Research on scientific thinking has often focused on a specific set of structured, almost stereotyped, thinking strategies. In particular, consciously formulating and testing hypotheses has been seen as a central aspect of scientific reasoning (Kuhn, 1989; Klahr, 2000; Schauble, 1996; Zimmerman, 2000). Recently, however, several experts have argued that these reasoning processes are only a subset of those needed in science. Focusing only on hypothesis testing leaves out a vast array of other forms of thinking that are also crucially important for science (Gleason and Schauble, 2000; National Research Council, 2007).

Paradoxically, while the reasoning skills involved in science are sometimes seen in the thinking of very young children, there is also evidence that many adults are less than proficient in some of these skills (Kuhn, 1996; Tversky and Kahneman, 1986). Do individuals somehow lose scientific reasoning skills as they age? We consider this paradox in this section, exploring how research on everyday thinking may clarify the discrepant results. Our focus is on two issues in scientific reasoning that are particularly relevant to everyday thinking: causality and context.

Gopnik and colleagues (2004) argue, in fact, that seeking causes is a basic human drive. Controlled studies of causal thinking have shown that young children can process complex causal relations (Gopnik, Sobel, Shulz, and Glymour, 2001; Kushnir and Gopnik, 2005). In classroom studies, ways of supporting students' causal thinking have been implemented and evaluated (Lehrer and Schauble, 2006). Everyday causal thinking is more ambiguous than laboratory or classroom tasks, and there is less of a chance that true causes can be determined. That doesn't change the fact that cause-seeking is a major preoccupation in everyday life. People are often trying to figure out causes for events in both the natural world and the social world.

Research has also demonstrated that various contextual factors have enormous impact on how effectively children and adults test claims and evaluate evidence. For example, Tschirgi (1980) showed that how people value an outcome will influence how likely they are to use scientific strategies in testing its cause. For example, if testing which ingredient made a cake turn out poorly, children and adults were likely to systematically control the variables they tested. However, if testing which ingredient made a cake turn out well, the participants tended to want to hold variables constant rather than test them systematically—presumably because it made sense to them to try and recreate the good cake. Similarly, there is an extensive body of research demonstrating that adults, even experts, often use logic that does not match the "scientific" approach (Kuhn, 1996; Tversky and Kahneman, 1986; Wason, 1960), and often these differences in logic or scientific reasoning are related to a variety of everyday heuristics for making sense of the world. Using controlled tasks in a laboratory setting, Amsterlaw and Meltzoff (2007) have recently documented that children develop more scientific ways of reasoning and making decisions over their elementary school years. Children exhibit less of an outcome bias and identify the crucial role of evidence in reasoning. These laboratory-based studies may shed light on a phenomenon that is pervasive in everyday learning situations.

Throughout much of the research on scientific reasoning, a pervading assumption about the "right" way to do science is apparent. This assumption is often overly simplistic, suggesting that the scientific method of testing hypotheses by controlling variables is the correct way to do science, when in fact there are many different methods involved in carrying out scientific work (Gleason and Schauble, 2000), and the way that scientists really go about their work can be quite different from the stereotype (Dunbar, 1999; Knorr-Cetina, 1999; Latour and Woolgar, 1986). One example is that not all sciences use hypothesis-testing in the same way. Paleontology and astronomy, for example, also make important use of putting together patterns or sequences into a plausible narrative. Erickson and Gutiérrez (2002) argue that it is crucial for understanding science learning to recognize the variety of methods required for rigorous scientific work. They describe an example in which qualitative observational research was necessary in order to make sense of an anomalous finding obtained with quantitative methods. We return to these issues later when discussing learners' engagement with the practices of science.

## Strand 4: Reflecting on Science

A large body of education research shows that, when asked questions about the nature of science, children and adults are likely to express some beliefs that contradict the notions of science held by most scientists and those espoused by philosophical and empirical accounts of scientific practice. This discrepancy in understanding the nature of science has been argued to hamper students' attempts to learn science (Bell and Linn, 2002; Driver, Leach, Millar, and Scott, 1996; Lederman, 1992; Sandoval, 2005). Reflecting on science and its processes, as well as reflecting on one's own science thinking, are crucial parts of everyday science thinking.

One of the most consistent ways that nonscientists' perceptions of science seems to differ from those of scientists is that many children and adults tend to perceive science as a set of established facts rather than as an ongoing process of knowledge construction (Songer and Linn, 1991). And many people show little awareness of the variety of methods used in science, and they tend to misunderstand the crucial role of evidence in the science community's process of reaching conclusions.

For example, Sandoval (2005) summarizes the literature on students' understanding of science assessed in school settings, arguing that they have difficulty with the following four aspects of science:

1. Science as constructed by people—rather than seeing science as a body of knowledge constructed through interpretation of evidence, Sandoval argues that students often seem to see science as a set of objectively true facts.
2. Science as varying in certainty—students often see science as certain knowledge. It is difficult for them to understand that because they use evidence to come to conclusions, scientists often change their conclusions when presented with new evidence.
3. Diversity of methods of science—not realizing that there are a variety of different methods involved in science, students often see science as based only on experiments. And they often tend to have trouble understanding how methods link to evidence and how evidence is used to answer questions.
4. Forms of scientific knowledge—students often are confused about the nature of different types of scientific knowledge; in particular, they see hypothesis, theory, and law as a linear sequence, going from less certain to more certain. Students are also sometimes confused about how theories differ from evidence and how models relate to real phenomena.[1]

The misunderstandings about science that Sandoval (2005) and others describe in classroom settings also have much in common with how many adults think about science. They frequently struggle with the four aspects of science listed above. In fact, one of the most powerful findings in the literature on scientific thinking is that adults as well as children have considerable difficulty taking their own thinking as an object of thought. Kuhn (1996) argues that scientific thinking is really not very different from everyday thinking, in that the difficulty of reflecting on one's thought leads to less than good reasoning in either domain. Sandoval (2005) argues that children have better access to their own reasoning in everyday settings than they do in classroom settings. Yet the persistent observations that children and adults typically misconstrue aspects of the goals, processes, and norms

---

[1] Of course there is great debate among historians, sociologists, and philosophers of science about the meanings of and relationships among hypothesis, theory, and law.

of science seem to imply that everyday learning does not typically build a strong basis for understanding science as a way of knowing.

## Strand 5: Engaging in Scientific Practices

A key challenge in the study of science learning in informal environments is to identify what counts as "doing science." Traditionally, scientific endeavors make use of specialized language, equipment, and representations, and the practice of science is typically seen as following a structured set of principles in particular laboratory-like or field-specific settings. However, research focused on everyday settings has highlighted that some features of scientific practice can often be found in routine activities (Nasir, Rosebery, Warren, and Lee, 2006). At the same time, studies of scientists in their actual daily practices have shown that the processes of science do not always follow the structured procedures taught as the scientific method (Latour and Woolgar, 1986; Knorr-Cetina, 1999; Collins, 1985).

The goal of pointing out that science-relevant practices appear in everyday activity is not to suggest that they are a replacement for the more traditional activities of the science lab or field site. But, as Nasir et al. (2006) argue, recognition of the overlap between everyday activities and the "official" activities of science can highlight some valuable access points to science for learners who might not otherwise engage in scientific activities. One example they describe is the research of Warren, Rosebery, and their colleagues (e.g., Warren and Rosebery, 1996) in which teachers working with Haitian youth help them to see that a common discourse practice of argumentation that they use in their community (called *bay odyans*) has much in common with the kind of persuasive language that scientists use to convince one another of their interpretations of findings. Encouraging these young people to see that this comfortable style of arguing has something in common with the practice of science has been shown to positively impact their science achievement (Warren and Rosebery, 1996). Many other researchers who investigate learning in informal settings have pointed to other spontaneous activities in these settings that can also be seen as part of the practice of science, including aspects of inquiry, analogy, imagining, and argumentation (Allen, 2002).

As noted above, in addition to the claim that everyday practices include components of doing science, a large body of research on scientists' practice shows that their work does not faithfully follow a single strict scientific method. Many specific examples have been documented. For example, Ochs, Gonzales, and Jacoby (1996) described the imaginative "theory talk" that took place in a physics research group. In working out hypothetical possibilities, scientists were often projecting themselves into the conceptual terrain of their subject matter and producing anthropomorphic talk about the entities that they study. Such talk seems to have much in common with children's imaginative talk. In a typical science classroom, such talk might be at risk of being labeled unscientific. Yet while this imaginative first-person talk isn't part of the stereotype of science, Ochs and colleagues show that it is indeed part of science.

Another aspect of science practice that has often been overlooked in studies of science learning is the social nature of the enterprise. There are two relevant versions of the argument that science is social. In one, science is social in that it involves groups of people working together to build explanations of the natural world. They communicate to identify scientific problems; to gather, analyze, and interpret evidence; to build explanations that account for the broadest set of observations; and to critique and improve on these accounts. In the other, the social dimension of science refers to the specialized norms and commitments that scientists share. They learn to talk in ways that build on other people's ideas or that criticize ideas. They learn to parse the evidence from the theory in ways that allow for careful analysis.

Studies of everyday science learning that do address social issues focus more on the former notion of science as a broadly social process, rather than on the sociological description of science as a specialized, normed form of interacting. Researchers have documented the fact that science is not the isolated activity that it is often perceived to be. Latour and Woolgar (1986) and Dunbar (1999) studied the process of scientists' work and documented the importance of the social aspects of the process. Similarly, children and adults reason about issues that are important to them while interacting with other people. Studies of dinner table conversations, visits to the zoo, and other everyday activities have uncovered rich conversations on a myriad of scientific topics and using scientific forms of discourse (Blum-Kulka, 1997; Callanan, Shrager, and Moore, 1995). Families with a working-class background as well as middle-class families engage in everyday conversations about a rich range of topics, including physics, biology, religion, and metaphysics (Blum-Kulka, 2002; Tenenbaum and Callanan, 2008). Researchers (Goodwin, 1994; Stevens and Hall, 1998) have documented how scientists and doctors learn to perceive in specialized ways, often through their use of technical tools and equipment or by recognizing the meaning of something (such as a bump on the skin or a particular flower in a meadow) that nonexperts see as normal or uninteresting. That research indicates that doctors- and scientists-in-training learn these specialized modes of perception through guided participation, or apprenticeship, as part of a deliberative practice (Prentice, 2004, 2005). Yet not as much is known specifically about how children learn to perceive the world in scientific ways, although there is no reason to doubt the utility of the apprenticeship and disciplined practice learning model.

Caregivers and other people around them interpret the world for children and guide them in learning about scientific topics (Gelman et al., 2004; Harris and Koenig, 2006; Harris et al., 2006). For example, one very powerful way that parents impart knowledge about the natural world to children is in their use of generic language—phrases that imply general rules, such as "pandas eat bamboo" and "stars come out at night." Gelman and her colleagues have found that even very young children are sensitive to the subtle differences between these generic statements and more specific statements, such as "that panda is eating bamboo," and that they make more inferences after hearing generic sentences (Gelman and Raman, 2003). By about the age of 3, children are likely to engage in causal conversations about mechanisms of change (Hood and Bloom, 1979; Callanan and Oakes, 1992; Tizard and Hughes, 1984). Similarly, in terms of scientific thinking, children engage with others in questioning, explaining, making predictions, and evaluating evidence (Callanan and Oakes, 1992; Chouinard, 2007). Thus, in a variety of ways, including family social activity and conversation, children may begin to learn about the content of science domains (at least in the middle-class, Western families where most of these studies have been done). Although young children and their parents may not think of any of these routine activities as relevant to later science classrooms, there is evidence that they are, in fact, important building blocks for later understanding of the domains of science (National Research Council, 2007). And the vast majority of this early learning about science occurs in settings in which there is no deliberate goal of teaching particular content or skills to the child. Instead, these are first and foremost everyday social activities in which children are motivated to participate, and in which learning is occurring as a result of that participation (Rogoff, 2003).

The particular things that children and adults learn are likely to vary depending on the physical environment as well as the community and particular family in which they are living (Heath, 2007). Variation in styles of everyday conversations across families in different communities has been noted. While explanation has been a focus of much of the work on scientific reasoning, narrative or story-telling is another example of a verbal form that researchers argue is relevant to science thinking (and crucially important for learning in general) (Bruner, 1996). Aukrust (2002), for example, found more focus on explanatory talk in American family conversations about their child's school day and more narrative conversation in Norwegian families (although both types of conversations occurred in both communities).

A number of studies have focused on whether families emphasize accountability, factuality, or evidence. While talk about accountability or factuality is emphasized by parents in many cultures, there is also evidence of some variation. Heath (2007) suggests that communities with literate traditions are more likely to expect children to be accountable to facts. For example, Valle (2007) found variation in middle-class, highly educated European American parents in their conversations during a homework-like task—whether they focused on evidence in evaluating conflicting claims (e.g., whether food additives are good or bad) was related to their major field of study in college. Blum-Kulka (1997) reports that Jewish American families in her study focused more on factuality in dinner conversations than did Israeli families (who gave equal emphasis to factuality, relevance, and politeness). Although the particular ways in which families talk about science topics vary widely, it may be that discussions of concepts and causal connections in the natural world are part of the experience of most children and adults.

The social nature of learning science also has consequences for how children interpret information from different adults. A growing field of study in cognitive development focuses on children's evaluation of sources of information. For example, quite young children are able to distinguish between adults who are knowledgeable on a topic and those who are not (Harris and Koenig, 2006; Lutz and Keil, 2002). Sabbagh and Baldwin (2001) found that when hearing a new word from a speaker who admits to not being certain, children did not learn the word as well as when hearing it from a knowledgeable speaker. Children also understand, at a young age, that there are experts on particular topics to whom one can turn for clarification of the true nature of things (Lutz and Keil, 2002).

More research is needed on exactly how children and their social partners negotiate new understandings of science in informal settings. Ash (2002) provides vivid descriptions of families making sense of natural phenomena in museum and aquarium settings. Crowley and Jacobs (2002) explore how children develop "islands of expertise" through interactions with parents in such settings as museums. One study found that children who became experts on some science topic (as defined by keeping a particular sustained interest for at least two years) were likely to have parents who focused more on supporting their children's curiosity and providing materials to support their interests (Leibham et al., 2005).

We are aware that throughout this chapter, and in the discussion of Strand 5 in particular, we may have left the sense that all social interactions are good and positive and move the learner forward. Certainly this is not always true. Gleason and Schauble (2000) have argued, for example, that parents often miss opportunities to support their children's learning. Goodnow (1990) argued that more sensitive attention is needed to the value judgments that parents make to support children's learning or steer them away from topics. Current debates about evolution and creationism provide a rich example of variation in the goals and choices that parents make in how they talk with children (Evans, 2001). In adolescence and adulthood, too, what is likely to be learned may or may not be consistent with the goals of science educators. These are issues for future research to address in more depth.

## Strand 6: Identifying with the Scientific Enterprise

Most people's everyday activities include experiences and social interactions that have the potential to engage them with science thinking or science content, although people differ in the extent to which they take such opportunities. Even more variation is apparent in the likelihood that one develops an identity as a science learner. Involvement in science has continued to be less common for girls and for students from nondominant groups, who must navigate a number of complex influences on their participation (Margolis and Fisher, 2002; Tate and Linn, 2005).

In Brown's (2004) work on discursive identities, for example, he discusses evidence that the same child may talk about contradictory religious and scientific beliefs in different contexts. Brown shows that African-American students may talk in scientific ways when focused on how they are viewed by teachers, but the same students may use a very different discourse style when their peers tease them about this scientific style of talk. There is clearly a complex set of issues surrounding learners' desire and willingness to see themselves as capable in science, and these issues vary depending on the age and life circumstances of the people involved.

There is significant evidence that a person's social network has a strong influence on their development of sustained interest (Barron, 2006; Lave and Wenger, 1991; Nasir, 2002). Developing and sustaining a personal, motivating connection to science (that is, an identity as a science learner) is influenced by one's social interactions and supports. As many personal anecdotes as well as systematic evidence show, parents, peers, mentors, and teachers can help to sustain the efforts of the learner, helping them to increase their competence, especially through difficult or trying periods (Barron, 2006; Nasir, 2002).

In considering these findings regarding cultural variability, however, it is important to keep in mind Heath's (2007) cautions regarding how best to think about culture (see also Gutiérrez and Rogoff, 2003). The key issue is socialization of children into ways of thinking about science and scientific topics, according to Heath. The important variations are based on cultural practices (including ways of talking), not cultural membership. The schooling of parents is often used as an important variable, but Heath brings up some major concerns about how schooling is viewed—including power issues. While an interest and effort in supporting children to succeed in school are broadly shared among parents across socioeconomic and other cultural groups, these factors play a powerful role in mediating how parents express their support and how they navigate the education system on behalf of their children.

Some approaches have suggested that nondominant students need help in bridging their everyday practices and ideas about science with more scientific ways of thinking. For example, Lee and Fradd (1996) found consistent but distinct patterns of discourse around science topics in different groups of students—bilingual Spanish, bilingual Haitian Creole, and monolingual English speakers. As noted above, some researchers disagree (e.g., Warren et al., 2001; Warren and Rosebery, 1996), arguing that all students' everyday ways of thinking include scientific skills, such as argument.

Gender is another factor that is related to identity as a science learner. A vast research literature documents gender differences in science achievement (Lawler, 2002; Mervis, 1999; National Science Foundation, 2002, 2007; Sax, 2001), and although the gap is narrowing in many areas (e.g., high school and college achievement, numbers entering fields like biology), some areas in science are still male dominated (e.g., highest levels of the academy, disciplines such as physics and engineering). Research on the everyday learning of science shows that parents provide more explanations to boys than to girls (Crowley, Callanan, Tenenbaum, and Allen, 2001), mirroring research on classroom settings (American Association of University Women, 1995; Jones and Wheatley, 1990; Tenenbaum and Leaper, 2003), suggesting that the gender gap may be at least partly encouraged by social influence.

Developing an identity as a science learner should also be understood as concurrent with and contingent on developing other identities. Tate and Linn (2005), for example, explored how identity influences the experiences of female engineering students of color using a multiple-identities framework. Their study design allows for analysis of gender and ethnic identity as semi-independent and interrelated. They explored three identity constructs: (1) academic identity, or how students engage in academic environments through such activities as help-seeking, tutoring, and mentoring; (2) social identity, or who they affiliate with, in what ways, for what purposes; and (3) intellectual identity, or understanding the knowledge base, driving questions, and operating practices of the field. They found that students tended to distinguish between their social and academic peer groups. Social groups tended to consist of individuals from the same ethnic background, whereas academic groups did not and reflected the predominant ethnic background of the program. Similarly, academic groups were infrequently used for non-academic (i.e., leisure) goals. They also found that the roles that students' identities play are context-dependent. In academic contexts, these women manifested very strong academic identities: they participated and sought help actively. At the same time, their social identity in academic contexts is characterized by a feeling of difference and not belonging.

Identity can be viewed as both a critical factor in shaping educational experiences and a goal into which a broad range of learning experiences can feed. And it is an important element for all learners. While discussions of identity, including many of the studies discussed in this section, draw on widely recognized ethnic and cultural identities, promoting identification with science learning is an important issue for learners from all backgrounds.

## CONCLUSION

Learners in everyday and family settings exemplify in their thinking and their practice the kinds of learning described in all the strands of science learning. We think the literature justifies and requires acknowledgment of the ways in which everyday science learning activities often overlap with more traditional science learning in labs and classrooms. Recognizing these links has particularly important promise for learners who have been outside the practice and identity of science, whether as children or as adults. More attention to everyday practices that are related to science may provide valuable tools for moving toward equity in access to science.

We recognize that the evidence for contributions from everyday science learning venues toward Strand 4 suggests less contribution than for other strands. The literature focuses more on learners' epistemic commitments and views of science (whether the learners are young or old) than on the ways that the everyday settings contribute to those commitments and views. The research, in fact, tends to focus on the limitations of learners' capabilities vis-à-vis reflection. We think further examination is warranted.

We acknowledge that everyday science cannot replace the kind of systematic and cumulative pedagogy that science educators have developed. For example, the concept of learning progressions has attracted substantial attention among science educators and researchers. Learning progressions call for the K-12 curriculum to build a small number of core scientific constructs across the curriculum. These major ideas are revisited recurrently from year to year with increasing depth and sophistication. Informed by developmental research, learning progressions also build on a broad range of science knowledge and skills, such as those reflected in the strands. Everyday learning cannot replace such systematic building of knowledge and experiences toward particular goals. However, everyday learning can augment and complement this and other curricular approaches to science learning. For example, they may be well suited to sparking early interest and for providing opportunities for deeper exploration of particular ideas.

A major challenge is to find more productive ways for everyday experiences with science to connect with more formal science learning. It is difficult to know how best to connect the pure moments of informal inquiry and exploration to the longer term goals of deeper scientific education. For example, creative use of spaces where the talk and practices of both science and everyday life can come together have shown particular promise in this arena (Barton, 2008).

Finally, we call attention to the disagreement in the literature as to the role of everyday experiences in children's developing scientific thinking. Some researchers are optimistic that everyday settings can be powerful, productive sources for (eventual) sophisticated, mature scientific knowledge. Others are more guarded and focus on how formal instruction should elicit and often correct scientific or science-like ideas that are developed in everyday settings. Further research is needed to illuminate the subtleties of the interaction between thinking about science in everyday and in school settings.

# REFERENCES

Agan, L., and Sneider, C. (2004). Learning about the Earth's shape and gravity: A guide for teachers and curriculum developers. *Astronomy Education Review, 2*(2), 90-117.

Allen, S. (2002). Looking for learning in visitor talk: A methodological exploration. In G. Leinhardt, K. Crowley, and K. Knutson (Eds.), *Learning conversations in museums* (pp. 259-303). Mahwah, NJ: Lawrence Erlbaum Associates.

American Association of University Women. (1995). *Growing smart: What's working for girls in school.* Researched by S. Hansen, J. Walker, and B. Flom at the University of Minnesota's College of Education and Human Development. Washington, DC: Author.

Amsterlaw, J., and Meltzoff, A.N. (2007, March). Children's evaluation of everyday thinking strategies: An outcome-to-process shift. In C.M. Mills (Chair), *Taking a critical stance: How children evaluate the thinking of others.* Symposium conducted at the Society for Research in Child Development, Boston.

Ash, D. (2002). Negotiation of biological thematic conversations about biology. In G. Leinhardt, K. Crowley, and K. Knutson (Eds.), *Learning conversations in museums* (pp. 357-400). Mahwah, NJ: Lawrence Erlbaum Associates.

Aukrust, V. (2002). What did you do in school today? Speech genres and tellability in multiparty family mealtime conversations in two cultures. In S. Blum-Kulka and C. Snow (Eds.), *Talking to adults: The contribution of multiparty discourse to language acquisition* (pp. 55-83). Mahwah, NJ: Lawrence Erlbaum Associates.

Azevedo, F.S. (2006). *Serious play: A comparative study of engagement and learning in hobby practices.* Unpublished dissertation, University of California, Berkeley.

Baillargeon, R. (2004). How do infants learn about the physical world? *Current Directions in Psychological Science, 3,* 133-140.

Ballantyne, R., and Bain, J. (1995). Enhancing environmental conceptions: An evaluation of cognitive conflict and structured controversy learning units. *Studies in Higher Education, 20*(3), 293-303.

Barron, B. (2006). Interest and self-sustained learning as catalysts of development: A learning ecology perspective. *Human Development, 49*(4), 153-224.

Barton, A.C. (2008). Creating hybrid spaces for engaging school science among urban middle school girls. *American Educational Research Journal, 45*(1), 68-103.

Bell, P., and Linn, M.C. (2002). Beliefs about science: How does science instruction contribute? In B. Hofer and P. Pintrich (Eds.), *Personal epistemology: The psychology of beliefs about knowledge and knowing* (pp. 321-346). Mahwah, NJ: Lawrence Erlbaum Associates.

Bell, P., Bricker, L.A., Lee, T.F., Reeve, S., and Zimmerman, H.H. (2006). Understanding the cultural foundations of children's biological knowledge: Insights from everyday cognition research. In A. Barab, K.E. Hay, and D. Hickey (Eds.), *7th international conference of the learning sciences, ICLS 2006* (vol. 2, pp. 1029-1035). Mahwah, NJ: Lawrence Erlbaum Associates.

Blum-Kulka, S. (1997). *Dinner talk: Cultural patterns of sociability and socialization in family discourse.* Mahwah, NJ: Lawrence Erlbaum Associates.

Blum-Kulka, S. (2002). Do you believe that Lot's wife is blocking the road (to Jericho)? Co-constructing theories about the world with adults. In S. Blum-Kulka and C.E. Snow (Eds.), *Talking to adults: The contribution of multiparty discourse to language acquisition* (pp. 85-116). Mahwah, NJ: Lawrence Erlbaum Associates.

Bricker, L.A., and Bell, P. (no date). *Evidentiality and evidence use in children's talk across everyday contexts.* Everyday Science and Technology Group, University of Washington.

Brodie, M., Hamel, E.C., Altman, D.E., Blendon, R.J., and Benson, J.M. (2003). Health news and the American public, 1996-2002. *Journal of Health Politics, Policy and Law, 28*(5), 927-950.

Brown, A.L., and Campione, J.C. (1996). Psychological theory and the design of innovative learning environments: On procedures, principles and systems. In L. Schauble and R. Glaser (Eds.), *Innovations in learning: New environments for education* (pp. 289-325). Mahwah, NJ: Lawrence Erlbaum Associates.

Brown, B.A. (2004). Discursive identity: Assimilation into the culture of science and its implications for minority students. *Journal of Research in Science Teaching, 41*(8), 810-834.

Bruner, J. (1996). *The culture of education.* Cambridge, MA: Harvard University Press.

Bullock, M., Gelman, R., and Baillargeon, R. (1982). The development of causal reasoning. In W.J. Friedman (Ed.), *The developmental psychology of time* (pp. 209-254). New York: Academic Press.

Callanan, M.A., and Oakes, L. (1992). Preschoolers' questions and parents' explanations: Causal thinking in everyday activity. *Cognitive Development, 7,* 213-233.

Callanan, M., Perez-Granados, D., Barajas, N., and Goldberg, J. (no date). *Everyday conversations about science: Questions as contexts for theory development.* Unpublished manuscript, University of California, Santa Cruz.

Callanan, M.A., Shrager, J., and Moore, J. (1995). Parent-child collaborative explanations: Methods of identification and analysis. *Journal of the Learning Sciences, 4,* 105-129.

Carey, S. (1985). *Conceptual change in childhood.* Cambridge, MA: MIT Press.

Chi, M.T.H., and Koeske, R.D. (1983). Network representation of a child's dinosaur knowledge. *Developmental Psychology, 19*(1), 29-39.

Chi, M.T.H., Hutchinson, J.E., and Robin, A.F. (1989). How inferences about novel domain-related concepts can be constrained by structured knowledge. *Merrill-Palmer Quarterly, 35*(1), 27-62.

Chouinard, M.M. (2007). Children's questions: A mechanism for cognitive development. *Monographs of the Society for Research in Child Development, 72*(1), 1-121.

Cohen, L.B., and Cashon, C.H. (2006). Infant cognition. In W. Damon and R.M. Lerner (Series Eds.) and D. Kuhn and R.S. Siegler (Vol. Eds.), *Handbook of child psychology: Cognition, perception, and language* (vol. 2, 6th ed., pp. 214-251). New York: Wiley.

Cole, M. (1996). *Cultural psychology: A once and future discipline.* Cambridge, MA: Harvard University Press.

Cole, M. (2005). Cross-cultural and historical perspectives on the developmental consequence of education. *Human Development, 48*(4), 195-216.

Collins, H.M. (1985). *Changing order: Replication and induction in scientific practice.* Beverley Hills, CA: Sage.

Crowley, K., and Galco, J. (2001). Everyday activity and the development of scientific thinking. In K. Crowley, C.D. Schunn, and T. Okada (Eds.), *Designing for science: Implications from everyday, classroom, and professional settings* (pp. 123-156). Mahwah, NJ: Lawrence Erlbaum Associates.

Crowley, K., and Jacobs, M. (2002). Islands of expertise and the development of family scientific literacy. In G. Leinhardt, K. Crowley, and K. Knutson (Eds.), *Learning conversations in museums* (pp. 333-356). Mahwah, NJ: Lawrence Erlbaum Associates.

Crowley, K., Callanan, M.A., Tenenbaum, H.R., and Allen, E. (2001). Parents explain more often to boys than to girls during shared scientific thinking. *Psychological Science, 12*(3), 258-261.

Csikszentmihalyi, M. (1996). *Creativity: Flow and the psychology of discovery and invention.* New York: HarperCollins.

Csikszentmihalyi, M., and Larson, R. (1984). *Being adolescent.* New York: Basic Books.

Dickerson, S., Reinhart, A.M., Feeley, T.H., Bidani, R., Rich, E., Garg, V.K., and Hershey, C.O. (2004). Patient Internet use for health information at three urban primary care clinics. *Journal of American Medical Information Association, 11*, 499-504.

diSessa, A. (1988). Knowledge in pieces. In G. Forman and P. Pufall (Eds.), *Constructivism in the computer age* (pp. 49-70). Mahwah, NJ: Lawrence Erlbaum Associates.

Driver, R., Leach, J., Millar, R., and Scott, P. (1996). *Young people's images of science.* Buckingham, England: Open University Press.

Dunbar, K. (1999). The scientist in vivo: How scientists think and reason in the laboratory. In L. Magnanai, N. Nersessian, and P. Thagard (Eds.), *Model-based reasoning in scientific discovery* (pp. 89-98). New York: Plenum.

Eccles, J.S., Lord, S., and Buchanan, C.M. (1996). School transitions in early adolescence: What are we doing to our young people? In J. Graber, J. Brooks-Gunn, and A. Petersen (Eds.), *Transitions through adolescence: Interpersonal domains and context* (pp. 251-284). Mahwah, NJ: Lawrence Erlbaum Associates.

Epstein, S. (1996). *Impure science: AIDS, activism, and the politics of knowledge.* Berkeley: University of California Press.

Erickson, F., and Gutiérrez, K. (2002). Culture, rigor, and science in educational research. *Educational Researcher, 31*(8), 21-24.

Evans, E.M. (2001). Cognitive and contextual factors in the emergence of diverse belief systems: Creation versus evolution. *Cognitive Psychology, 42*(3), 217-266.

Evans, E.M. (2005). Teaching and learning about evolution. In J. Diamond (Ed.), *The virus and the whale: Explore evolution in creatures small and large.* Arlington, VA: NSTA Press.

Falk, J.H., and Dierking, L.D. (2002). *Lessons without limit: How free choice learning is transforming education.* Walnut Creek, CA: AltaMira Press.

Falk, J.H., and Storksdieck, M. (2005). Using the contextual model of learning to understand visitor learning from a science center exhibition. *Science Education, 89*(5), 744-778.

Farenga, S.J., and Joyce, B.A. (1997). Beyond the classroom: Gender differences in science experiences. *Education, 117*, 563-568.

Flynn, K.E., Smith, M.A., and Freese, J. (2006). When do older adults turn to the Internet for health information? Findings from the Wisconsin longitudinal study. *Journal of General Internal Medicine, 21*(12), 1295-1301.

Fox, S. (2006). *Online health search.* Washington, DC: Pew Internet and American Life Project.

Gelman, R., and Baillargeon, R. (1983). A review of some Piagetian concepts. In J. H. Flavell and E. Markman (Eds.), *Cognitive development: Handbook of child development* (vol. 3, pp. 167-230). New York: Wiley.

Gelman, S.A. (2003). *The essential child: Origins of essentialism in everyday thought.* New York: Oxford University Press.

Gelman, S.A., and Gottfried, G.M. (1996). Children's causal explanations of animate and inanimate motion. *Child Development, 67*(5), 1970-1987.

Gelman, S.A., and Kalish, C.W. (2006). Conceptual development. In D. Kuhn and R. Siegler (Eds.), *Handbook of child psychology: Cognition, perception and language* (vol. 2, pp. 687-733). New York: Wiley.

Gelman, S.A., and Raman, L. (2003). Preschool children use linguistic form class and pragmatic cues to interpret generics. *Child Development, 74*(1), 308-325.

Gelman, S.A., Taylor, M.G., Nguyen, S., Leaper, C., and Bigler, R.S. (2004). Mother-child conversations about gender: Understanding the acquisition of essentialist beliefs. *Monographs of the Society for Research in Child Development, 69*(1), 145.

Gleason, M.E., and Schauble, L. (2000). Parents' assistance of their children's scientific reasoning. *Cognition and Instruction, 17,* 343-378.

Goodnow, J.J. (1990). The socialization of cognition: What's involved? In J.W. Stigler, R.A. Shweder, and G.H. Herdt (Eds.), *Cultural psychology: Essays on comparative human development* (pp. 259-286). New York: Cambridge University Press.

Goodwin, C. (1994). Professional vision. *American Anthropologist, 96*(3), 606-633.

Goodwin, M.H. (2007). Occasioned knowledge exploration in family interaction. *Discourse and Society, 18*(1), 93-110.

Gopnik, A. (1998). Explanation as orgasm. *Minds and Machines, 8*(1), 101-118.

Gopnik, A., and Wellman, H.M. (1992). Why the child's theory of mind really is a theory. *Mind and Language, 7*(1-2), 145-171.

Gopnik, A., Glymour, C., Sobel, D., Schulz, L., Kushnir, T., and Danks, D. (2004). A theory of causal learning in children: Causal maps and Bayes nets. *Psychological Review, 111,* 1-31.

Gopnik, A., Meltzoff, A., and Kuhl, P. (1999). *The scientist in the crib.* New York: Morrow.

Gopnik, A., Sobel, D.M., Schulz, L., and Glymour, C. (2001). Causal learning mechanisms in very young children: Two, three, and four-year-olds infer causal relations from patterns of variation and covariation. *Developmental Psychology, 37*(5), 620-629.

Gutiérrez, K., and Rogoff, B. (2003). Cultural ways of learning: Individual traits or repertoires of practice. *Educational Researcher, 32*(5), 19-25.

Halford, G.S., and Andrews, G. (2006). Reasoning and problem solving. In D. Kuhn and R. Siegler (Eds.), *Handbook of child psychology: Cognitive, language and perceptual development* (6th ed., vol. 2, pp. 557-608). Hoboken, NJ: Wiley.

Hall, R., and Schaverien, L. (2001). Families' participation in young children's science and technology learning. *Science Education, 85*(4), 454-481.

Harris, P.L., and Koenig, M.A. (2006). Trust in testimony: How children learn about science and religion. *Child Development, 77*(3), 505-524.

Harris, P.L., Pasquini, E.S., Duke, S., Asscher, J.J., and Pons, F. (2006). Germs and angels: The role of testimony in young children's ontology. *Developmental Science, 9*(1), 76-96.

Heath, S.B. (1983). *Ways with words: Language, life and work in communities and classroom.* New York: Cambridge University Press.

Heath, S.B. (1999). Dimensions of language development: Lessons from older children. In A.S. Masten (Ed.), *Cultural processes in child development: The Minnesota symposium on child psychology* (vol. 29, pp. 59-75). Mahwah, NJ: Lawrence Erlbaum Associates.

Heath, S.B. (2007). *Diverse learning and learner diversity in "informal" science learning environments.* Commissioned paper prepared for the National Research Council Committee on Science Education for Learning Science in Informal Environments. Available: http://www7.nationalacademies.org/bose/Brice%20Heath_Commissioned_Paper.pdf [accessed February 2009].

Hidi, S., and Renninger, K.A. (2006). The four-phase model of interest development. *Educational Psychologist, 41*(2), 111-127.

Hood, L., and Bloom, L. (1979). What, when, and how about why: A longitudinal study of early expressions of causality. *Monographs of the Society for Research in Child Development, 44*(6), 1-47.

Howe, C., McWilliam, D., and Cross, G. (2005). Chance favours only the prepared mind: Incubation and the delayed effects of peer collaboration. *British Journal of Psychology, 96*(1), 67-93.

Howe, C., Tobmie, A., and Rodgers, C. (1992). The acquisition of conceptual knowledge in science by primary school children: Group interaction and the understanding of motion down an incline. *British Journal of Psychology, 10*(2), 113-130.

Inagaki, K., and Hatano, G. (1996). Young children's recognition of commonalities between animals and plants. *Child Development, 67*(6), 2823-2840.

Ioannides, C.H., and Vosniadou, S. (2002). The changing meaning of force. *Cognitive Science Quarterly, 2*(1), 5-62.

Irwin, A., and Wynne, B. (Eds.). (1996). *Misunderstanding science? The public reconstruction of science and technology.* New York: Cambridge University Press.

Jipson, J.L., and Gelman, S.A. (2007). Robots and rodents: Children's inferences about living and nonliving kinds. *Child Development, 78*(6), 1675-1688.

Johnson, K., Alexander, J., Spencer, S., Leibham, M., and Neitzel, C. (2004). Factors associated with the early emergence of intense interests within conceptual domains. *Cognitive Development, 19*(3), 325-343.

Jones, G., and Wheatley, J. (1990). Gender differences in teacher-student interactions in science classrooms. *Journal of Research in Science Teaching, 27*(9), 861-874.

Kelly, L., Savage, G., Landman, P., and Tonkin, S. (2002). *Energised, engaged, everywhere: Older Australians and museums.* Canberra: National Museum of Australia.

Klahr, D. (2000). *Exploring science: The cognition and development of discovery processes.* Cambridge, MA: MIT Press.

Knorr-Cetina, K.D. (1999). *Epistemic cultures: How the sciences make knowledge.* Cambridge, MA: Harvard University Press.

Korpan, C.A., Bisanz, G.L., Bisanz, J., and Lynch, M.A. (1998). *Charts: A tool for surveying young children's opportunities to learn about science outside of school.* Ottawa: Canadian Social Science and Humanities Research Council.

Krist, H., Fieberg, E.L., and Wilkening, F. (1993). Intuitive physics in action and judgment: The development of knowledge about projectile motion. *Journal of Experimental Psychology: Learning, Memory, and Cognition, 19*(4), 952.

Kuhn, D. (1989). Children and adults as intuitive scientists. *Psychological Review, 96*(4), 674-689.

Kuhn, D. (1996). Is good thinking scientific thinking? In D. Olson and N. Torrance (Eds.), *Modes of thought: Explorations in culture and cognition* (pp. 261-281). New York: Cambridge University Press.

Kushnir, T., and Gopnik, A. (2005). Young children infer causal strength from probability and intervention. *Psychological Science, 16*, 678-683.

Latour, B., and Woolgar, S. (1986). *Laboratory life: The social construction of scientific facts.* Princeton, NJ: Princeton University Press.

Lave, J., and Wenger, E. (1991). *Situated learning: Legitimate peripheral participation.* New York: Cambridge University Press.

Lawler, A. (2002). Engineers marginalized, MIT report concludes. *Science, 295*(5563), 2192.

Layton, D. (1993). *Inarticulate science? Perspectives on the public understanding of science and some implications for science education.* Driffield, England: Studies in Education.

Lederman, N.G. (1992). Students' and teachers' conceptions of the nature of science: A review of the research. *Journal of Research in Science Teaching, 29*(4), 331-359.

Lee, O., and Fradd, S. (1996). Interactional patterns of linguistically diverse students and teachers: Insights for promoting science learning. *Linguistics and Education, 8*(3), 269-297.

Lehrer, R., and Schauble, L. (2006). Scientific thinking and scientific literacy. In W. Damon, R. Lerner, K.A. Renninger, and E. Sigel (Eds.), *Handbook of child psychology* (6th ed., vol. 4, pp. 153-196). Hoboken, NJ: Wiley.

Leibham, M.E., Alexander, J.M., Johnson, K.E., Neitzel, C., and Reis-Henrie, F. (2005). Parenting behaviors associated with the maintenance of preschoolers' interests: A prospective longitudinal study. *Journal of Applied Developmental Psychology, 26*(4), 397-414.

Lumpe, A. (1995). Peer interaction in science concept development and problem solving. *School Science and Mathematics, 6*, 302-310.

Luo, Y., and Baillargeon, R. (2005). When the ordinary seems unexpected: Evidence for incremental physical knowledge in young infants. *Cognition, 95*, 297-328.

Lutz, D.R., and Keil, F.C. (2002). Early understanding of the division of cognitive labor. *Child Development, 73*(4), 1073-1084.

Madden, M., and Fox, S. (2006). *Finding answers online in sickness and in health.* Washington, DC: Pew Internet and American Life Project.

Margolis, J., and Fisher, A. (2002). *Unlocking the clubhouse: Women in computing.* Cambridge, MA: MIT Press.

McDermott, R.P., Goldman, S.V., and Varenne, H. (1984). When school goes home: Some problems in the organization of homework. *Teachers College Record, 85*(3), 391-409.

Mervis, J. (1999, April). High-level groups study barriers women face. *Science, 284*(5415), 727.

Nasir, N.S. (2002). Identity, goals, and learning: Mathematics in cultural practice. *Mathematical Thinking and Learning, 4*(2-3), 213-248.

Nasir, N.S., Rosebery, A.S., Warren B., and Lee, C.D. (2006). Learning as a cultural process: Achieving equity through diversity. In R. Keith Sawyer (Ed.), *The Cambridge handbook of the learning sciences* (pp. 489-504). New York: Cambridge University Press.

National Research Council. (2000). *How people learn: Brain, mind, experience, and school* (expanded ed.). Committee on Developments in the Science of Learning, J.D. Bransford, A.L. Brown, and R.R. Cocking (Eds.), and Committee on Learning Research and Educational Practice, M.S. Donovan, J.D. Bransford, and J.W. Pellegrino (Eds.), Commission on Behavioral and Social Sciences and Education. Washington, DC: National Academy Press.

National Research Council. (2007). *Taking science to school: Learning and teaching science in grades K-8*. Committee on Science Learning, Kindergarten Through Eighth Grade. R.A. Duschl, H.A. Schweingruber, and A.W. Shouse (Eds.). Board on Science Education, Center for Education, Division of Behavioral and Social Sciences and Education. Washington, DC: The National Academies Press.

National Science Foundation. (2002). *Gender differences in the careers of academic scientists and engineers*. (NSF 04-323.) Arlington, VA: Author.

National Science Foundation. (2007). *Women, minorities, and persons with disabilities in science and engineering* (NSF 07-315.) Arlington, VA: Author. Available: http://www.nsf.gov/statistics/wmpd [accessed October 2008].

Nussbaum, J., and Novak, J.D. (1976). An assessment of children's concepts of the earth utilizing structured interviews. *Science Education, 60*(4), 535-555.

Ochs, E., Gonzales, P., and Jacoby, S. (1996). When I come down I'm in the domain state: Grammar and graphic representation of the interpretive activity of physicists. In E. Ochs, E.A. Schegloff, and S.A. Thompson (Eds.), *Interaction and grammar* (pp. 328-369). New York: Cambridge University Press.

Ochs, E., Smith, R., and Taylor, C.E. (1996). Detective stories at dinnertime: Problem solving through co-narration. In C.L. Briggs (Ed.), *Disorderly discourse: Narrative, conflict and inequality* (pp. 95-113). New York: Oxford University Press.

Palmquist, S., and Crowley, K. (2007). From teachers to testers: How parents talk to novice and expert children in a natural history museum. *Science Education, 91*(5), 783-804.

Pereira, J.L., Koski, S., Hanson, J., Bruera, E.D., and Mackey, J.R. (2000). Internet usage among women with breast cancer: An exploratory study. *Clinical Breast Cancer, 1*(2), 148-153.

Prentice, R. (2004). *Bodies of information: Reinventing bodies and practice in medical education*. Unpublished doctoral dissertation, Massachusetts Institute of Technology.

Prentice, R. (2005). The anatomy of a surgical simulation: The mutual articulation of bodies in and through the machine. *Social Studies of Science, 35*(6), 837-866.

Rogoff, B. (2003). *The cultural nature of human development*. New York: Oxford University Press.

Rogoff, B., Paradise, R., Mejía Arauz, R., Correa-Chávez, M., and Angelillo, C. (2003). Firsthand learning by intent participation. *Annual Review of Psychology, 54,* 175-203.

Ross, N., Medin, D., Coley, J.D., and Atran, S. (2003). Cultural and experiential differences in the development of folkbiological induction. *Cognitive Development, 18*(1), 35-47.

Sabbagh, M.A., and Baldwin, D.A. (2001). Learning words from knowledgeable versus ignorant speakers: Links between preschoolers' theory of mind and semantic development. *Child Development, 72*(4), 1054-1070.

Sachatello-Sawyer, B. (2006). *Adults and informal science learning.* Presentation to the National Research Council Committee on Learning Science in Informal Environments, Washington, DC. Available: http://www7.nationalacademies. org/bose/Learning_Science_in_Informal_Environments_Commissioned_Papers. html [accessed November 2008].

Sachatello-Sawyer, B., Fellenz, R.A., Burton, H., Gittings-Carlson, L., Lewis-Mahony, J., and Woolbaugh, W. (2002). *Adult museum programs: Designing meaningful experiences.* American Association for State and Local History Book Series. Blue Ridge Summit, PA: AltaMira Press.

Samarapungavan, A., Vosniadou, S., and Brewer, W.F. (1996). Mental models of the earth, sun and moon: Indian children's cosmologies. *Cognitive Development, 11*(4), 491-521.

Sandoval, W.A. (2005). Understanding students' practical epistemologies and their influence on learning through inquiry. *Science Education, 89*(4), 634-656.

Sax, L.J. (2001). Undergraduate science majors: Gender differences in who goes to graduate school. *Review of Higher Education, 24*(2), 153-172.

Saxe, R., Tzelnic, T., and Carey, S. (2007). Knowing who dunnit: Infants identifying the casual agent in an unseen casual interaction. *Developmental Psychology, 43*(1), 149-158.

Schauble, L. (1996). The development of scientific reasoning in knowledge-rich contexts. *Developmental Psychology, 32*(1), 102-119.

Simon, H.A. (2001). "Seek and ye shall find": How curiosity engenders discovery. In K. Crowley, C. Schunn, and T. Okada (Eds.), *Designing for science: Implications from everyday, classroom, and professional settings* (pp. 5-20). Mahwah, NJ: Lawrence Erlbaum Associates.

Smith, J.P., diSessa, A.A., and Roschelle, J. (1993). Misconceptions reconceived: A constructivist analysis of knowledge in transition. *Journal of the Learning Sciences, 3*(2), 115-163.

Snir, J., Smith, C.L., and Raz, G. (2003). Linking phenomena with competing underlying models: A software tool for introduction students to the particulate model of matter. *Science Education, 87,* 794-830.

Snyder, C.I., and Ohadi, M.M. (1998). Unraveling students' misconceptions about the earth's shape and gravity. *Science Education, 82*(2), 265-284.

Songer, N.B., and Linn, M.C. (1991). How do students' views of the scientific enterprise influence knowledge integration? *Journal of Research in Science Teaching, 28*(9), 761-784.

Spelke, E.S. (2002). Developmental neuroimaging: A developmental psychologist looks ahead. *Developmental Science, 5*(3), 392-396.

Spradley, J.P. (1980). *The ethnographic interview*. New York: Holt, Rinehart, and Winston.

Springer, K., and Keil, F. (1991). Early differentiation of causal mechanisms appropriate to biological and nonbiological kinds. *Child Development, 62*, 767-781.

Stevens, R., and Hall, R. (1998). Disciplined perception: Learning to see in techno-science. In M. Lampert and M.L. Blunk (Eds.), *Talking mathematics in school: Studies of teaching and learning* (pp. 107-149). New York: Cambridge University Press.

Tardy, R.W., and Hale, C.L. (1998). Bonding and cracking: The role of informal, interpersonal networks in health care decision making. *Health Communication, 10*(2), 151-173.

Tarlowski, A. (2006). If it's an animal it has axons: Experience and culture in preschool children's reasoning about animates. *Cognitive Development, 21*(3), 249-265.

Tate, E.D., and Linn, M.C. (2005). How does identity shape the experiences of women of color engineering students? *Journal of Science Education and Technology, 14*(5-6), 483-493.

Tenenbaum, H.R., and Callanan, M.A. (2008). Parents' science talk to their children in Mexican-descent families residing in the United States. *International Journal of Behavioral Development, 32*(1), 1-12.

Tenenbaum, H.R., and Leaper, C. (2003). Parent-child conversations about science: The socialization of gender inequities? *Developmental Psychology, 39*(1), 34-47.

Tharp, R.G., and Gallimore, R. (1989). *Rousing minds to life: Teaching and learning in social context*. New York: Cambridge University Press.

Tizard, B., and Hughes, M. (1984). *Young children learning*. Cambridge, MA: Harvard University Press.

Treagust, D.F. (1988). Development and use of diagnostic tests to evaluate students' misconceptions in science. *International Journal of Science Education, 10*(2), 159-169.

Tschirgi, J.E. (1980). Sensible reasoning: A hypothesis about hypotheses. *Child Development, 51*, 1-10.

Tversky, A., and Kahneman, D. (1986). Rational choice and the framing of decisions. *Journal of Business, 59*(4), 251-278.

Valle, A. (2007, April). *Developing habitual ways of reasoning: Epistemological beliefs and formal bias in parent-child conversations*. Poster presented at biennial meeting of the Society for Research in Child Development, Boston.

Valle, A., and Callanan, M.A. (2006). Similarity comparisons and relational analogies in parent-child conversations about science topics. *Merrill-Palmer Quarterly, 52*(1), 96-124.

von Hofsten, C. (2004). An action perspective on motor development. *Trends in Cognitive Sciences, 8*(6), 266-272.

Vosniadou, S., and Brewer, W.F. (1992). Mental models of the earth: A study of conceptual change in childhood. *Cognitive Psychology, 24*(4), 535-585.

Warren, B., and Rosebery, A.S. (1996). This question is just too, too easy! Students' perspectives on accountability in science. In L. Schauble and R. Glaser (Eds.), *Innovations in learning: New environments for education* (pp. 97-126). Mahwah, NJ: Lawrence Erlbaum Associates.

Warren, B., Ballenger, C., Ogonowski, M., Rosebery, A., and Hudicourt-Barnes, J. (2001). Rethinking diversity in learning science: The logic of everyday sense-making. *Journal of Research in Science Teaching, 38,* 529-552.

Wason, P.C. (1960). On the failure to eliminate hypotheses in a conceptual task. *Quarterly Journal of Experimental Psychology, 12*(4), 129-140.

Waxman, S.R. (2005). Why is the concept "living thing" so elusive? Concepts, languages, and the development of folkbiology. In W. Ahn, R.L. Goldstone, B.C. Love, A.B. Markman, and P. Wolff (Eds.), *Categorization inside and outside the laboratory: Essays in honor of Douglas L. Medin.* Washington, DC: American Psychological Association.

Waxman, S., and Medin, D. (2007). Experience and cultural models matter: Placing firm limits on anthropocentrism. *Human Development, 50,* 23-30.

Zimmer-Gembeck, M.J., and Collins, W.A. (2003). Autonomy development during adolescence. In G.R. Adams and M.D. Berzonsky (Eds.), *Blackwell handbook of adolescence* (pp. 175-204). Malden, MA: Blackwell.

Zimmerman, C. (2000). The development of scientific reasoning skills. *Developmental Review, 20*(1), 99-149.

5

# Science Learning in Designed Settings

This chapter describes informal environments that are intentionally designed for learning about science and the physical and natural world. Designed settings include institutions such as museums, science centers, aquariums, and environmental centers, and the smaller components contained within these settings, such as exhibits, exhibitions, demonstrations, and short-term programs. Like everyday learning, learning in designed settings is highly participant structured, but also reflects the intended communicative and pedagogical goals of designers and educators. And in important ways, designed spaces are unlike science learning programs. Science learning programs serve a subscribed group and recur over time, whereas learning in designed spaces tends to be more fluid and sporadic. An important feature for structuring learning in these environments is that they are typically experienced episodically, rather than continuously.

Another defining characteristic of designed spaces is that they are navigated freely, with limited or often no direct facilitation from institutional actors. Visitors may freely choose which of the exhibits to interact with, and they receive little guidance as to which path they should follow as they explore. This design is typical, and reflects the learner's personal choice about learning in these settings. Should the learner choose to design their own systematic study of a given topic, the option is available. Institutions typically shy away from directing a particular course, opting instead for multiple entry levels and possible navigational paths through the public space. Whereas classrooms have teachers and Cub Scouts have den leaders, designed settings rely primarily on objects, labels, spaces, recorded messages, brief interpretive guides, and occasionally docents or interpreters to facilitate learner engagement. They are designed to serve a diverse public in the myriad social configurations they assemble. Thus, individuals, families, and teen peer groups are all understood as participants whose needs and interests should be accommodated in designed spaces.

Individual learners and groups play an important role in determining their own learning outcomes in designed spaces (Moussouri, 2002). Contemporary views of learning as an active, constructive process have led to increased attention to learners' motivations, prior experiences, tacit knowledge, and cultural identity (National Research Council, 2007). While professional educators—designers, facilitators, teachers, curators—have scientific, social, practical, or other goals for participants, these are achieved only in partnership with learners. This is particularly salient in designed spaces, where learners are not assumed to operate under strong cultural pressures to participate or achieve a particular goal, as they may be pressured to do in schools, educational programs, and workplace settings. Participants in designed science learning settings control their own learning agenda.

The science learning that takes place in designed settings is shaped by elements of intentional design, personal interpretation and choice, and chance. The environment—both large-scale characteristics of the institution and small-scale features of exhibits and programs—helps to guide or mediate the visitors' attitudes or perspectives, their relationship with the content and the institution, the meaning of their activity there, and how the institution views them. Learners typically participate of their own volition and at their own pace. They may be scientific experts or novices, or anyone in between.

Not surprisingly, experiences in these spaces are often designed to elicit participants' emotions or sensory responses to scientific and natural phenomena. For example, zoos and aquariums may develop conservation themes linking plant, animal, and human well-being. Science centers use multimedia to engage multiple senses, or build larger-than-life models that make phenomena visible and inspire participants' awe. Emotional and interactive sensory experiences are design priorities, though they are typically accompanied by particular informational or cognitive goals as well.

From the perspective of science learning, a key educational challenge for designed spaces is to link emotional and sensory responses with science-specific phenomena. Associating scientific thinking with engaging and enjoyable events and real-world outcomes can create important connections on a personal level. Promoting or supporting a variety of emotional responses (surprise, puzzlement, awe) and a variety of processing modes (observation, discovery, contemplation) increases the likelihood of connecting with a greater variety of people and encouraging them as learners (Jacobson, 2006).

## LEARNING IN DESIGNED SPACES

Although the process of learning itself is not necessarily different in designed settings than it is in everyday settings or in programs for science learning, designed spaces do use special methods for structuring, teaching, guiding, and prompting learning.

The scale of designed learning spaces varies, and so does the way that the public interacts with these spaces. At the institutional level, there are distinctions among the types of materials and objects housed or collected. Zoos, aquariums, and nature centers, for example, typically maintain live collections. Traditional museums and science centers typically (though not always) organize nonliving collections that may include scientific artifacts (e.g., mineral specimens), tools employed in scientific inquiry (e.g., telescopes), and pedagogical exhibits (e.g., a supersized panpipe designed to explore vibration and pitch). The substantive focus of a particular institution has important implications for its goals. For example, designed spaces with live animal collections may focus primarily on conservation goals—goals with observable behavioral implications (e.g., participants may make unique consumer choices that reflect a conservation ethic). Science centers may pursue somewhat broader or less easily observable goals, such as supporting future inquiry and inspiring curiosity.

Research on learning in designed spaces has provided evidence of learning across the strands. Some studies focus on the importance of developing scientific ideas and processes of science, in interaction with others (Ash, 2003; Crowley and Jacobs, 2002; Tunnicliffe, 2000). Other studies have described science learning in informal settings as an opportunity to appropriate the language or participate in the "culture" of science (Borun et al., 1998; Crowley and Callanan, 1998; Ellenbogen, 2003). Still others have explored the idea that learning involves a change in identity—specifically, how people view or present themselves, and how others see them (Holland, Lachicotte, Skinner, and Cain, 1998; Wenger, 1999).

Before delving into the specific strands, we should not lose sight of the fact that individuals choose to spend their time in these settings and that this choice in itself can be seen as an indication of their participation in science (as indicated in Strand 5) and at least a weak proxy for learning. As mentioned in Chapter 1, the scale of participation in designed settings, though crudely estimated, is certainly vast: U.S. museums and science centers tally hundreds of millions of visits each year. While counting heads is no substitute for careful analysis of how learners participate and what they learn, and there are significant biases in terms of the cultural and demographic characteristics of individuals and families that tend to participate in designed settings, nevertheless the fact that large numbers of people choose to attend, often paying for admission, is an important measure for a field that is predicated on learner choice. In addition, attendance records and many large-scale visitor surveys show that the public has a positive view of informal environments for science learning, seeks them out during leisure time (Hilke, 1987; Ivanova, 2003; Briseno-Garzon, Anderson, and Anderson, 2007; Moussouri, 1998), and values both the entertainment and learning aspects that these institutions offer. This suggests that such institutions are viewed positively on a broad scale. Some contend that they are part of the nation's science education infrastructure (St. John and Perry, 1993), one measure of system-wide impact. Although we focus primarily on designed settings, we also note that schools and field trips play an important role; Box 5-1 is a summary of the relevant research on field trips.

## Strand 1: Developing Interest in Science

Some key assumptions about learning in informal environments are that exciting experiences lead to intrinsically motivated learning, and that these experiences are personally meaningful, providing experiential foundations for more advanced structured, science learning. Perry (1994), for example, proposes that curiosity, confidence, challenge, and play are among the essential elements of intrinsically motivating experiences in museums. This is an area of tremendous interest to informal science educators and has been documented extensively in evaluations and the accounts of practitioners. To provide an inclusive summary here, we integrate conventional forms of published, peer-reviewed literatures with anecdotes and excerpts from evaluation reports.

### *Excitement*

Numerous evaluation studies show that visitors to informal environments report feeling excitement as a result of their experiences. For example, consider the following from Tisdal (2004, p. 24):

> Another visitor noted the pleasure he took in watching children get excited about science: "I was talking to the mother of the other boy that was there and just kind of—not necessarily small talk, but talking about the objects and how you could see how he was really excited when he was playing with it. And we had some jokes going on about (inaudible) when he had the football up in the air, and he got a little excited about the whole thing. It was cool to see him light up over something that—you know, science isn't normally fun for those kids. So I thought that was kind of cool, that we were having a good time over there" (Case 6, male, age 18).

Researchers also often observe signs of positive excitement among visitors. They cite expressions of joy, delight, awe, wonder, appreciation, surprise, intrigue, interest, caring, inspiration, satisfaction, and meaningfulness. For example:

"The size of animals that you have in there . . . I was just flabbergasted. But they are all extremely well maintained. I can tell by looking that everything is thriving. It's not just living" (120404-3) (Beaumont, 2005, p. 14).

"I think [the exhibition] is inspirational—that regular people can invent things. That is how I felt [when I read] about the lady [who invented] Kevlar [Stephanie Kwolek]" (National Museum of American History; female, age 42) (Korn, 2004, p. 44).

"It was fun. It was beautiful. The ice crystals, the colors in the ice crystals were beautiful. I think it is a great exhibit. It's the only time I've seen that kind of exhibit—it's sort of, each crystal is different, each time you do it will be different" (Tisdal, 2004, p. 29).

Allen (2002) notes that affective responses (defined as verbal expressions of feeling) were one of the three most common forms of "learning talk" in visitors' conversations while viewing an exhibition on frogs. Visitors expressed their feelings at 57 percent of all exhibit elements at which they stopped. The most common subcategories were surprise/intrigue (37 percent) and pleasure (36 percent).

Some evidence from experimental social psychology and neuropsychology suggests a link between excitement and other forms of learning (e.g., Steidl, Mohi-uddin, and Anderson, 2006). Models of the relation of mood to substantive cognitive processing, as well as studies of operant conditioning, have predicted and demonstrated that mood states or internal responses influence the information used during processing in laboratory situations (Bower, 1981; Eich et al., 2000). The precise relationship is not yet well understood, and the influence of excitement can alternately enhance or detract from learning. Specific connections between affect, thinking, and activity settings, moreover, have not been studied and are clearly needed.

## Interest

The construct of interest takes one deeper into the question of what people learn from experiences in informal environments. Hidi and Renninger (2006) distinguish between situational interest (short-lived, typically evoked by the environment) and individual interest (more stable and specific to an individual). Based on a number of studies, they propose a four-phase model of interest development: (1) triggered situational interest, typically sparked by such environmental features as incongruous/surprising information or personal relevance; (2) maintained situational interest, sustained through the meaningfulness of tasks and personal involvement; (3) emerging individual interest; and (4) well-developed individual interest, in which the individual chooses to engage in an extended pursuit using systematic approaches to questioning and seeking answers. Interestingly, this sequence of increasing

## BOX 5-1 Field Trips

School groups make up a large proportion of the visitors to science learn-ing institutions. Several studies have pointed to possible long-term impacts of field trips—typically, memories of specific experiences (Anderson and Piscitelli, 2002; Falk and Dierking, 1997). In fact, all of the elementary and middle school students and adults interviewed by Falk and Dierking (1997), in a study of students who visited a museum on a field trip, were able to recall at least one thing they had learned on a field trip. The nature and more immediate impact of schoolchildren's visits vary widely, however (Kisiel, 2006; Orion and Hofstein, 1994; Price and Hein, 1991; Storksdieck, 2006). Although results are mixed regarding the impact of field trips to informal institutions on children's attitudes, interest, and knowledge of science, the majority of studies that have measured knowledge and attitudes have found positive changes (Koran, Koran, and Ellis, 1989). Most of the work on interpreted visits to museums looks at the structure of field trips and how their effectiveness can be improved.

In general, the impact of field trips made to such institutions as museums, zoos, and nature centers is dependent on several critical factors: advance content preparation (Anderson, Kisiel, and Storksdieck, 2006; Falk and Balling, 1982; Griffin and Symington, 1997; Kubota and Olstad, 1991), active participa-tion in activities (Griffin, 1994; Griffin and Symington, 1997; Price and Hein, 1991), teacher involvement (Griffin, 1994; Price and Hein, 1991), and follow-up activities (Anderson, Lucas, Ginns, and Dierking, 2000; Griffin, 1994; Koran, Lehman, Shafer, and Koran, 1983).

### Advance Preparation

Advance field trip preparation activities give students the framework for how to interpret what they will see and guide what they should pay attention to during the visit. Students who receive appropriate advance preparation from their teachers, in such forms as previsit activities and orientation, have been noted, via observational studies and pre-post survey-based studies, to concen-trate and learn more from their visits (Griffin, 1994; Griffin and Symington, 1997; Anderson, Lucas, Ginns, and Dierking, 2000; Orion and Hofstein, 1994).

Advance preparation is most effective when it reduces the cognitive, psychological, and geographical novelty of the field trip experience (Kubota and Olstad, 1991; Orion and Hofstein, 1994). Such preparation has been linked to students spending more time interacting with exhibits (Kubota and Olstad, 1991) and learning from their visits (Orion and Hofstein, 1994). Studies have shown, however, that teachers spend very little time preparing students for field trips (Anderson, Kisiel, and Storksdieck, 2006; Griffin, 1994; Griffin and Symington, 1997).

**Active Participation in Museum Activities**

A review of over 200 evaluations of field trips to informal institutions (Price and Hein, 1991) indicates that effective ones include both hands-on activities and time for more structured instruction (e.g., viewing films, listening to presentations, participating in discussions with facilitators and peers). In general, children who were able to handle materials, engage in science activities, and observe animals or objects were excited about and enjoyed their field trip experience and displayed cooperative learning strategies. Similarly, Koran and colleague's review of earlier field trip studies—from 1939 to 1989—revealed that hands-on involvement with exhibits results in more changes in attitudes and interest than passive experiences (1989). At the same time, Griffin and Symington (1997) argued for the inclusion of structured activities to help keep students engaged throughout their field trip experience. Observing 30 unstructured classroom visits to museums, they noted that very few students continued purposefully exploring the museum after the first half hour of hands-on activities. Instead, most students were observed talking in the coffee shop, sitting on gallery benches, copying each other's worksheets, or moving quickly from exhibit to exhibit.

**Involvement by Teachers and Chaperones**

Classroom teacher involvement is a key ingredient to successful field trips, yet studies have consistently found that teachers often play a very small role or no role in the planning or execution of excursions and that institution staff are responsible for connecting exhibits to classroom content (Anderson and Zhang, 2003; Griffin, 1994; Griffin and Symington, 1997; Tal, Bamberger, and Morag, 2005).

There is wide variation in the amount and level of teacher involvement

*continued*

**BOX 5-1 Continued**

during field trips (Griffin, 1994; Griffin and Symington; 1997; Kisiel, 2006; Price and Hein, 1991). Price and Hein (1991) found a range of teacher involvement, from cases in which teachers congregated in such areas as the cafeteria and were not involved in the field trip activities, to cases in which teachers remained with the students and were actively involved in all phases of the trip. This review indicates that teacher involvement in various aspects of field trip planning and implementation is important. For example, a correlation was found between involvement in planning field trip activities and greater buy-in by teachers. When teachers are involved in planning, it is more likely that the activities will align with classroom curriculum and be viewed as valuable experiences by the teachers. Furthermore, alignment of classroom and field trip content and teacher buy-in are important, because they have been connected with student learning from field trips (Price and Hein, 1991; Griffin and Symington, 1997).

**Reinforcement After the Field Trip**

Teachers often plan to do follow-up after visiting informal institutions but in fact do little more than collect and mark student worksheets completed

investment and meaningfulness has parallels with work done by a group of museum professionals (e.g., Serrell, 2006) in generating criteria for exhibition excellence based on principles from the visitor studies literature. This group defined an "excellent exhibition" as one that is (1) comfortable—opening the door to other positive experiences; (2) engaging—enticing visitors to attend; (3) reinforcing—providing reinforcing experiences and supporting visitors to feel competent; and (4) meaningful—providing personally relevant experiences that change visitors cognitively and affectively (Serrell, 2006).

Research in various settings has shown that interest is in fact a gateway to deeper and sustained forms of learning. For example, when participants have a more developed interest for science, they pose curiosity questions and are also more inclined to learn and/or to use systematic approaches to seek answers (Engle and Conant, 2002; Kuhn and Franklin, 2006; Renninger, 2000). Interested people are also more likely to be motivated learners, to seek out challenge and difficulty, to use effective learning strategies, and to make use of feedback (Barron, 2006; Csikszentmihalyi, Rathunde, and Whalen, 1993; Lipstein and Renninger, 2006; Renninger and Hidi, 2002).

during the field trip (Griffin, 1994; Griffin and Symington, 1997). In Griffin's (1994) study of field trips taken by students in 13 Australian schools, about half of the teachers reported they planned to do follow-up activities, but only about a quarter of the teachers reported doing so. Furthermore, no students expected to receive meaningful follow-up, which may indicate that this was a common experience for them.

Developing productive post-visit activities is often complicated by the fact that the topics being covered in the classroom do not align with the field trip (Griffin and Symington, 1997). This can make it difficult to plan follow-up activities without disrupting regular classroom activities. However, even when the topics covered in the classroom align with the field trip content, connections between field trip experiences and classroom topics are often not made (Griffin, 1994). In addition, when post-visit activities do occur, they are often not designed to have any lasting impact. For example, a study of 36 field trips revealed that only 9 of the 18 teachers who reported conducting post-visit activities did more than ask students if they enjoyed the experience (Storksdieck, 2001). However, when well-designed examples of classroom follow-up have been noted, they are associated with positive educational impacts (Anderson et al., 2000; Griffin, 1994).

Another aspect of Strand 1 is motivation. Some researchers distinguish between intrinsic motivation, in which people do activities that interest them or provide spontaneous enjoyment, and extrinsic motivation, in which people do activities as a means to desired ends (such as good grades or career advancement). Deci and Ryan (2002) argue that intrinsic motivation is key for learning throughout the life span, because much of what people learn stems from spontaneous interests, curiosity, and their desire to master problems and affect their surroundings. They point to a body of work that documents the advantages of this type of learning in various settings. For example, Grolnick and Ryan (1987) conducted an experiment with 91 fifth graders who read material after they were told either that they would be tested on it or that they would be asked questions about how interesting and difficult they found it. The results showed that students in the second group had both higher interest and understanding in the material, and that, overall, students with more self-determined learning styles showed greater conceptual learning.

A meta-analysis by Utman (1997) showed that both intrinsic and extrinsic motivation was effective for simple tasks, but that intrinsic motivation led to greater success on creative or complex performance tasks. Of particular relevance, Zuckerman and colleagues (1978) found that intrinsic motivation was enhanced when problem-solvers could choose the activities and amounts of time they spent on them. More recently, research on motivation for learning has emphasized a broader set of constructs in "goal-orientation theory," which includes needs, values, and situated meaning-making processes (reviewed by Kaplan and Maehr, 2007). However, this theory has yet to be applied to informal environments.

### Comfort

Finally, while Strand 1 focuses primarily on arousing emotions, such as excitement, many studies have shown the importance of comfort, both physical and intellectual, as a prerequisite to learning in designed settings. For example, Maxwell and Evans (2002) link the physical environment to learning through psychological processes, such as cognitive fatigue, distraction, motivation, and anxiety, and they offer some evidence that learning is enhanced in quieter, smaller, better differentiated spaces. Physical and conceptual orientation (using maps, guides, and films) has also been shown to contribute to learners' comfort, presumably by reducing cognitive overwhelm and allowing them to make more informed choices about what to attend to. Much of this literature is summarized in Serrell (2006) and Crane, Nicholson, Chen, and Bitgood (1994).

## Strand 2: Understanding Scientific Knowledge

There is some research demonstrating that people gain understanding of scientific concepts, arguments, explanations, models, and facts, even after single museum visits. For example, Guichard (1995) studied the effect of an interactive exhibit designed to help visitors understand the form and function of the human skeleton. The exhibit consisted of a stationary bicycle that a visitor could ride, next to a large reflecting pane of glass. When the visitor pedaled the bicycle, the exhibit was arranged so that an image of a moving skeleton appeared inside the pedaling person's reflection. The movements of the legs and skeleton attracted the visitor's attention to the role and structure of the lower part of the skeleton.

Even without any additional mediation, this exhibit experience seemed to transform children's understanding. Children ages 6-7 were given an outline of a human body and asked to "draw the skeleton inside the silhouette" after the cycling experience. Of the 93 children in the sample, 96 percent correctly drew skeletons whose bones began or ended at the joints of the body; this result was in sharp contrast to the figure of 3 percent for a sample of children of similar age in a previous study who did not experience the exhibit. Even more impressively, the children's understanding persisted over time, with

92 percent of them retaining the idea of bones extending between places where the body bends 8 months after their museum visit and without any additional schooling, practice, or warning that they would be tested.

Multifaceted cognitive learning of this type has also been documented over a collection of exhibits. For example, Falk, Moussouri, and Coulson (1998) used the technique of personal meaning mapping, in which visitors complete pre- and post-exhibit diagrams, to record the deepening and broadening of their understanding of a science topic as a result of visiting an exhibition.

Typically, exhibition evaluations include self-reports from visitors that they have learned some content knowledge, usually small-scale, counterintuitive facts rather than large-scale abstractions or principles. For example:

> More than one-half of interviewees said they learned something new about plants while visiting the Conservatory. While learning was highly individualized and personal, all of these interviewees consistently referred to topics presented in the Conservatory exhibits and text. Several mentioned carnivorous plants, for example, and being surprised about the Venus flytrap's small size or the pitcher plant's feeding mechanism. A few expressed amazement by the water lily pollination story, while a few others appreciated experiencing a bog firsthand. Other topics mentioned by a few interviewees were: epiphytes ("plants can grow on top of other plants"), the co-evolution of plant nectar and pollinators ("different concentrations of nectar attract different animals"), the precipitation level of Los Angeles compared with a rain forest, and elephants as seed dispersers. The remaining responses were idiosyncratic; for example, one interviewee learned that "leaves have holes" and another that orchids are the source of vanilla beans (Jones, 2005, p. 8).

Most visitors' conceptual understanding was articulated as surprise at a counterintuitive phenomenon, that is, objects floating on a stream of air:

> "Oh, yeah. I was like, oh, I didn't know that. I didn't know it could stay up for so long. I thought eventually it would just die down and the weight would overcome the air pressure and stuff. But it just kept on floating. Like the football kept on doing misties and stuff. It was pretty cool" (Case 6, male, age 13) (Tisdal, 2004, p. 28).

> "[The exhibition is about] all the different life forms that we have on our planet and how there's a possibility that these life forms can exist on other planets. I just learned about the vents in the ocean. I never knew there were those kinds of things. And now I can understand how maybe there is life on Mars underneath all that ice. It's something I never understood before so I think it kind of expanded my world" (Adult) (Korn, 2006, p. 18).

Occasionally an exhibit experience may be powerful enough to challenge a common conception held by visitors. In a classic visitor study of the impact of short-term exposure to exhibits, Borun, Massey, and Lutter (1993) documented that, at least in the short term, many visitors changed their mistaken belief that gravity needs air in order to work after interacting with an exhibit showing a ball in a tube that could be evacuated.

For children, play may result in science learning of this kind, although some kinds of play seem more fruitful than others. Rennie and McClafferty (1993), working with schoolchildren using interactive exhibits, showed that they were more likely to learn the scientific ideas and principles that the curators intended if they were engaged in investigatory rather than fantasy forms of play. This may be because older children are familiar with school routines and expectations, and so benefit more from experiences structured around those kinds of expectations.

Conceptual change over the long term has not been studied in great depth in informal settings. There is certainly evidence that visitors can synthesize the big ideas of an exhibition or program and recall them or elaborate on them over time, although memories fade or change depending on many subject and condition variables (for a review of the museum memory literature, see Anderson, Storksdieck, and Spock, 2007). Measuring the long-term impact of museum visits is problematic because of the many variables at play (see discussion in Chapter 3). But as an example of a positive finding, Stevenson (1991) visited British families at home six months after their museum visits and interviewed 79 adults and children. The study found that each person was able to remember spontaneously, on average, 5 of the 15 exhibits in the exhibition, often clearly and in detail. Furthermore, over one-quarter of the memories were classified as "thoughts" (rather than feelings or exhibit descriptions), providing evidence of thinking or reflection about the exhibit in some way.

A commonly reported outcome from exhibition evaluations is that learners self-report a deeper understanding of a concept by virtue of having a direct sensory or immersive experience. For example, Korn (2006) collected the following observation from an adult participant following a visit to the *Search for Life* exhibition:

> I think the water exhibit is really brilliant. I can read something in a paragraph and not really have a sense of how much water 16 gallons is. It was just beautifully illustrated and really surprising. I had no idea that that much water is in our body. I think the [New York Hall of Science staff] do a great job of taking abstract contents and making it concrete so you can touch it and see it. That's why I like to bring my kids. You're going to absorb something somehow, even if you're not really trying at all (p. 17).

In addition to the key role of direct experiences, there is evidence that interpretive materials, such as labels, signs, and audio-guides, contribute significantly to this strand of science learning. For example, controlled experimental studies of exhibits in various science and natural history museums have shown that visitors showed significantly greater cognitive gains when objects were accompanied by interpretive labels than when they were experienced purely as sensory phenomena (Allen, 1997; Borun and Miller, 1980; Peart, 1984).

There is also substantial literature on the environmental design of informal science learning settings, including architectural and interior design, exhibit arrangement, label design and positioning, graphical and textual design, lighting, and other physical characteristics. Much of this involves recommendations based on practice, although there have also been many experimental studies, summarized in such reviews as Bitgood (2002), Bitgood and Loomis (1993), and Screven (1992). Given the scope of this report, the committee did not review this literature in detail because, although it has contributed significantly to practice, it mostly emphasizes such outcomes as visitor movements and behaviors rather than direct assessments of learning, as described in the strands.

Another commonly reported outcome from the evaluation literature is that learners self-report being reminded of learning experiences earlier in their lives. Since rehearsal is key to memory (Belmont and Butterfield, 1971), we regard this as a significant form of activity in Strand 2. For example, Jones (2005) heard the following from an adult male participant in a botanical conservatory:

> (What did you like most about the Conservatory?) "One place that I particularly liked and was pleased with was the Plant Lab because it showed me the way plants come to form life and the microscopes show you the different shapes of the seeds, the leaves, the roots—so many things that I didn't know before. . . . I came here and many of them refreshed my memory of when I was a child and took classes at school" (male, age 28; translated from Spanish) (p. 5).

In sum, there are documented cases showing that people who participate in a designed educational experience can generate, explain, and apply new knowledge to new examples and think in generalities (abstractions) about phenomena both familiar and new. Conceptual understanding and mental models of phenomena on which knowledge is built, however, take time to form (National Research Council, 1999, 2007; Lehrer and Schauble, 2000) and seem to depend on a person's existing knowledge base (Inagaki and Hatano, 2002; Carey, 1985) and cultural practices (Rogoff, 2003). Determining whether designed environments support more elaborated forms of conceptual knowledge development would probably entail longer time scales and close analysis of learning across settings.

## Strand 3: Engaging in Scientific Reasoning

The investigatory processes of science, often clustered under the title "scientific inquiry," are seen as a vital part of science literacy by educators and researchers alike (e.g., American Association for the Advancement of Science, 1993; Minstrell and van Zee, 2000; National Research Council, 1996, 2000). Designed environments provide opportunities to engage in many of these processes, and visitors have been observed to manipulate, test, explore, observe, predict, and question, as well as to make sense of the natural and physical world. The most studied environments in which visitors engage in these processes are physically interactive exhibits at science centers, which typically support a broader range of investigatory behaviors than animals or living ecosystems. The most common audience studied has been family groups, which is the largest single audience (numerically and economically) at many science centers.

## Interactivity

A key finding from the field is that learners are engaged by experiences that offer interactivity, which is defined by McLean (1993) in terms of reciprocity: "The visitor acts upon the exhibit, and the exhibit does something that acts upon the visitor" (p. 92). The field of practice is committed to this idea which, in a generic sense, has strong support from research. Learning—whether viewed in a purely mental or more broadly social perspective—is essentially interactive.

Summative evaluations of museum exhibitions frequently show evidence that learners, particularly parents, are aware of interactivity as a design feature of these environments and embrace it, although they often use related terms, such as "hands-on," to express this idea. For example:

> "[The exhibition] is trying to get kids involved in science [by] letting them know that it is fun. It is not all [about] some boring book somewhere. There are really fun, hands-on things that you can do. [It is] trying to give them opportunities to learn more complex principles with hands-on materials" (National Museum of American History; female, age 33) (Korn, 2004, p. 45).

It is well established that interactive exhibits tend to attract more visitors and engage them for longer times than static exhibits (e.g., Allen, 2007; Brooks and Vernon, 1956; Borun, 2003; Korn, 1997; Rosenfeld and Terkel, 1982; Serrell, 2001). At the same time, the specific impact of interactivity tends to be difficult to determine because authentic interactive exhibits usually differ in multiple design properties from noninteractive ones, and also because it is difficult to separate the effect of longer time spent from intellectual stimulation (Lucas, 1983). Koran, Koran, and Longino (1986) did find that simply removing the plexiglass cover from an exhibit case of seashells increased the number of visitors who stopped there and the amount of time they spent, even though only 38 percent of those who stopped actually picked up a shell.

Even in institutions with live animals, visitors seek out interactivity in particular. For example, Taylor (1986) found that families sought out interactions with live aquarium creatures. Goldowsky (2002) studied this experimentally by comparing the learning experiences of visitors to an exhibit on penguins. This was an experimental study in which the control condition used a typical aquarium exhibit, including live penguins, naturalistic habitat, and graphics. The interactive condition added a device designed to mediate interaction between participants and penguins, which allowed participants to move a light beam across the bottom of the pool, which the penguins would chase. Videotaped data were analyzed for 301 visitor groups (756 individuals). Goldowsky found that those who interacted with the penguins were significantly more likely to reason about the penguins' motivations.

Apart from supporting interaction with the physical world, interactive exhibits may also create a broader temporal space in which additional learning can transpire, including stimulating constructive exchanges between parents and children more frequently than static exhibits (Blud, 1990). Visitors self-report a variety of outcomes from interactives, including learning knowledge and skills, gaining new perspectives, and generating enthusiasm and interest (Falk et al., 2004).

While interactive experiences are prevalent across designed settings, they are not uniformly desirable in all exhibits and may be overutilized. For example, Allen and Gutwill (2004) documented several examples of exhibit designs that incorporated too many interactive features, leading to participant misunderstandings or to their feeling overwhelmed. Problematic design features included multiple undifferentiated options, features that allow multiple users to interfere with one another, options that encourage users to disrupt the phenomenon being displayed, features that make the critical phenomenon difficult to find, and secondary features that obscure the primary feature.

## Doing and Seeing

Given the widespread embracing of interactivity in designed spaces, it is unsurprising that the most frequently observed processes of science are those that Randol (2005) characterized as "do and see": Visitors manipulate an exhibit to explore its capabilities and observe what happens as a result. Randol conducted very detailed studies of visitors' inquiry behaviors at eight interactive exhibits from three science centers. The exhibits were selected to optimize the possibilities for scientific inquiry processes, as well as family learning as defined by Borun, Chambers, and Cleghorn (1996) in an influential study. In particular, there were many possible outcomes, so families were able to conduct a range of investigations of their own choosing. Randol discovered that visitors used the exhibits purposefully and successfully, and that their main interactions were focused on doing what the exhibit afforded (turning a dial, rolling a wheel) and watching what happened. These two actions, coded as "manipulate variable" and "observe," accounted for more than half of visitors' inquiry-related actions at the exhibits (see Figure 5-1). A similar pattern of typical family behaviors was reported by Diamond (1986).

Interestingly, Rennie and McClafferty (2002) found that these same inquiry activities did in fact lead to conceptual learning of science by children, supporting the notion that the strands are mutually reinforcing. Using Hutt's (1981) distinction between symbolic or fantasy play ("What can *I* do with this object?") and investigation ("What can this *object* do?"), they studied children using an interactive science exhibit. The exhibit, *Magnetic Maze*, was designed to support a range of learning experiences: enjoyment, mystery, role-playing, and development of hand-eye coordination, in addition to the goal of understanding that magnets can attract some objects, even at a distance and through materials. Rennie and McClafferty found that this latter science content goal was reached almost exclusively by children who took an investigatory approach to the exhibit. In other words, the "do and see" approach observed so frequently by Randol did enhance children's understanding of the intended science content.

Another common form of observation is pointing out to others a feature of particular interest. Allen (2002) calls this kind of spoken observation "perceptual talk" and regards it as a significant process measure of learning because it is an act of identifying and sharing what is significant in a complex environment. She defined four subcategories: identification ("Oh, look

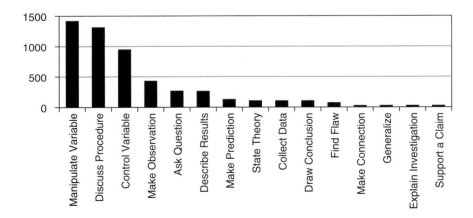

FIGURE 5-1   *Frequency of visitor actions at interactive exhibits.*
SOURCE: Randol (2005).

at this guy"), naming ("It's a Golden Frog"), pointing out a feature ("Check out the bump on his head"), and quoting from a label. Audio-recorded visitors engaged in perceptual talk at 70 percent of the exhibit elements they stopped at, the most common category of talk in Allen's scheme. Similarly, Callanan and Jipson (2001) argue that this kind of talk is an important way in which adults help guide children's scientific literacy. They propose that adults who point out salient features of the environment are helping children guide their attention, interpret their experiences, and frame them in terms of relevant domains of knowledge. This same benefit need not be limited to parents talking with children but could also apply to the contributions of any member of a group communicating with other members.

## Meaning-Making and Explanation

Meaning-making (i.e., interpreting experiences to give them personal significance) has become so central to descriptions of learning in informal environments that it is sometimes regarded as the essential learning behavior (e.g., Silverman, 1995; Hein, 1998; Ansbacher, 1999; Rounds, 1999). Callanan and Jipson (2001) make the point that visitors vary not only in terms of how they interpret experience, but also in terms of what they find worthy of interpretation. The degree and quality of sense-making have been the basis of a number of systems for coding learning. For example, Borun and colleagues (1998) defined three levels of family learning in informal environments: identifying, describing, and interpreting/applying. They found that 88 percent of families fell within the first two levels. Similarly, Leinhardt and Knutson (2004) list four levels of interpretation: listing, analysis, synthesis, and explanation.

Explanation has also been the subject of extensive study. The research consensus seems to be that explanations in designed spaces tend to be concrete, local, and incomplete. In studying the parents' explanations to their children in a museum context, Callanan and Jipson (2001) defined three types of explanation: (1) abstract scientific principles (e.g., "It's because of the gravitational attraction") were used in only 12 percent of explanations; (2) causal connections (e.g., "Each of those pictures is a little different pose on the horse, and it makes it look like it is galloping") constituted 54 percent of the explanations; and (3) connections to prior experience (e.g., "Remember the stethoscope at the doctor's? We can listen to your heart beat") made up a further 25 percent of the adults' explanations. The authors argue that the connections to prior experience served the purpose of contextualizing the experience for children by linking it to their previous knowledge and history, giving weight to a design strategy that has been used for over a century by practitioners to help visitors find personal meaning in exhibitions and programs. Similarly, studies in various designed settings (e.g., Crowley and Jacobs, 2002; Taylor, 1986) have shown that parents tend to focus on helping children to understand the particular event at hand, rather than learning more abstract principles.

In related work, Callanan, Jipson, and Soennichsen (2002), studying families' use of representational objects, such as maps and globes, found that parents tended to explain to children by using specific referents as if they were the real objects (e.g., "There's your school!"), rather than explaining the more abstract relationships between the representation and real world it shows. The authors point out that it is not just children who learn from fragments of scientific reasoning; adults and even scientists can learn this way as well.

Gleason and Schauble (1999) show that parents may not coach their children equally in all aspects of scientific inquiry at an exhibit; they may in some ways limit children's access to cognitively complex tasks. The researchers asked 20 highly educated parent-child pairs to design a complex experiment at an interactive exhibit in which a boat was towed down a small canal. Specifically, each parent-child pair was asked to spend 45 minutes designing and interpreting a series of experimental trials to determine the features influencing how quickly the boat would be towed. The researchers found that the parent-child pairs spent considerably more time on experimentation with materials than on interpretation of results. Parents did support and advance their children's reasoning, but they tended to do the more challenging conceptual parts of the activity themselves (such as looking up the results of previous trials and drawing conclusions aloud) and only rarely encouraged their children to take these on, even over time. By comparison, children did the logistical or mechanical aspects (such as releasing the boat in the canal, operating the stopwatch). At the end of the 45 minutes, it was the parents rather than the children who made gains in understanding regarding the true causal features of the boat and canal system. The findings from this study raise intriguing questions about how designed settings might better support parents and other adult care providers to take advantage of these opportunities.

## Questioning and Predicting

Questioning and predicting are typically inquiry behaviors that involve articulating ideas to others prior to physical experimentation. The study by Randol (2005) shows that, while visitors did engage in questioning and predicting at interactive exhibits, these were approximately 10 times rarer than manipulating and observing. Even lower frequencies of prediction (3 percent) were found by Allen (2002) in her analysis of visitors' conversations in a multidisciplinary exhibition about frogs. She noted, however, that this figure may be particularly low because many of the elements were live animals rather than interactive exhibits.

Questioning is widely regarded by educators as one of, if not the, central inquiry behaviors that support learning in informal environments. Borun et al. (1996) found that asking and answering questions were some of the key behaviors that discriminated among levels of family learning as defined in their study: families that asked and answered questions were more likely to engage in the processes of "describing" (including making connections between an exhibit and their personal experience) rather than the lower level of "identifying." So it is perhaps surprising that these behaviors, too, appeared relatively infrequently in Randol's (2005) study. One explanation for this is that the asking and answering of questions may be taking place implicitly, rather than being spoken by participants. For example, if it is true that the most common approach to interactive exhibits is "What can this object do?" this already frames an implicit question that need not be publicly stated. Similarly, visitors' common expressions of surprise and intrigue (a mainstay of the "counterintuitive" genre of exhibit design) suggest that some form of implicit prediction must have been made to evoke a surprised response. Callanan and Jipson (2001) report that in contrast to other settings explanatory conversations at museum exhibits were started only rarely by a "why" question from a child. The elements of physical interactivity and novel phenomena available in a museum may encourage a form of discourse that is more of an implicit "what if" than a "why."

Humphrey and Gutwill (2005) showed that the number and kinds of questions visitors ask depends in part on the design of the exhibits they are using. Their team created and studied a class of interactive exhibits that supported active prolonged engagement (APE), a combination of inquiry behaviors that included visitors staying at an exhibit for an extended time, asking and then answering their own questions. These exhibits took several forms, based on the primary form of activity they supported: exploration, investigation, observation, and construction. The APE exhibits were compared with more traditional "planned discovery" exhibits, in which visitors are surprised by a single intriguing phenomenon that is explained in a label. The researchers found that, in interactions with APE exhibits, the number and type of participants' questions varied. Visitors asked more questions overall, and more of them related to using or understanding the exhibit, rather than questions about the logistical aspects of working the exhibit or about what others were experiencing. Also, the team found that visitors using APE exhibits were more likely to answer their own questions by using or discussing the exhibit rather than reading the label. Related studies by Hein, Kelley, Bailey, and Bronnenkant (1996) showed that a series of open-ended exhibits at the Boston Museum of Science also encouraged visitors to ask questions, although no quantitative comparisons were made with other exhibits on the floor.

Drawing conclusions, generalizing, and argumentation are much less frequently observed inquiry behaviors in designed settings. Randol's study of eight interactive exhibits found that, although the exhibits were selected for their ability to support a range of inquiry actions, some actions were very rare. The rarest observed were supporting a claim, explaining an investigation, generalizing, and making a generalized connection between an exhibit phenomenon and a situation outside the museum. These behaviors were observed roughly 100 times less often than manipulation of variables. After looking at the inquiry-related actions through several different theoretical lenses, Randol concludes that most visitors in his study did not engage in what experts consider to be high-level inquiry behaviors, such as drawing conclusions or making generalizations. Nor did they tend to engage in such actions as presenting alternatives or supporting claims, key aspects of building and testing theories in science. Randol attributed this latter finding to visitors' reluctance to do anything that might seem confrontational in a situation focused on leisure and social interaction with companions.

## Strand 4: Reflecting on Science

A number of designed environments have created exhibitions and programs that focus specifically on issues in science or on the processes of science from a social and historical perspective. Such exhibitions give visitors the opportunity to reflect on science as a human endeavor and to think about the nature and generation of scientific knowledge.

Perhaps the best known example is *A Question of Truth*, created at the Ontario Science Center, which invites visitors to consider the cultural and political influences that affect scientific activity. The three main themes of the exhibition are (1) frames of reference (e.g., sun-centered versus earth-centered); (2) bias (e.g., concepts of race, eugenics, and intelligence testing); and (3) science and community (e.g., interviews with diverse groups of scientists). Pedretti (2004) conducted interviews with casual visitors, as well as students on school field trips, and found that the exhibition contributed to their understanding of science and society by considering science and social responsibility, controversy and debate, decision making, and ethics. She found that 84 percent of the comment cards left by visitors were overwhelmingly positive, "applauding the science center's efforts to demystify and deconstruct the practice of science while providing a social cultural context" (Pedretti, 2004, p. S43). For example, a visiting student commented, "The exhibit makes us think a lot about our beliefs and why we think in certain ways. . . . I didn't think that the gene that affects the color of your skin was so small and unimportant. Most people don't think of things like that." Another student challenged the view of science as being amoral: "We view science as often being separate from morals, and it's kind of negative because it allows them to do all sorts of things like altering human life, and it may not necessarily be beneficial to our society. . . . Some scientists are saying, should we actually be doing this?"

Pedretti and colleagues (2001) argued that such exhibitions encourage visitors to reflect on the processes of science, politics, and personal beliefs, and they achieve this by personalizing the subject matter, evoking emotion, and stimulating debate by presenting material from multiple perspectives.

## Self-Reflections on Learning

Designed environments also provide opportunities for visitors to reflect on their own learning processes, although this has been less frequently studied than other inquiry-related actions. Randol (2005) found that visitors using eight interactive exhibits at science museums frequently made some kind of self-reflective comment, typically with a focus on the way they were using the particular exhibit they were engaged with. Specifically, he reported that over 70 percent of the groups observed made at least one statement regarding the group's progress toward their goal (e.g., "Okay, just two more") or a comment regarding possible problems in procedures (e.g., "Wait, wait—they have to start at the same time").

By contrast, Allen (2002), in her recorded conversations with pairs of visitors at an exhibition on frogs, reported much lower frequencies of self-reflective comments. She distinguished among three subcategories of such talk. (1) Metacognitive comments, in which visitors talked about their own state of current or previous knowledge, were heard at 9 percent of the elements visitors engaged with. Of the 66 elements in the exhibition (exhibits or other components), the element that most frequently evoked metacognitive comments was *Mealtime*, a compilation of video clips of frogs catching and eating their food. Visitors reflected on their surprise at the variety and nature of what frogs ate: "I never would have believed . . ." or "I didn't realize they got them with their tongue." (2) Comments about exhibit use were heard at 16 percent of the stops, for example: "You have to start from here, and then jump as far as you can." (3) Evaluative comments, in which visitors judged their performance or actions, were heard at 8 percent of the exhibit stops. The element that evoked most comments in the latter two categories was *Croak Like a Frog*, an audio-based multimedia exhibit in which visitors could listen to a variety of prerecorded frog calls and record their own imitations. Visitors' comments included: "You have to do it before the red line disappears or it doesn't record," and "This was right, except I made it too long." Allen proposed that several exhibit features probably accounted for the high frequency of evaluative talk: high overall appeal of the exhibit, a challenging interface to problem-solve, and computer-generated graphs that supported visitors' efforts to visually compare their vocalizations with the standard frog calls.

A large body of evidence also shows that visitors are able to reflect on their own learning if asked. Many exit interviews used in summative evaluations of exhibitions ask visitors whether there was anything that they had not previously known, realized, or appreciated. While these are cued reflections rather than spontaneous ones, they provide evidence that visitors can and do reflect on their own learning in designed settings. For example:

> You learn—it's amazing. . . . I'm going on 74 and . . . and you're learning something new everyday. And when you see a statement like scientists still don't agree about algae whether they're plants. You know they work a little like a plant but then they don't and so some say, "yes it is" and some say "no it isn't." I'm looking at the spores—amazing tiny little specimens underneath the microscope—the variety. It's quite intriguing. I think anyone would find it interesting (male, age 73 years) (Jones, 2005, p. 6).

## Strand 5: Engaging in Scientific Practices

> By the end, [my son] was working collaboratively with four other kids, which was very nice. They were total strangers. That is how it happens in the lab sometimes when you are working on one thing and your colleagues get together and you start working on something together. . . . He would try something, and then another kid would try something. When it did not work, they would try a different way (National Museum of American History; female, age 43, with male, age 7) (Korn, 2004, p. 42).

In informal settings, participation in science is expected and deliberately designed into the experiences. Children do projects with each other, their parents, or other adults, such as group leaders and museum staff; adults on nature trails or families in zoos and botanical gardens walk and observe together. They use tools and instruments like microscopes or rulers that may be helpful for learning (Jones, 2003; Ma, 2002) but are not necessarily scientific equipment.

Verbal communication, or discourse, is a particularly prevalent and well-studied form of scientific practice in designed settings. In fact, the importance of discourse in learning is broadly acknowledged across a range of subject areas and settings (e.g., Cazden, 2001; National Research Council, 2007) and is of considerable interest to classroom-based science education. Researchers have found that successful science education depends on the learners' involvement in forms of communication and reasoning that models that of scientific communities (Gee, 1994; Lemke, 1990; National Research Council, 2007). There is increasing interest in designing programs and exhibitions that explicitly support social mediation and conversation (e.g., Morrissey, 2002; Schauble and Bartlett, 1997).

Leinhardt and Knutson (2004) combined into a single learning model the notion of conversation as both an outcome and a means of learning. After studying exhibitions at five different types of museums, they listed four levels of visitors' interpretation: listing, analysis, synthesis, and explanation.

They then studied how these contributed to overall learning, defined as a combination of holding time and frequency with which visitors mentioned the exhibition's intended themes during an interview after their visit. While their measure of learning is unorthodox, they found that it was slightly higher for visitors who had higher levels of interpretation while in the exhibition space.

The nature of participant explanation and commentary observed in designed settings varies according to many factors, including gender (Crowley, Callanan, Tenenbaum, and Allen, 2001a), age of the children (Gleason and Schauble, 1999), educational approach or goal (Schauble et al., 2002; Ellenbogen, 2002), available resources, and the skill and background of the leader and of the participants as well as situational demands (e.g., summer camp versus school group field trip). Gelman, Massey, and McManus (1991), for example, found it very difficult to design a stand-alone exhibit that promoted scientific observations and experimenting, as did Schauble and Bartlett (1997).

### Parent-Child Interactions

Much of the research on language use has focused on interactions in family groups in museums and science centers. This work dates back over two decades (e.g., Hensel, 1987; McManus, 1987; Taylor, 1986) and is largely comprised of detailed descriptive studies characterizing how adults and children behave and talk while visiting aquariums, museums, and the like. A common emphasis of this work is parent-child interactions.

One critical finding in this literature is that the participation of a parent improves the quality of child engagement with exhibits. For example, Crowley and colleagues (2001a) observed 91 families with children ages 4-8 as they interacted with a zoetrope exhibit in the Children's Discovery Museum in San Jose, California. They found that children who participated with their parents discussed evidence over longer periods of time and in a more focused manner than children who participated without their parents. Parents, they observed, played an important role in helping children select appropriate evidence and identify it as such. When using interactive exhibits with their children, parents tend to focus their explanations on the functions and mechanics of the exhibit, connecting the exhibit with real phenomena, and making connections to formal science ideas (Crowley and Callanan, 1998). Such explanations are often brief and fragmented—Crowley and Galco (2001) call them "explanatoids"—but they seem well targeted to a moment of authentic, collaborative parent-child activity. When parents explain a feature in an exhibit, children are more likely to talk about their experiences with the exhibit. Similarly, as previously noted, Gleason and Schauble (1999) found the educational potential of exhibits in a science gallery depended on mediation by parents. Parents tended to assume the most difficult conceptual tasks, delegating manual tasks to the children. Furthermore, there was little emphasis on science talk or thinking by parents and staff. Callanan, Jipson, and Soennichsen (2002), observing dyads at a science exhibit, noted that parents focus on specific events rather than general principles. They suggest, but do not explore empirically, that this may set the stage for more complex thinking. Parents who have experience in science may be comfortable enough to use the exhibits as props for sharing their knowledge on a particular topic. For example:

> One father described how he used Pulley Table to explain and demonstrate to his son: "Well, mostly I was explaining to my son what it was doing. Showing him that—for instance, there was one pulley that powered and the difference in putting the string on the smaller wheel as compared to the larger wheel, what it does to the other wheels. . . . Another boy walked up as well, and so I showed them the faster you turn it, the faster it plays, depending on the size of the pulley you use will also determine the power" (Case 24, male, early 40s) (Tisdal, 2004, p. 12).

Parents' conversations also depend on what they believe about the setting in relation to their children's learning. For example, Schauble et al. (2002) observed 94 parents of children ages 6-10, as well as 16 museum staff interacting with children at an exhibit in a science gallery. The researchers also conducted interviews to find out the beliefs about learning of each group and what each thought would help children's learning at the exhibit. Nearly half the parents believed that activity, observing, and fun with hands-on materials would lead to learning through sensory experience and excitement; many of these parents sat back and watched their children play, believing that the best assistance was to keep out of their way. Other parents seemed to distinguish play from learning and wondered about how learning could be enriched by resources in the museum. They tended to be less sure about how to assist their children. The museum staff were more likely than parents to value adult mediation (getting involved in children's activities), and some were critical of parents they perceived as passive. The staff talked about a variety of different ways to help children learn and emphasized asking them provocative questions or explaining how things work, compared with the parents' more frequent focus on logistical forms of help. The researchers point out that the staff's larger repertoire of assistance techniques presumably results from their experience in deciding how to mediate visitor experiences on a daily basis in the gallery.

### Specialized Science Talk

Not all forms of talk are equally effective supports for science learning in designed settings. For example, Crowley and Jacobs (2002) showed that higher levels of certain kinds of talk by parents were associated with children's learning. In particular, they found that children of parents who read artifact labels aloud (in this case, for fossils) and helped children make connections to shared family history were better at identifying fossils after their museum visit.

Another form of specialized talk, of great interest in classrooms, is argumentation (National Research Council, 2007; Bell and Linn, 2000; Driver, Newton, and Osborne, 2000; Duschl and Osborne, 2002). While there is little research on science-specific modes of argumentation and discussion in informal environments, most observers agree that typical presentations of science in such institutions are built on everyday language in order to engage a general public. Studies of classroom-based science argumentation have found that such discourse generally requires extensive instruction and support, intentional development of shared norms, and long-term practices of reflection (National Research Council, 2007). Thus, even in the cases in which inquiry and scientific talk are encouraged in designed settings, it may be that the experiences are not extended enough to be internalized by the learner. And as noted previously, it also seems plausible that scientific argumentation can be perceived as threatening to the social interactions and leisure goals learners have for their visit. There is no immediate reward for challenging the conceptual structures of others in the group, especially in multigenerational groups in which power is unequally shared. Thus, it is not yet clear whether scientific argumentation can be incorporated into these settings without jeopardizing defining properties of informal environments.

There are documented examples of the use of scientific terminology and language on occasions when museum visitors read labels aloud (Crowley and Jacobs, 2002; Borun et al., 1996) and explain or comment on exhibit features to each other (Tunnicliffe, 1996; Ash, 2002). The time frames of such studies are generally too short to assess whether learners internalize such scientific terminology and use it in other settings. There has not been sufficient work analyzing participation in designed settings over months or years to explore how the use of scientific language might deepen over time in designed spaces.

## Scientific Tools

Research on learning broadly employs varied notions of tools, which include not only conventional scientific tools (e.g., laboratory equipment), but also a broader range of representational tools, such as language, graphs, and mathematical formulas. This broader notion of tools is evident in the growing body of research on classrooms (National Research Council, 2007). However, there has been very little emphasis on tools in research on learning science in designed settings. Furthermore, how people are introduced to conventional scientific tools in informal environments has not been directly evaluated, but at least some programs involve participants in inquiries that go beyond communicating scientific language and ideas and require them to use lab equipment, research tools, and measurement tools. For example, in the Cell Lab Exhibition at the Science Museum of Minnesota, participants use a number of tools as they visit the exhibition and its seven wet lab experiment benches. Visitors have the opportunity to use a number of scientific instruments and tools, including microscopes, cameras, monitors, glass slides, test tubes, incubators, dry baths, and UV detectors (National Science Foundation, 2006).

Again, available resources, the skill and background of the leader and the participants, and situational demands are likely to determine the depth of contact and talk, rather than the design of the space or materials alone (Gelman et al., 1991; Gleason and Schauble, 1999; Schauble and Bartlett, 1997). Gleason and Schauble (1999) found that the educational potential of exhibits in a science gallery depended on the mediation, which may be particularly important the more individual exhibits or stations require participants to use scientific tools.

### Social Group Influences

There is some evidence that use of scientific language may be influenced by gender (see Chapter 7). One body of work looks at the ways in which parents and facilitators interact with boys and girls. Several studies (Crowley et al., 2001a; Tenenbaum and Leaper, 1998; Tenenbaum, Snow, Roach, and Kurland, 2005) have found that parents engage in modes of discourse associated with higher cognitive demands at higher rates with boys, than with girls. Crowley and colleagues (2001b), for example, examined 298 naturally occurring conversations among parents and their children at interactive exhibits in a science museum. They observed interactions of families with boys and girls, girls only, and boys only and with one, two, or no parents present. They found that parents, both fathers and mothers, tend to provide causal explanations of phenomena to boys more frequently than to girls. Although families seemed not to make gender-based distinctions in bringing children to museums, engaging them in interactive science activities, talking about what exhibits do, or talking about what to perceive in an exhibit, they placed significantly greater emphasis on explaining science to boys. This subtle distinction could have consequences for girls' science learning, raising concerns for parents and educators who design and facilitate learning in designed settings. In *Alice's Wonderland*, an exhibit designed with a theme that parents would think of as interesting to girls, no gender differences in explanations were found, suggesting that modifications to exhibits could influence parents' tendency to engage girls with science (Callanan et al., 2002).

Level of expertise is another factor that may shape group learning processes in designed settings. The varied expertise of group members can influence learning interactions. For example, an individual with a lot of information, even a child, may play an important role in facilitating the learning of others by pointing out critical elements or information and by providing input and structure for a more focused discussion of science (Moll, Amanti, Neff, and Gonzalez, 2005; Palmquist and Crowley, 2007). In a small study of an exhibition about glass, Fienberg and Leinhardt (2002) found that adults with high prior knowledge and interest in glass tended to engage in more explanatory talk (discussing how or why something happened or worked), than those with less prior knowledge or interest.

Vom Lehn, Heath, and Hindmarsh (2001) reported that visitors' activities at an exhibit could be significantly affected by the behavior of other visitors, either companions or strangers. Meisner and colleagues (2007) showed that visitors sometimes turned their interactions with interactive exhibits into spontaneous performances with a theatrical flavor, which allowed them to be shared with other family members or even strangers. And Koran, Koran, and Foster (1988) documented that visitors can learn exhibit-related behavior from strangers, even without any conversation taking place. They found that museum visitors, especially adults, were more likely to engage in such behaviors as touching a manipulative exhibit, listening to headphones, or attending to an exhibit for an extended period if they had previously witnessed a person silently modeling these behaviors.

## Strand 6: Identifying with the Scientific Enterprise

Informal environments for science learning, like all educational institutions, can be seen as places of enculturation (Bruner, 1996; Martin and Toon, 2005; Pearce, 1994). Enculturation is about developing identity as a part of a community, and informal settings include different environments that may influence people's identities as science learners (Ivanova, 2003).

Personal identity, viewed as "the cluster of knowledge, dispositions, and activities brought with the visitor" (Leinhardt and Knutson, 2004, p. 50), highly influences museum visitors' conversations (Fienberg and Leinhardt, 2002) and can shape learning experiences more broadly (Ellenbogen, Luke, and Dierking, 2004; Leinhardt and Gregg, 2002; Falk et al., 1998; Leinhardt, Tittle, and Knutson, 2002; Anderson, 2003; Anderson and Shimizu, 2007). For example, Falk, Heimlich, and Bronnenkant (2008) used the following categories to classify 1,555 visitors to a group of four zoos and aquariums:

1.  Explorers are curiosity-driven and seek to learn more about whatever they might encounter at the institution.
2.  Facilitators are focused primarily on enabling the experience and learning of others in their accompanying social group.
3.  Professional/hobbyists feel a close tie between the institution's content and their professional or hobbyist passions.

4. Experience seekers primarily derive satisfaction from the fact of visiting this important site.
5. Spiritual pilgrims are primarily seeking a contemplative and/or restorative experience.

The researchers reported that 55 percent of the visitors showed one dominant kind of motivation under this scheme. The motivations accounted for about a quarter of the variation in visitors' conservation-related attitudes and also correlated with aspects of visitors' long-term memories, suggesting that aspects of identity served as a framework for visitors to make sense of their experience.

Identities such as these may be drivers for what participants do and learn in designed settings. For example, parents who want to develop a particular family identity are able to quickly adapt the general museum experience, as well as specific content, to reinforce the desired identity. Everything from expectations ("We don't bang on the computer screen like that") to personal narrative history ("Do you remember the last time we saw one like that?") can be used to reinforce the values and identity of the family (Ellenbogen, 2003).

## *Agenda*

One aspect of identity is the learners' agenda, that is, the cognitive, affective, or social expectations and goals the individual expects to pursue or satisfy during the event. For example, families tend to see visits as social events (Laetsch, Diamond, Gottfried, and Rosenfeld, 1980) and pursue an identity-related agenda as they generate their own pathway through museums (Cohen, Winkel, Olsen, and Wheeler, 1977; Falk, 2008). For example, Falk tells of Frank, a 40-year-old father whose agenda in museum visits is closely tied to his own childhood experiences. Frank's father, a busy academic, spent little time with him as a child, although he valued science. Similarly, Frank sought to explore science with his own daughter and, at the same time, to play a more active role in his daughter's life. Museum experiences gave him occasion to pursue deep, identity-building experiences. The goals of individuals and groups may be multiple (e.g., pursuing learning, enjoyment, and socialization in a single event) and may incorporate additional practical agendas as well, such as providing tours for out-of-town visitors and for entertaining young children.

Several researchers have interpreted their data to argue that learners act purposefully to meet their individual family's learning goals. Hilke (1989), for example, concluded from her detailed analysis of family behavior in an exhibition that families are pursuing an agenda to learn during their visits to museums.

Anderson and colleagues (Anderson, 2003; Anderson and Shimizu, 2007) have found that the degree to which a learner's agenda is satisfied or frustrated can greatly affect his or her memories of learning experiences. In interviewing participants about world expos they attended after almost two decades of elapsed time, they found that visitors' "social context," which includes the participants' agenda, "dominated their recall of their . . . experiences [15 to 17 years] after the event—more than any other encounter or episode they were able to report" (Anderson, 2003, p. 417).

The participants' agendas and the pedagogical or communicative goals of a particular designed setting may coincide, conflict, or simply fail to connect. Dierking, Burtnyk, Buchner, and Falk (2002), for example, conducted a literature review on participation and learning in zoos and aquariums. They observed frequent disconnects between the agendas of zoo visitors and the ecological goals of zoos. The staff and institutional commitments of zoos typically espouse ecological conservation. However, individuals who visit zoos may fail to perceive ecological principles and conservation commitments in their visit. In fact, they found that even individuals who were zoo-goers and who also made financial contributions to nonzoo ecological organizations may fail to link their ecology and conservation interests to zoo visits. Rather, zoos were seen simply as places to see animals up close.

Science-rich institutions have historically varied in the degree to which they take seriously the agenda of visitors (Doering, 1999). Viewing the visitor as "stranger" reflects a tradition in which the personal collections of gentry were used for their own individual investigations of natural history. When visitors are seen as strangers, the institution focuses primarily on its responsibility and interest in its collection or subject matter and not on the interests or needs of the visiting public. When visitors are viewed as "guests," the institution is inclined to attend to their interests through educational and entertainment activities. Objects and ideas are still central to the institution's values and work, but they also give significant credence to their visitors. For example, a visitor who reads a meteorite exhibit label may choose not to focus on fundamental scientific aspects of the text—the origin of the meteorite, evidence of impact, what it tells us about the universe. Instead, the visitor may focus on an aspect of the label that is personally meaningful to herself or to her family (e.g., the meteorite was found in Alberta, Canada) and use this observation as an opportunity to explore family identity (e.g., recalling that a family member was once in Alberta) rather than a strictly scientific meaning (Ellenbogen, 2003).

Designers and researchers have explored various ways to embrace visitors' agendas, such as supporting visitors to write, speak, or draw their own ideas (McLean and Pollack, 2007) or by changing their scientific labels to embody a conversational tone more compatible with visitors' own (McManus, 1989; Rand, 1990; Serrell, 1996). Both techniques have been shown, at least in some cases, to increase visitors' engagement with scientific material.

## *Prior Knowledge and Experience*

Another aspect of identity is prior knowledge and experience, long recognized as critical to learning by cognitive and sociocultural theories of learning. More recently, analyses of conversations in museum exhibitions have shown that families regularly make verbal connections between prior experiences and novel observations (Callanan and Jipson, 2001; Allen, 2002). Crowley and Jacobs (2002) have suggested that young children learn science by building "islands of expertise," or topics in which they become interested and knowledgeable about over a period of weeks, months, or years. These topics become integrated into family activities, such as field trips, reading books, and dinnertime conversations. In this sense, previously developed ideas and interests influence the trajectory of learning activities over a sustained period of time, becoming a focal point of activity for individuals, peers, and family members.

Practitioners have long been aware of visitors' desire to make links from their sensory experiences in designed settings to their prior knowledge and experiences. Typically, practitioners talk about this as creating "hooks" from science content to everyday life and familiar activities. Such connections are commonly reported in evaluations, as suggested in the following excerpts (Korn, 2004, p. 59):

> "I am very interested in the way scientists are working to map cells and create tools to diagnose [disease]. At my age I'm very interested in health protection" (female, age 55).

> "[I feel connected to the invention process] in everyday life, especially as a mother, when I am called upon [to] solve some kind of problem. When I don't have the right materials [to solve the problem], I have to look at what I have [around me] and try to be creative and come up with some solution" (female, age 48).

Anderson et al. (2002) found that, for children, experiences that were embedded in familiar sociocultural contexts of the child's world, such as play, story, and familiar objects, acted as powerful mediators and supported children's recollections and reflections about their activities. Facilitator-led narrative discussions were particularly memorable. Interestingly, the children's memories were very idiosyncratic. Still, the most memorable aspects of their experience tended to be those that took a familiar form (e.g., play and storytelling). The researchers concluded: "exhibits and programmatic museum experiences that provide context and links with children's own culture . . . will provide greater impact and meaning than [those] that are decontextualized in nature" (p. 229).

At the same time, the effect of prior knowledge on learning is not fully understood in informal settings. Leinhardt and Knutson (2004), in a study of learning in an art museum, found that the background knowledge visitors bring to the museum was the best single predictor of how long they spend and what they learned from an exhibit: the more people already knew, the longer they stayed and the more they tend to talk about and learn from the curatorial themes. These findings, however, conflict with the results of a study by Falk and Storksdieck (2005). They found that while prior knowledge was the most potent predictor of learning in museums, in this case, the more a visitor knew about life science when entering the exhibition, the less they gained, suggesting a ceiling effect or a limitation in the type of gains that could be measured (rather than a disavowal of the importance of prior knowledge). It seems possible that the role prior knowledge plays could depend on many factors, including the domain in question, exact nature of the museum offerings, particular visitors studied, and assessment methods. Clearly, more research is needed to determine how to interpret these findings.

### Personal Commitment to Action

Another aspect of personal identity in relation to science is the gradual understanding of the implications of one's own actions on the world and the potential to change those actions in light of scientific evidence.

Many exhibitions and programs at aquariums and zoos focus on this aspect in particular, emphasizing conservation and stewardship, and some have seen results. For example, Falk et al. (2007) studied visitors to two museums and two zoos. They concluded that such visits prompted 54 percent of individuals to reconsider their role in conservation action and to see themselves as part of the solution to environmental problems. Other studies have shown less success in promoting this aspect of identity. For example, Dierking et al. (2004) found that visitors to Disney's Animal Kingdom Conservation Station showed significant short-term increase in their level of planned action, but follow-up phone calls two months later revealed that they had not initiated the intended activities.

Schneider and Cheslock (2003) reviewed studies from a number of fields related to behavior change, including visitor studies and environmental education. They concluded that the most successful programs were those that targeted actions, tailored interventions to the particular audience, built self-efficacy, and used prompts or tools to trigger action. Hayward (1998, in an aquarium exhibition study reported by Yalowitz, 2004) showed the importance of suggesting specific behaviors visitors can engage in to ameliorate environmental problems; without these, visitors left more disillusioned and less empowered than a control group.

One embodiment of this principle is Monterey Bay Aquarium's Seafood Watch, a program that has been operating for over a decade on a national scale. Seafood Watch offers visitors wallet-sized cards containing information about the environmental impact of various fishing practices and makes recommendations about which types of seafood to avoid purchasing. Findings from a large-scale evaluation, using surveys and focus groups, found that participation in Seafood Watch was correlated with changes not only in the purchasing patterns of visitors but also in the selling practices of seafood restaurants across the country (Quadra Planning Consultants, Ltd., 2004). Although linking environmental knowledge to behavior has often proven elusive, the Seafood Watch evaluation found evidence that increased knowledge strengthened pro-environmental attitudes and behavior.

Several studies indicate that an individual's prior interest and involvement in conservation may serve as a better predictor of their responses and actions than typical demographic variables, such as age, gender, ethnicity, or education. For example, visitors with high interest in conservation stopped at more of the exhibits in a conservation-themed aquarium exhibition (Yalowitz, 2004; Hayward, 1997), and zoo visitors' emotional responses to animals were more closely associated with emotional or personality variables (Myers, Saunders, and Birjulin, 2004) than demographic variables.

A common assumption in the field is that affective responses, such as caring for individual animals, will provide a basis for future behavior change. Carol Saunders at the Brookfield Zoo is developing the notion of "conservation psychology" to describe an emerging field that studies how humans behave toward nature, in particular how they come to value and care for it (Saunders, 2003). For example, a detailed study of zoo visitors' self-reported emotional responses showed that certain emotions, including love, sense of connection, and amusement, related powerfully to their interest in the animals' subjective feelings and to their desires to preserve the animals. Such emotions tended to be selectively felt, evoked by some types of animals more than by others. At the same time, the emotions of wonder and respect were also correlated with a desire to save the animal concerned, and these were "equal opportunity" emotions that were experienced at high levels by visitors watching a range of types of animals (Myers et al., 2004). Interestingly, emotions related to love and caring were elicited more frequently by active animals than by passive ones, and a visitor's sense of connection to an animal was particularly enhanced if the person perceived the animal to be attending to them or to other people.

Several evaluation studies suggest that a range of designed settings for science learning afford learners opportunities to experience this kind of wonder and respect toward the natural world. For example:

> "I learned all about plants—where they come from and how they live—so that makes me respect them [plants] more" (male, age 50; translated from Spanish) (Jones, 2005, p. 9).

> "[I think the main purpose of this Africa Savanna exhibit is . . .] to make people aware of the problems regarding the Savanna; it helps personalize it so if you hear about problems regarding the Savanna one is more likely to help" (Meluch, 2006, pp. 16-21).

### Building Science Identity Across Age and Background

One of the most common underlying agendas of informal environments is not only to interest people in science, but also possibly to propel children into science careers and engagement in lifelong science learning through hobbies and other everyday pursuits. Compelling stories from leading scientists and science educators often point to museums and similar settings as a contributing influence on their lifelong passion for science (Csikszentmihalyi, 1996; Spock, 2000). Such experiences may serve as a general or specific impetus for a brilliant career, for example:

> A fairly typical childhood is one recalled by Isabella Karle, one of the leading crystallographers in the world, a pioneer in new methods of electron diffraction analysis and X-ray analysis. Her parents were Polish immigrants with minimal formal education and limited means. Yet even during the worst years of the Great Depression Isabella's mother saved from her housekeeping money so that the family could take two-week vacations to explore the East Coast. The parents took their children to the library, to museums, and to concerts. . . . So even though a child need not develop an early interest in a domain in order to become creative in it later, it does help a great deal to become exposed early to the wealth and variety of life (Csikszentmihalyi, 1996, p. 163).

> [E.O.] Wilson wanted to be an entomologist by age ten; some issues of the National Geographic and a visit with a friend to the Washington zoo confirmed that what he wanted most to do in life was to become an explorer and a naturalist (Csikszentmihalyi, 1996, p. 267).

A study by Sachatello-Sawyer and colleagues (2002) shows that adults seeking learning experiences in their midlives often turn to subjects that were of interest to them around the age of 10. These studies highlight the impact of experiences in informal environments at an early age on later life decisions for some, offering evidence of ongoing learning progressions in science. Interestingly, these progressions may falter and stall, especially without continuing involvement. For example, Jarvis and Pell (2005) interviewed children ages 10-11 two months after their visit to a space center, including a mediated group experience at a Challenger Center simulation. They found that 20 percent of the students were more interested in science careers after their visit than before, but that this interest declined over a 4-5 month period following the experience.

Some attempts have been made by practitioners to extend the learning trajectories of participants over space and time. For example, Schauble and Bartlett (2002) designed an extended trajectory for science learning by using the notion of a funnel, in which the outermost, largest physical space is designed to invite learners through easily accessible, compelling, and loosely structured experiences. The outer edge of the funnel would serve all learners, and those who chose to continue to pursue the big idea in question would move further into the funnel. The second level of the funnel was a series of quieter, restricted areas that they called Discovery Labs. One example of these was the Dock Shop, where participants could explore boat design, including the design of different types of hulls tested for carrying capacity and various sail types tested with a wind machine. The deepest portion of the funnel was designed for repeat visitors, such as members and children from the local neighborhood. The activities in this portion of the gallery were designed to build on children's prior experiences in the museum, at home, and at school. Visitors would borrow kits that were housed in the museum and also distributed through local libraries. These kits contained materials that allowed children to extend their explorations in more detailed, sustained studies and to send in their results to the museum through Science Postcards. For learners who wished to pursue a particular topic in depth, they would need to find ways to extend their learning over time, perhaps over the course of a 90-minute visit or for return visits and for additional activities (e.g., future reading, watching educational television). Similarly, many institutions have created systems for lending visitors objects and interpretive materials, such as books, for a period of time, and some (e.g., Science North in Canada) have borrowed or bought reciprocal contributions from visitors, which they have developed doing science outside the institution.

Some examples of longer term identity development come from studies of youth interns at science centers (Beane and Pope, 2002; Gupta and Siegel, 2008). Such studies suggest that the combination of appropriate mentorship, support, responsibility, and resources provided by these internships can support the personal learning and empowerment that lead a young person to choose a science-related career. An example of this is the New York Hall of Science Explainer Program, which has created an institutionalized career path for its young docents, providing them direct access to a science teaching training program. Since 1987 that program has followed approximately 400 young people using various forms of communication and involved them in four formal evaluations. The museum staff found that the program builds knowledge and teaching skills, skills for careers in a variety of professions, social bonds, and leadership (Gupta and Siegel, 2008).

Intensive programming for science learning has also been shown to have a long-lasting impact on children's identities as learners. A longitudinal study of young women from urban, low-income, single-parent families who participated in an after-school science museum program found that more than 90 percent of the participants went on to attend college (Fadigan and Hammrich, 2004). For those young women, careers in medical or health-related fields, followed by careers related to science, mathematics, and technology, were the highest ranking chosen career paths four to nine years after initial participation in the program. The young women pointed to three characteristics—having staff to talk to, learning job skills, and having the museum as a safe place to go—as most influential on their chosen educational and career paths. Extended programs are discussed in greater depth in Chapter 6.

As is discussed in Chapter 7, institutions tend to represent or reflect the dominant culture, which may present a conflict for those from nondominant groups (Ivanova, 2003). In order to manage these differences, a child from a marginalized culture may temporarily adopt an identity for science learning experiences (Heath, 1982). If one can better understand how children come to integrate science into their home cultures, rather than temporarily adopting an identity, such knowledge can be used to create science learning environments that are more accessible and meaningful (Warren, Rosebery, and Conant, 1994).

To find out whether individual learners integrate experiences in informal environments with their personal and community-related identities, further study and models are needed to explore the long-term impacts of these experiences. It may be that temporarily adopting a science-specific identity does not advance a long-term or permanent sense of oneself as a science learner. It may be, however, that experimenting with identities in informal environments is an important form of creative play in a low-stakes situation. Further work is needed, then, on identity development and sustainability in relation to learning science in informal environments over time, focused on learners' multiple identities and how exploring a new identity or integrating multiple identities can lead to greater participation in science.

## CONCLUSION

The literature on designed settings for science learning provides considerable evidence of learning across the strands. For Strand 1 there is evidence of learner excitement and strong positive emotional responses to experiences of science and the natural world. This may lead to other forms of valuable learning (sustained interest, flexible reasoning, etc.), although the evidence on this is less clear and the research is limited. There is also clear evidence for learning science content (Strand 2), in the form of factual recall after experiences in designed settings. Recollection seems to be supported by experiential linkages that ground abstractions in sensory experiences. It is unclear how learners draw from these experiences to assemble broader conceptual knowledge. This is an issue for future research, which is likely to require tracking participants over time and across settings. Strand 3 has strong support as learners engage in exploration and interaction, "doing and seeing," questioning, explaining, and making sense of the natural and designed world. There is some evidence for aspects of Strand 4, reflecting on science, in designed settings. Although analysis of visitor behavior suggests that reflection is limited, in the context of interviews, researchers and evaluators have found that participants can reflect on the enterprise of science and on their own thinking about science in the context of designed settings. Facilitation appears to be critical to supporting reflection. However, in designed settings, extensive facilitation by professional staff may not be feasible. And it may not always be desirable, as it can interfere with leisure experiences and interrupt other important developments in the participant experience.

Strand 5, engaging in science, is also strongly supported, especially in the general form of social interaction, in which learners jointly explore and interpret the natural world. Social interaction is a notable strong tendency in multigenerational group visits. However, participating in practices such as scientific argumentation as is often studied in school settings is not explored here. Further, it is likely not an appropriate goal for most designed settings for science learning which do not afford for facilitated, longer term investigations within a community of learners.

For Strand 6, there is evidence of learners' attempts to personalize and integrate science learning experiences with their values and identity. This lends support to the educational practice of adjusting science content and learning experiences to be compatible with learner agendas.

## REFERENCES

Allen, S. (1997). Using scientific inquiry activities in exhibit explanations. *Science Education, 81*(6), 715-734.

Allen, S. (2002). Looking for learning in visitor talk: A methodological exploration. In G. Leinhardt, K. Crowley, and K. Knutson (Eds.), *Learning conversations in museums* (pp. 259-303). Mahwah, NJ: Lawrence Erlbaum Associates.

Allen, S. (2007). *Secrets of circles summative evaluation report.* Report prepared for the Children's Discovery Museum of San Jose. Available: http://www.informalscience.org/evaluation/show/115 [accessed October 2008].

Allen, S., and Gutwill, J. (2004). Designing science museum exhibits with multiple interactive features: Five common pitfalls. *Curator, 47*(2), 199-212.

American Association for the Advancement of Science. (1993). *Benchmarks for science literacy.* New York: Oxford University Press.

Anderson, D. (2003). Visitors' long-term memories of world expositions. *Curator, 46*(4), 400-420.

Anderson, D., and Piscitelli, B. (2002). Parental recollections of childhood museum visits. *Museum National, 10*(4), 26-27.

Anderson, D., and Shimizu, H. (2007). Factors shaping vividness of memory episodes: Visitors' long-term memories of the 1970 Japan world exposition. *Memory, 15*(2), 177-191.

Anderson, D., and Zhang, Z. (2003). Teacher perceptions of fieldtrip planning and implementation. *Visitor Studies Today, 6*(3), 6-12.

Anderson, D., Kisiel, J., and Storksdieck, M. (2006). Understanding teachers perspectives on field trips: Discovering common ground in three countries. *Curator, 49*(3), 365-386.

Anderson, D., Lucas, K.B., Ginns, I.S., and Dierking, L.D. (2000). Development of knowledge about electricity and magnetism during a visit to a science museum and post-visit activities. *Science Education, 84*(5), 658-679.

Anderson, D., Piscitelli, B., Weier, K., Everett, M., and Tayler, C. (2002). Children's museum experiences: Identifying powerful mediators of learning. *Curator, 45*(3), 213-231.

Anderson, D., Storksdieck, M., and Spock, M. (2007). The long-term impacts of museum experiences. In J. Falk, L. Dierking, and S. Foutz (Eds.), *In principle, in practice: New perspectives on museums as learning institutions* (pp. 197-215). Walnut Creek, CA: AltaMira Press.

Ansbacher, T. (1999). Experience, inquiry, and making meaning. *Exhibitionist, 18*(2), 22-26.

Ash, D. (2002). Negotiations of thematic conversations about biology. In G. Leinhardt, K. Crowley, and K. Knutson (Eds.), *Learning conversations in museums* (pp. 357-400). Mahwah, NJ: Lawrence Erlbaum Associates.

Ash, D. (2003). Dialogic inquiry in life science conversations of family groups in museums. *Journal of Research in Science Teaching, 40*(2), 138-162.

Barron, B. (2006). Interest and self-sustained learning as catalysts of development: A learning ecology perspective. *Human Development, 49*(4), 153-224.

Beane, D.B., and Pope, M.S. (2002). Leveling the playing field through object-based service learning. In S.G. Paris (Ed.), *Perspectives on object-centered learning in museums* (pp. 325-349). Mahwah, NJ: Lawrence Erlbaum Associates.

Beaumont, L. (2005). *Summative evaluation of wild reef-sharks at Shedd.* Report for the John G. Shedd Aquarium. Available: http://www.informalscience.com/download/case_studies/report_133.doc [accessed October 2008].

Bell, P., and Linn, M.C. (2000). Scientific arguments as learning artifacts: Designing for learning from the web with KIE. *International Journal of Science Education, 22*(8), 797-817.

Belmont, J.M., and Butterfield, E.C. (1971). Learning strategies as determinants of memory deficiencies. *Cognitive Psychology, 2*, 411-420.

Bitgood, S. (2002). Environmental psychology in museums, zoos, and other exhibition centers. In R.B. Bechtel and A. Churchman (Eds.), *The environmental psychology handbook* (2nd ed.). New York: John Wiley.

Bitgood, S., and Loomis, C. (1993). Introduction: Environmental design and evaluation in museums. *Environment and Behavior, 26*(6), 683-697.

Blud, L.M. (1990). Social interaction and learning among family groups visiting a museum. *Museum Management and Curatorship, 9*(1), 43-51.

Borun, M. (2003). *Space Command summative evaluation.* Philadelphia: Franklin Institute Science Museum. Available: http://www.informalscience.org/evaluations/report_24.pdf [accessed October 2008].

Borun, M., and Miller, M. (1980). *What's in a name? A study of the effectiveness of explanatory labels in a science museum.* Philadelphia: Franklin Institute Science Museum.

Borun, M., Chambers, M., and Cleghorn, A. (1996). Families are learning in science museums. *Curator, 39*(2), 123-138.

Borun, M., Dritsas, J., Johnson, J.I., Peter, N.E., Wagner, K.F., Fadigan, K., Jangaard, A., Stroup, E., and Wenger, A. (1998). *Family learning in museums: The PISEC perspective.* Philadelphia: Franklin Institute.

Borun, M., Massey, C., and Lutter, T. (1993). Naive knowledge and the design of science museum exhibits. *Curator, 36*(3), 201-219.

Bower, H. (1981). Mood and memory. *American Psychologist, 36*(2), 129-148.

Briseno-Garzon, A., Anderson, D., and Anderson, A. (2007). Entry and emergent agendas of adults visiting an aquarium in family groups. *Visitor Studies, 10*(1), 73-89.

Brooks, J.A.M., and Vernon, P.E. (1956). A study of children's interests and comprehension at a science museum. *British Journal of Psychology, 47*, 175-182.

Bruner, J. (1996). *The culture of education.* Cambridge, MA: Harvard University Press.

Callanan, M.A., and Jipson, J.L. (2001). Explanatory conversation and young children's developing scientific literacy. In K. Crowley, C.D. Schumm, and T. Okada (Eds.), *Designing for science: Implications from everyday, classroom, and professional settings.* Mahwah, NJ: Lawrence Erlbaum Associates.

Callanan, M., Esterly, J., Martin, J., Frazier, B., and Gorchoff, S. (2002). *Family conversations about science in an "Alice's Wonderland" exhibit.* Paper presented at the meetings of the American Educational Research Association, New Orleans, LA, April.

Callanan, M.A., Jipson, J.L., and Soennichsen, M. (2002). Maps, globes, and videos: Parent-child conversations about representational objects. In S.G. Paris (Ed.), *Perspectives on object-centered learning in museums* (pp. 261-283). Mahwah, NJ: Lawrence Erlbaum Associates.

Carey, S. (1985). *Conceptual change in childhood.* Cambridge, MA: MIT Press.

Cazden, C.B. (2001). *Classroom discourse: The language of teaching and learning* (2nd ed.). Westport, CT: Heinemann.

Cohen, M., Winkel, G., Olsen, R., and Wheeler, F. (1977). Orientation in a museum: An experimental visitor study. *Curator, 20*(2), 85-97.

Crane, V., Nicholson, H.J., Chen, M., and Bitgood, S. (1994). *Informal science learning: What the research says about television, science museums and community-based projects.* Dedham, MA: Research Communications.

Crowley, K., and Callanan, M.A. (1998). Identifying and supporting shared scientific reasoning in parent-child interactions. *Journal of Museum Education, 23*, 12-17.

Crowley, K., and Galco, J. (2001). Family conversations and the emergence of scientific literacy. In K. Crowley, C. Schunn, and T. Okada (Eds.), *Designing for science: Implications from everyday, classroom, and professional science* (pp. 393-413). Mahwah, NJ: Lawrence Erlbaum Associates.

Crowley, K., and Jacobs, M. (2002). Building islands of expertise in everyday family activity. In G. Leinhardt, K. Crowley, and K. Knutson (Eds.), *Learning conversations in museums* (pp. 333-356). Mahwah, NJ: Lawrence Erlbaum Associates.

Crowley, K., Callanan, M.A., Jipson, J., Galco, J., Topping, K., and Shrager, J. (2001a). Shared scientific thinking in everyday parent-child activity. *Science Education, 85*(6), 712-732.

Crowley, K., Callanan, M.A., Tenenbaum, H.R., and Allen, E. (2001b). Parents explain more often to boys than to girls during shared scientific thinking. *Psychological Science, 12*(3), 258-261.

Csikszentmihalyi, M. (1996). *Creativity: Flow and the psychology of discovery and invention.* New York: HarperCollins.

Csikszentmihalyi, M., Rathunde, K.R., and Whalen, S. (1993). *Talented teenagers: The roots of success and failure.* New York: Cambridge University Press.

Deci, E.L., and Ryan, R.M. (2002). The paradox of achievement: The harder you push, the worse it gets. In J. Aronson (Ed.), *Improving academic achievement: Impact of psychological factors on education* (pp. 61-87). Orlando, FL: Academic Press.

Diamond, J. (1986). The behavior of family groups in science museums. *Curator, 29*(2), 139-154.

Dierking, L.D., Adelman, L.M., Ogden, J., Lehnhardt, K., Miller, L., and Mellen, J.D. (2004). Using a behavior changes model to document the impact of visits to Disney's Animal Kingdom: A study investigating intended conservation action. *Curator, 47*(3), 322-343.

Dierking, L.D., Burtnyk, K., Buchner, K.S., and Falk, J.H. (2002). *Visitor learning in zoos and aquariums: A literature review.* Silver Spring, MD: American Zoo and Aquarium Association.

Doering, Z.D. (1999). Strangers, guests, or clients? Visitor experiences in museums. *Curator, 42*(2), 74-87.

Driver, R., Newton, P., and Osborne, J. (2000). Establishing the norms of scientific argumentation in classrooms. *Science Education, 84*(3), 287-312.

Duschl, R.A., and Osborne, J. (2002). Supporting and promoting argumentation discourse in science education. *Studies in Science Education, 38,* 39-72.

Eich, E., Kihlstrom, J., Bower, G., Forgas, J., and Niedenthal, P. (2000). *Cognition and emotion.* New York: Oxford University Press.

Ellenbogen, K.M. (2002). Museums in family life: An ethnographic case study. In G. Leinhardt, K. Crowley, and K. Knutson (Eds.), *Learning conversations in museums* (pp. 81-101). Mahwah, NJ: Lawrence Erlbaum Associates.

Ellenbogen, K.M. (2003). From dioramas to the dinner table: An ethnographic case study of the role of science museums in family life. *Dissertation Abstracts International, 64*(3), 846-847.

Ellenbogen, K.M., Luke, J.J., and Dierking, L.D. (2004). Family learning research in museums: An emerging disciplinary matrix? *Science Education, 88*(S1), S48-S58.

Engle, R.A., and Conant, F.C. (2002). Guiding principles for fostering productive disciplinary engagement: Explaining an emergent argument in a community of learners classroom. *Cognition and Instruction, 20*(4), 399-483.

Fadigan, K.A., and Hammrich, P.L. (2004). A longitudinal study of the educational and career trajectories of female participants of an urban informal science education program. *Journal of Research on Science Teaching, 41*(8), 835-860.

Falk, J.H. (2008). Calling all spiritual pilgrims: Identity in the museum experience. *Museum* (Jan./Feb.).

Falk, J.H., and Balling, J.D. (1982). The field trip milieu: Learning and behavior as a function of contextual events. *Journal of Educational Research, 76*(1), 22-28.

Falk, J.H., and Dierking, L.D. (1997). School field trips: Assessing their long-term impact. *Curator, 40,* 211-218.

Falk, J.H., and Storksdieck, M. (2005). Using the "contextual model of learning" to understand visitor learning from a science center exhibition. *Science Education, 89,* 744-778.

Falk, J.H., Heimlich, J., and Bronnenkant, K. (2008). Using identity-related visit motivations as a tool for understanding adult zoo and aquarium visitors' meaning-making. *Curator, 51*(1), 55-79.

Falk, J.H., Moussouri, T., and Coulson, D. (1998). The effect of visitors' agendas on museum learning. *Curator, 41*(2), 107-120.

Falk, J.H., Reinhard, E.M., Vernon, C.L., Bronnenkant, K., Deans, N.L., and Heimlich, J.E. (2007). *Why zoos and aquariums matter: Assessing the impact of a visit.* Silver Spring, MD: Association of Zoos and Aquariums.

Falk, J.H., Scott, C., Dierking, L., Rennie, L. and Jones, M.C. (2004). Interactives and visitor learning. *Curator, 47*(2), 171-192.

Fienberg, J., and Leinhardt, G. (2002). Looking through the glass: Reflections of identity in conversations at a history museum. In G. Leinhardt, K. Crowley, and K. Knutson (Eds.), *Learning conversations in museums* (pp. 167-211). Mahwah, NJ: Lawrence Erlbaum Associates.

Gee, J.P. (1994). First language acquisition as a guide for theories of learning and pedagogy. *Linguistics and Education, 6*(4), 331-354.

Gelman, R., Massey, C., and McManus, P.M. (1991). Characterizing supporting environments for cognitive development: Lessons from children in a museum. In L.B. Resnick, J.M. Levine, and S.D. Teasley (Eds.), *Perspectives on socially shared cognition* (pp. 226-256). Washington, DC: American Psychology Society.

Gleason, M.E., and Schauble, L. (1999). Parents' assistance of their children's scientific reasoning. *Cognition and Instruction, 17*(4), 343-378.

Goldowsky, N. (2002). *Lessons from life: Learning from exhibits, animals, and interaction in a museum.* UMI#3055856. Unpublished doctoral dissertation, Harvard University.

Griffin, J. (1994). Learning to learn in informal science settings. *Research in Science Education, 24*(1), 121-128.

Griffin, J., and Symington, D. (1997). Moving from task-oriented to learning-oriented strategies on school excursions to museums. *Science Education, 81*(6), 763-779.

Grolnick, W.S., and Ryan, R.M. (1987). Autonomy in children's learning: An experimental and individual difference investigation. *Journal of Personality and Social Psychology, 52*(5), 890-898.

Guichard, H. (1995). Designing tools to develop the conception of learners. *International Journal of Science Education, 17*(2), 243-253.

Gupta, P., and Siegel, E. (2008). Science career ladder at the New York Hall of Science: Youth facilitators as agents of inquiry. In R.E. Yaeger and J.H. Falk (Eds.), *Exemplary science in informal education settings: Standards-based success stories.* Arlington, VA: National Science Teachers Association.

Hayward, J. (1997). *Conservation study, phase 2: An analysis of visitors' perceptions about conservation at the Monterey Bay Aquarium.* Northampton, MA: People, Places, and Design Research.

Hayward, J. (1998). *Summative evaluation: Visitors' reactions to "Fishing for Solutions."* Northampton, MA: People, Places, and Design Research.

Heath, S.B. (1982). What no bedtime story means: Narrative skills at home and school. *Language Socialization, 11*(1), 49-76.

Hein, G.E. (1998). *Learning in the museum.* New York: Routledge.

Hein, G.E., Kelley, J., Bailey, E., and Bronnenkant, K. (1996). *Investigate: Summative evaluation report.* Unpublished report, Leslie College Program Evaluation Group.

Hensel, K.A. (1987). Families in a museum: Interactions and conversations at displays. *Dissertation Abstracts International, 49*(9). University Microfilms No. 8824441.

Hidi, S., and Renninger, K.A. (2006). The four-phase model of interest development. *Educational Psychologist, 41*(2), 111-127.

Hilke, D.D. (1987). Museums as resources for family learning: Turning the question around. *Museologist, 50*(175), 14-15.

Hilke, D.D. (1989). The family as a learning system: An observational study of families in museums. In B.H. Butler and M.B. Sussman (Eds.), *Museum visits and activities for family life enrichment* (pp. 101-129). New York: Haworth Press.

Holland, D., Lachicotte, W., Skinner, D., and Cain, C. (1998). *Identity and agency in cultural worlds.* Cambridge, MA: Harvard University Press.

Humphrey, T., and Gutwill, J.P. (Eds.). (2005). *Fostering active prolonged engagement: The art of creating APE exhibits.* San Francisco: The Exploratorium.

Hutt, C. (1981). Toward a taxonomy and conceptual model of play. In H.I. Day (Ed.), *Advances in intrinsic motivation and aesthetics.* New York: Plenum Press.

Inagaki, K., and Hatano, G. (2002). *Young children's naive thinking about the biological world.* New York: Psychology Press.

Ivanova, E. (2003). Changes in collective memory: The schematic narrative template of victimhood in Kharkiv museums. *Journal of Museum Education, 28*(1), 17-22.

Jacobson, W. (2006). *Why we do what we do: A conceptual model for the work of teaching and learning centers.* Presentation for the International Symposium on Excellence in Teaching and Learning (ISETL), National Taiwan University, November.

Jarvis, T., and Pell, A. (2005). Factors influencing elementary school children's attitudes toward science before, during, and after a visit to the UK National Space Centre. *Journal of Research on Science Teaching, 42*(1), 53-83.

Jones, J. (2003). *Cell lab summative evaluation.* St. Paul: Science Museum of Minnesota. Available: http://www.informalscience.org/evaluation/show/57 [accessed October 2008].

Jones, J. (2005). *Huntington Botanical Gardens summative evaluation conservatory for botanical science.* Report for Huntington Botanical Gardens. Available: http://www.informalscience.org/evaluation/show/105 [accessed October 2008].

Kaplan, A., and Maehr, M.L. (2007). The contributions and prospects of goal orientation theory. *Educational Psychology Review, 19*(2), 141-184.

Kisiel, J. (2006). An examination of fieldtrip strategies and their implementation within a natural history museum. *Science Education, 90*(3), 434-452.

Koran, J.J., Koran, M.L., and Ellis, J. (1989). Evaluating the effectiveness of field experiences: 1939-1989. *Scottish Museum News, 4*(2), 7-10.

Koran, J.J., Koran, M.L., and Foster, J.S. (1988). Using modeling to direct attention. *Curator, 31*(1), 36-42.

Koran, J.J., Koran, M.L., and Longino, S.J. (1986). The relationship of age, sex, attraction, and holding power with two types of science exhibits. *Curator, 29*(3), 227-235.

Koran, J.J., Lehman, J., Shafer, L.D., and Koran, M.L. (1983). The relative effects of pre- and post attention directing devices on learning from a "walk-through" museum exhibit. *Journal of Research in Science Teaching, 20*(4), 341-346.

Korn, R. (1997). *Electric space: A summative evaluation*. Boulder, CO: Space Science Institute. Available: http://www.informalscience.org/evaluation/show/19 [accessed October 2008].

Korn, R. (2004). *Invention at play: Summative evaluation*. New York: New York Hall of Science. Available: http://www.informalscience.org/evaluation/show/66 [accessed October 2008].

Korn, R. (2006). *Search for life: Summative evaluation*. New York: New York Hall of Science. Available: http://www.informalscience.org/evaluations/report_151.pdf [accessed October 2008].

Kubota, C.A., and Olstad, R.G. (1991). Effects of novelty-reducing preparation on exploratory behavior and cognitive learning in a science museum setting. *Journal of Research in Science Teaching, 28*(3), 225-234.

Kuhn, D., and Franklin, S. (2006). The second decade: What develops (and how)? In D. Kuhn and R. Siegler (Eds.), *Handbook of child psychology: Cognition, perception and language* (vol. 2, 6th ed., pp. 953-993). New York: Wiley.

Laetsch, W.M., Diamond, J., Gottfried, J.L., and Rosenfeld, S. (1980). Children and family groups in science centres. *Science and Children, 17*(6), 14-17.

Lehrer, R., and Schauble, L. (2000). The development of model-based reasoning. *Journal of Applied Developmental Psychology, 21*(1), 39-48.

Leinhardt, G., and Gregg, M. (2002). Burning buses, burning crosses: Pre-service teachers see civil rights. In G. Leinhardt, K. Crowley, and K. Knutson (Eds.), *Learning conversations in museums* (pp. 139-166). Mahwah, NJ: Lawrence Erlbaum Associates.

Leinhardt, G., and Knutson, K. (2004). *Listening in on museum conversations*. Walnut Creek, CA: AltaMira Press.

Leinhardt, G., Tittle, C., and Knutson, K. (2002). Talking to oneself: Diaries of museum visits. In G. Leinhardt, K. Crowley, and K. Knutson (Eds.), *Learning conversations in museums* (pp. 103-133). Mahwah, NJ: Lawrence Erlbaum Associates.

Lemke, J.L. (1990). *Talking science: Language, learning and values*. Norwood, NJ: Ablex.

Lipstein, R., and Renninger, K.A. (2006). "Putting things into words": The development of 12-15-year-old students' interest for writing. In S. Hidi and P. Boscolo (Eds.), *Writing and motivation* (pp. 113-140). Oxford, England: Elsevier.

Lucas, A.M. (1983). Scientific literacy and informal learning. *Studies in Science Education, 10*, 1-36.

Ma, J. (2002). *Outdoor exploratorium: Front end study. Open-ended exploration with a noticing toolkit at the Palace of Fine Arts*. San Francisco: The Exploratorium. Available: http://www.informalscience.org/evaluation/show/39 [accessed October 2008].

Martin, L., and Toon, R. (2005). Narratives in a science center: Interpretation and identity. *Curator, 48*(4), 407-426.

Maxwell, L.E., and Evans, G.W. (2002). Museums as learning settings: The importance of the physical environment. *Journal of Museum Education, 27*(1), 3-7.

McLean, K. (1993). *Planning for people in museum exhibitions*. Washington, DC: Association of Science-Technology Centers.

McLean, K., and Pollock, W. (Eds.). (2007). *Visitor voices in museum exhibitions*. Washington, DC: Association of Science-Technology Centers.

McManus, P.M. (1987). It's the company you keep: The social determination of learning-related behavior in a science museum. *International Journal of Museum Management and Curatorship, 6*(3), 260-270.

McManus, P.M. (1989). What people say and how they think in a science museum. In D. Uzzell (Ed.), *Heritage interpretation* (pp. 174-189). London: Bellhaven Press.

Meisner, R., vom Lehn, D., Heath, C., Burch, A., Gammon, B., and Reisman, M. (2007). Exhibiting performance: Co-participation in science centres and museums. *International Journal of Science Education, 29*(12), 1531-1555.

Meluch, W. (2006). *San Francisco Zoo African savanna exhibit summative evaluation*. San Francisco: San Francisco Zoo. Available: http://www.informalscience. org/evaluation/show/54 [accessed October 2008].

Minstrell, J., and van Zee, E.H. (Eds.). (2000). *Inquiring into inquiry learning and teaching in science*. Washington, DC: American Association for the Advancement of Science.

Moll, L., Amanti, C., Neff, D., and Gonzalez, N. (2005). Funding of knowledge for teaching: Using a qualitative approach to connect homes and classrooms. In L. Moll, C. Amanti, and N. Gonzalez (Eds.), *Funds of knowledge: Theorizing practices in households, communities, and classrooms* (pp. 71-88). London: Routledge.

Morrissey, K. (2002). Pathways among objects and museum visitors. In S.G. Paris (Ed.), *Perspectives on object-centered learning in museums* (pp. 285-299). Mahwah, NJ: Lawrence Erlbaum Associates.

Moussouri, T. (1998). Family agendas and the museum experience. *Museum Archeologist, 24*, 20-30.

Moussouri, T., (2002). *Researching learning in museums and galleries, 1990-1999: A bibliographic review*. Part of the Leicester Museum Studies Series. Leicester, England: Research Centre for Museums and Galleries, Department of Museum Studies.

Myers, G., Saunders, C.D., and Birjulin, A.A. (2004). Emotional dimensions of watching zoo animals: An experience sampling study building on insights from psychology. *Curator, 47*, 299-321.

National Research Council. (1996). *National science education standards*. National Committee on Science Education Standards and Assessment. Washington, DC: National Academy Press.

National Research Council. (1999). *How people learn: Brain, mind, experience, and school*. Committee on Developments in the Science of Learning. J.D. Bransford, A.L. Brown, and R.R. Cocking (Eds.). Washington, DC: National Academy Press.

National Research Council. (2000). *Inquiry and the national science education standards: A guide for teaching and learning*. Committee on the Development of an Addendum to the National Science Education Standards on Scientific Inquiry. S. Olson and S. Loucks-Horsley (Eds.). Washington, DC: National Academy Press.

National Research Council. (2007). *Taking science to school: Learning and teaching science in grades K-8*. Committee on Science Learning, Kindergarten Through Eighth Grade. R.A. Duschl, H.A. Schweingruber, and A.W. Shouse (Eds.). Washington, DC: The National Academies Press.

National Science Foundation. (2006). *Now showing: Science Museum of Minnesota "Cell Lab."* Available: http://www.nsf.gov/news/now_showing/museums/cell_lab. jsp [accessed October 2008].

Orion, N., and Hofstein, A. (1994). Factors that influence learning during a scientific field trip in a natural environment. *Journal of Research in Science Teaching, 31*(10), 1097-1119.

Palmquist, S., and Crowley, K. (2007). From teachers to testers: How parents talk to novice and expert children. *Science Education, 91*(5), 783-804.

Pearce, S. (Ed.). (1994). *Interpreting objects and collections.* London: Routledge.

Peart, B. (1984). Impact of exhibit type on knowledge gain, attitudes, and behavior. *Curator, 27*(3), 220-227.

Pedretti, E.G. (2004). *Perspectives on learning through research on critical issues-based science center exhibitions.* Hoboken, NJ: Wiley.

Pedretti, E.G., MacDonald, R.G., Gitari, W., and McLaughlin, H. (2001). Visitor perspectives on the nature and practice of science: Challenging beliefs through "a question of truth." *Canadian Journal of Science, Mathematics and Technology Education, 4,* 399-418.

Perry, D.L. (1994). Designing exhibits that motivate. In R.J. Hannapel (Ed.), *What research says about learning in science museums* (vol. 2, pp. 25-29). Washington, DC: Association of Science-Technology Centers.

Price, S., and Hein, G.E. (1991). More than a field trip: Science programmes for elementary school groups at museums. *International Journal of Science Education, 13*(5), 505-519.

Quadra Planning Consultants, Ltd., and Galiano Institute for Environmental and Social Research. (2004, June). *Seafood watch evaluation: Summary report.* Saltspring Island, BC, Canada: Author.

Rand, J. (1990). Fish stories that hook readers: Interpretive graphics at the Monterey Bay Aquarium. In *Proceedings of the 1986 American Association of Zoological Parks and Aquariums* (pp. 404-413). Columbus, OH: American Association of Zoological Parks and Aquariums.

Randol, S.M. (2005). *The nature of inquiry in science centers: Describing and assessing inquiry at exhibits.* Unpublished doctoral dissertation, University of California, Berkeley.

Rennie, L.J., and McClafferty, T.P. (1993). Learning in science centres and science museums: A review of recent studies. *Research in Science Education, 23*(1), 351.

Rennie, L.J., and McClafferty, T.P. (2002). Objects and learning: Understanding young children's interaction with science exhibits. In S.G. Paris (Ed.), *Perspectives on object-centered learning in museums* (pp. 191-213). Mahwah, NJ: Lawrence Erlbaum Associates.

Renninger, K.A. (2000). Individual interest and its implications for understanding intrinsic motivation. In C. Sansone and J.M. Harackiewicz (Eds.), *Intrinsic motivation: Controversies and new directions* (pp. 373-404). San Diego: Academic Press.

Renninger, K.A., and Hidi, S. (2002). Student interest and achievement: Developmental issues raised by a case study. In A. Wigfield and J.S. Eccles (Eds.), *Development of achievement motivation* (pp. 173-195). New York: Academic Press.

Rogoff, B. (2003). *The cultural nature of human development.* New York: Oxford University Press.

Rosenfeld, S., and Terkel, A. (1982). A naturalistic study of visitors at an interactive mini-zoo. *Curator, 25*(3), 187-212.

Rounds, J. (1999). Meaning-making: A new paradigm for museum exhibits? *Exhibitionist, 18*(2), 5-8.

Sachatello-Sawyer, B., Fellenz., R.A., Burton, H., Gittings-Carlson, L., Lewis-Mahony, J.L., and Woolbaugh, W. (2002). *Adult museum programs: Designing meaningful experiences.* Walnut Creek, CA: AltaMira Press.

Saunders, C.D. (2003). The emerging field of conservation psychology. *Human Ecology Review, 10*(2), 137-149.

Schauble, L., and Bartlett, K. (1997). Constructing a science gallery for children and families: The role of research in an innovative design process. *Science Education, 81*(6), 781-793.

Schauble, L., Gleason, M., Lehrer, R., Bartlett, K., Petrosino, A.J., Allen, A., Clinton, C., Ho, E., Jones, M.G., Lee, Y.L., Phillips, J., Seigler, J., and Street, J. (2002). Supporting science learning in museums. In G. Leinhardt, K. Crowley, and K. Knutson (Eds.), *Learning conversations: Explanation and identity in museums* (pp. 425-452). Mahwah, NJ: Lawrence Erlbaum Associates.

Schneider, B., and Cheslock, N. (2003). *Measuring results: Gaining insight on behavior change strategies and evaluation methods from environmental education, museum, health and social marketing programs.* San Francisco: Coevolution Institute. Available: http://www.pollinator.org/Resources/Metrics%20Executive%20Summary.pdf [accessed October 2008].

Screven, C.G. (1992). Motivating visitors to read labels. *ILVS Review, 2*(2), 183-221.

Serrell, B. (1996). *Exhibit labels: An interpretive approach.* Walnut Creek, CA: AltaMira Press.

Serrell, B. (2001). *Marvelous molecules: The secret of life.* Corona Park: New York Hall of Science. Available: http://www.informalscience.org/evaluation/show/41 [accessed October 2008].

Serrell, B. (2006). *Judging exhibitions: A framework for assessing excellence.* Walnut Creek, CA: Left Coast Press.

Silverman, L.H. (1995). Visitor meaning-making in museums for a new age. *Curator, 38*(3), 161-170.

Spock, M. (Ed.) (2000). *Philadelphia stories: A collection of pivotal museum memories.* Video and study guide. Washington, DC: American Association of Museums.

St. John, M., and Perry, D.L. (1993). Rethink role, science museums urged. *ASTC Newsletter, 21*(5), 1, 6-7.

Steidl, S., Mohi-uddin, S., and Anderson, A (2006). Effects of emotional arousal on multiple memory systems: Evidence from declarative and procedural learning. *Learning & Memory, 13*(5), 650-658. Available: http://www.pubmedcentral.nih.gov/picrender.fcgi?artid=1783620&blobtype=pdf [accessed March 2009].

Stevenson, J. (1991). The long-term impact of interactive exhibits. *International Journal of Science Education, 13*(5), 521-531.

Storksdieck, M. (2001). Differences in teachers' and students' museum field-trip experiences. *Visitor Studies Today, 4*(1), 8-12.

Storksdieck, M. (2006). *Field trips in environmental education.* Berlin: Berliner Wissenschafts-Verlag.

Tal, R.T., Bamberger, Y., and Morag, O. (2005). Guided schools visits to natural history museums in Israel. *Science Education, 89*(6), 920-935.

Taylor, S.M. (1986). *Understanding processes of informal education: A naturalistic study of visitors to a public aquarium.* Unpublished doctoral dissertation, University of California, Berkeley.

Tenenbaum, H.R., and Leaper, C. (1998). Mothers' and fathers' responses to their Mexican-descent child: A sequential analysis. *First Language, 18*(53), 129-147.

Tenenbaum, H.R., Snow, C., Roach, K., and Kurland, B. (2005). Talking and reading science: Longitudinal data on sex differences in mother-child conversations in low-income families. *Journal of Applied Developmental Psychology, 26*(1), 1-19.

Tisdal, C. (2004). *Going APE (active prolonged exploration) at the Exploratorium: Phase 2 summative evaluation.* Available: http://www.informalscience.org/evaluation/show/67 [accessed October 2008].

Tunnicliffe, S.D. (1996). A comparison of conversations of primary school groups at animated, preserved, and live animal specimens. *Journal of Biological Education, 30*(3), 195-206.

Tunnicliffe, S.D. (2000). Conversations of family and primary school groups at robotic dinosaur exhibits in a museum: What do they talk about? *International Journal of Science Education, 22*(7), 739-754.

Utman, C.H. (1997). Performance effects of motivational state: A meta-analysis. *Personality and Social Psychology Review, 1*(2), 170-182.

vom Lehn, D., Heath, C., and Hindmarsh, J. (2001). Exhibiting interaction: Conduct and collaboration in museums and galleries. *Symbolic Interaction, 24*(2), 189-216.

Warren, B., Rosebery, A., and Conant, F. (1994). Discourse and social practice: Learning science in language minority classrooms. In D. Spener (Ed.), *Adult biliteracy in the United States* (pp. 191-210). McHenry, IL: Center for Applied Linguistics and Delta Systems.

Wenger, E. (1999). *Communities of practice: Learning, meaning, and identity.* New York: Cambridge University Press.

Yalowitz, S.S. (2004). Evaluating visitor conservation research at the Monterey Bay Aquarium. *Curator, 47*(3), 283-298.

Zuckerman, M., Porac, J., Lathin, D., Smith, R., and Deci, E.L. (1978). On the importance of self-determination for intrinsically motivated behavior. *Personality and Social Psychology Bulletin, 4*(3), 443-446.

# 6

# Programs for Young and Old

This chapter focuses on science learning programs for children, youth, and adults. These programs take place in many different environments—schools, community centers, universities, and a range of informal institutions. They are held indoors and out and in urban, suburban, and rural areas. Program schedules vary, with some taking place daily and others weekly or even monthly. The way in which participants spend their time also varies. Some programs mirror a traditional classroom structure, with program leaders teaching mini-lessons and students practicing skills. Some programs conduct projects off-site in the community, and others take place in a science lab or field study setting. Program goals may include developing basic scientific knowledge, advancing academic school goals, or applying knowledge to improve the quality of life for the participant or the community.

What these programs have in common is an organizational goal to achieve curricular ends—a goal that distinguishes them from everyday learning activities and learning in designed environments. Science learning programs are typically led by a professional educator or facilitator, and, rather than being episodic and self-organized, they tend to extend for a period of weeks or months and serve a prescribed population of learners. Ideally, the programs are informal in design—they are learner driven, identifying and building on the interests and motivations of the participant, and use assessment in constructive, formative ways to give learners useful, valued information. Yet as programs that retain much of the structure identified with schools—a curriculum that unfolds over time, facilitators or teachers, a consistent group of participants—and yet that occur in nonschool hours, they have a natural tension. Nowhere is this tension more evident than in discussions of after-school programs in which establishing learning goals, outcome measures, and accountability processes can be especially contentious (see Box 6-1).

In this chapter we organize the discussion of programs for science learning around three distinct age groups: children and youth in after-school and out-of-school programs; adults, including K-12 teachers; and older adults, who have unique developmental capabilities and life-course interests. The emphasis in this chapter on programs for school-age children, and specifically after-school programs, reflects several considerations:

- the committee's charge to examine the articulation between schools and informal settings;
- the scale and proliferation of out-of-school-time programs;
- the fact that there has been considerable research on this topic, much of it evoking controversy that the committee hopes to illuminate and address;
- the relative paucity of research on programs for adults (including senior citizens); and
- the promise of out-of-school time as a means of engaging a diverse population of children and youth in science (e.g., U.S. Department of Education, 2003).

## LEARNING SCIENCE IN OUT-OF-SCHOOL-TIME PROGRAMS

Out-of-school-time programs have existed for some time, first appearing at the end of the 19th century.[1] Throughout the years, they have changed and adapted to serve different purposes, needs, and concerns, including providing a safe environment, academic enrichment, socialization, acculturation, problem remediation, and play (Halpern, 2002). Diverging goals and the fact that multiple institutions and professional communities share claim to these programs has periodically caused tensions (see Box 6-2).

Today, out-of-school-time programs typically incorporate three blocks of time devoted to (1) homework help and tutoring, (2) enriched learning experiences, and (3) nonacademic activities, such as sports, arts, or play (Noam, Biancarosa, and Dechausay, 2003). Programs are also expanding, in large part due to strong federal and private support. They continue to be supported by various stakeholders with diverse goals for a broad range of student populations.

The bulk of the research on out-of-school-time programs has occurred in the past two decades in conjunction with a rise in governmental and public

---

[1]We use the term "out-of-school" to refer to the broad set of educational programs that take place before or after the school day and during nonschool periods, such as summer vacation.

support for them. Politicians, parents, and educators increasingly view these programs as an important developmental contributor in the lives of young people and a necessary component of public education. One indication of their importance is funding for the 21st Century Community Learning Centers (CCLCs), a federal program providing out-of-school-time care: it rose from zero in 1994, to $40 million in 1998, to $1 billion in 2002 (U.S. Department of Education, 2003). In 2007, the House of Representatives voted to increase funding to $1.1 billion (Afterschool Alliance, 2008).

Society has also witnessed changes in the workforce, resulting in a greater proportion of homes in which all adults are employed and an increase in student participation in out-of-school-time programs and other care arrangements. In 2005, 40 percent of all students in grades K-8 were in at least one weekly nonparental out-of-school-time care arrangement (National Center for Education Statistics, 2006). School-based or center-based programs were the most common care arrangement. Out-of-school-time programs have the potential to provide large-scale enrichment opportunities that were once reserved for wealthier families. In fact, at the 21st CCLCs, more than half the participants are of minority background and from low-income schools. The students who attend most frequently are more likely to be black, from single-parent homes, low-income, and on public assistance. This means that out-of-school-time programs often serve the most vulnerable populations. One consequence of this demographic structure is that much of the research on learning in out-of-school-time programs focuses on nondominant groups, a feature that will be seen in the evidence cited throughout this chapter.

## Evidence of Science Learning

Despite its long history, research on learning in general in out-of-school programs is controversial and inconclusive (Miller, 2003; Dynarski et al., 2004; Bissel et al., 2003). However, a range of evaluation studies show that out-of-school programs can have positive effects on participants' attitudes toward science, grades, test scores, graduation rates, and specific science knowledge and skills (Gibson and Chase, 2002; Building Science and Engineering Talent, 2004; Archer, Fanesali, Froschl, and Sprung, 2003; Project Exploration, 2006; Ferreira, 2001; Harvard Family Research Project, 2003; DeHaven and Weist, 2003; Jarman, 2005; Campbell et al., 1998, as cited in Fancsali, 2002; Brenner, Hudley, Jimerson, and Okamoto, 2001; Johnson, 2005; Fusco, 2001; Jeffers, 2003). Yet there is little evidence of a synthesized literature on out-of-school-time science programs.

Program goals, outcome measures used to evaluate them, and research methods vary tremendously in this area. Some researchers, drawing on social psychology and youth development traditions, are primarily concerned with the development of positive attitudes, skills, and social relationships. Other

## BOX 6-1 The Relationship Between School and Out-of-School Programs

Historically, relationships between school and out-of-school programs—particularly community-based out-of-school programs—have often been characterized by mutual mistrust and conflict. In a report based on 10 years of research studying approximately 120 youth-based community organizations throughout the United States, McLaughlin (2000) explains, "adults working with youth organizations frequently believe that school people do not respect or value their young people. Educators, for their part, generally see youth organizations as mere 'fun' and as having little to contribute to the business of schools. Moreover, educators often establish professional boundaries around learning and teaching, considering them the sole purview of teachers. If we want to better serve our youth, there is an obvious need for rethinking the relationship between schools and out-of-school programs, particularly for out-of-school programs that have an academic focus such as science."

In *Afterschool Education*, Noam, Biancarosa, and Dechausay (2003) outline different models of relationships between school and out-of-school programs in an effort to create better relationships, management connections, and interesting curricula and materials. At one extreme, there is the model of "unified" programs that are the equivalent of what is now called extended-day programming. Under this model, out-of-school programs can become essentially indistinguishable from school, since they take place in the same space and are usually under the same leadership (the school principal). At the other extreme lie "self-contained" programs, which intentionally choose to be separate from schools. Taking place in a different location, they often provide students with an entirely different experience from school. Between these two extremes lie three other models: "associated," "coordinated," and "integrated," each connecting out-of-school programs with schools at different levels of intensity. Noam and colleagues also outline the different ways these

researchers are more concerned with academic skills and improved academic achievement, as measured by standardized test scores, grades, graduation rates, and continued involvement in school science (Campbell et al., 1998; Building Engineering and Science Talent, 2004; Brenner et al., 2001). Given these different approaches (and the concerns we noted in Chapter 3 about relying on solely academic outcomes), we cannot provide definitive conclu-

connections can take place, dividing them into interpersonal, systemic, and curricular domains. The curricular domain is perhaps the most significant one in the discussion of relationships between out-of-school science and school science, although it is obviously influenced by such factors as physical location, philosophy, and interpersonal relationships. These models of relationships between out-of-school programs and school can be used as a foundation for more specific models describing the spectrum of relationships between out-of-school science and school science.

With the associated model, the out-of-school curriculum is closely connected to the school curriculum. Out-of-school coordinators and staff know on a week-by-week basis the material teachers are covering in class and can directly connect it to out-of-school activities. Out-of-school science is essentially an extension of school science, but with a more informal feel. The benefit of this model is that out-of-school and school science are connected, and the connection between the two is explicit.

In the coordinated model, out-of-school science programs connect their activities to the general school science curriculum and standards but not to what students are learning in class on a daily or weekly basis. This model avoids some of the conflicts between science in schools and out-of-school programs, while allowing out-of-school programs to support students' learning in schools. It also has logistical benefits, since it does not require the same level of planning and day-to-day communication between schoolteachers and out-of-school staff.

Finally, in the integrated model, out-of-school science is entirely disconnected from school science. Out-of-school programs make sure that participants are engaging in high-quality science experiences, but consider it undesirable for students to connect out-of-school science to school science. By keeping the two worlds separate, integrated out-of-school programs say they can provide students with an alternate entry point into science if they have already been turned off from school science.

sions about what learning outcomes can be achieved. Our goal here is to organize the evidence of science-specific learning outcomes in a way that can provide a foundation for exploring two questions more thoroughly in the future: To what extent are the types of science-specific goals described in this report reflected in the evidence base? How do the commitments of

## BOX 6-2 Learning Goals for Science Learning Programs

There is an ongoing debate about the goals of out-of-school programs and appropriate measures for evaluating them. On one side of the debate are those who view out-of-school time as an extension of regular school time. They argue that, in an age of accountability, when many students are failing to meet state and national academic standards, out-of-school time should be used to further the academic goals of schools.

On the other side of the debate are those who view out-of-school time as part of the broader realm of development. In their view, out-of-school programs should ensure healthy development and well-being for participants by developing personal and social assets in physical, intellectual, psychological, emotional, and social development domains (Institute of Medicine, 2002). The focus in programming is less on the acquisition of specific academic skills and knowledge and much more on providing a physically and psychologically safe environment with supportive relationships and a sense of belonging. Adding a science focus does not conflict with these nonacademic outcomes. Learners

informal education, such as learner choice and low-stakes assessment, shape the program and evaluation agenda in out-of-school settings?

In many cases, the dominance of a youth development, academic accountability, or science-specific perspective is evident in program goals and outcome measures. In an effort to integrate the findings and identify patterns of strong evidence with respect to science-specific outcomes across studies, we integrate the evidence across these varied perspectives. We examine evidence in light of the strands of science learning—some but not all of which are evident in the research base on out-of-school-time programs.

### Strand 1: Developing Interest in Science

Promoting interest in science is a common goal of out-of-school science programs (e.g., Brenner et al., 2001; Building Science and Engineering Talent, 2004; Gibson and Chase, 2002; Archer et al., 2003; Project Exploration, 2006). A number of evaluations that have examined this outcome suggest that sustained engagement in out-of-school science programs can promote science interest.

For example, a comparison study by Gibson and Chase (2002) exam-

must feel safe with science and find value in it if they are to make progress along the strands.

A science focus does call for careful attention to the specificity of socio-emotional and cognitive outcomes—that is, to the ways that out-of-school programs may contribute specifically to the learning outcomes described in the strands. In fact, out-of-school settings may provide a place where science learning can have a greater impact through higher "dosage" than incidental experiences in designed settings without losing an informal feel. Lucy Friedman, president of The After-School Corporation, writes, "While both the afterschool and science fields are at a crossroads, association with the other enhances the potential for each to flourish" (Friedman, 2005, p. 75).

It has also become more important to find new venues for science learning as time spent on science in schools decreases (Dorph et al., 2007; McMurrer, 2007). Schools classified as "in need of improvement" under the No Child Left Behind Act, in particular, have limited science instructional time; 43 percent of these schools have cut science to an average of 91 minutes per week (McMurrer, 2007).

ined the effects of a two-week summer science program for middle school students that employed inquiry-based instruction. Using stratified random sampling, the researchers selected a group of 158 students to participate in the summer program; of these 79 participated in the study. In addition, a group of 35 students who applied for the program but were not selected to participate in the summer program and a group of 500 students who did not apply to participate in the summer program served as comparison groups. Two surveys were used to gauge students' interest in science, and qualitative interview data were collected from 22 students who attended the summer program. There was complete longitudinal data for only 8 of the 35 students who applied for the program but were not selected.

By following these three groups over a five-year time period, the researchers were able to determine not only if the two-week program had an immediate effect on participants' attitudes toward science, but also if this interest was sustained over time. In all three groups, interest in science decreased, but students who participated in the two-week science program retained a more positive attitude toward science and higher interest in science careers than the other two groups. Although outcomes were not tightly linked to program features or components, the study focused on the role of an inquiry-based approach to teaching science in increasing students' long-term interest in science. In interviews conducted several years after completion of the program, program participants pointed to its hands-on, inquiry-based nature as what they best remembered and most enjoyed (Gibson and Chase, 2002).

A large number of other studies also indicate that participation in out-of-school programs focused on science and mathematics can support more positive attitudes toward science, particularly among girls. For example, several noncomparative studies of Operation SMART, an out-of-school-time program for girls ages 6-18, showed increased levels of confidence and comfort with mathematics and science immediately after the program (Building Science and Engineering Talent, 2004). Operation SMART's curriculum also consists of hands-on, inquiry-based activities.

Project Exploration, an out-of-school-time program that primarily serves students from groups that are typically underrepresented in the sciences—80 percent low income, 90 percent minority, and 73 percent female—has remarkable statistics on participants' sustained interest in science: 25 percent of all students and 35 percent of female students major in sciences in college (Archer et al., 2003; Project Exploration, 2006). Project Exploration serves students in the Chicago public schools, and an alliance with the school district appears to be strategic in allowing its services to reach a traditionally underserved population. When compared with the graduation rate of students attending the same schools, Project Exploration alumni graduate from high school at a rate 18 percent higher than their peers. These data suggest a positive result, but the basis for selection into the program is not explained in the evaluation reports, other than the statement that "academic achievement is not a requirement for selection into Project Exploration programs. . . . [I]t is not known whether the students are exactly representative of their respective schools. Additional data [are] needed to increase confidence in this measure" (Project Exploration, 2006, p. 6).

In a program in which African American middle school girls worked on projects with female engineers, participating girls held more positive attitudes toward science class and science careers after participation in the program (Ferreira, 2001). This study emphasized the importance of female mentors in changing the girls' attitudes toward science (with the caveat that, to be most successful, mentors must have subject matter expertise as well as pedagogical knowledge of cooperative learning strategies).

Two other studies of summer science programs for girls showed similar positive results. A three-year evaluation of Raising Interest in Science and Engineering, a program aimed at increasing middle school girls' confidence in mathematics and science and decreasing attrition in secondary-level mathematics and science classes reported that 86 percent of participants planned on pursuing careers in mathematics and science, and 52 percent had changed their career plans after participating in the program (Jarvis, 2002).

An important component of the program was that each participant was given a female mentor, most of whom were Latina and African American college students studying engineering. The Girls Math and Technology Program placed a similar emphasis on female role models for middle school girls and also showed increased confidence in mathematics based on pre- and post-test data (DeHaven and Weist, 2003).

These studies also support the observation made in Chapter 2 and elsewhere that the strands must be understood as interrelated. For example, here the evidence indicates that interest (Strand 1) can be sustained over many years. At some point, a sustained interest in science is likely to change the ways in which individuals understand the concepts in a domain (Strand 2) and how they view themselves in relation to science (Strand 6).

### Strand 2: Understanding Scientific Knowledge

Several studies have examined students' learning of science concepts and explanations by relying largely on academic outcome measures—test scores, grades, and graduation rates. One program exception is an evaluation of Kinetic City After School (Johnson, 2005). Kinetic City includes a variety of investigations, hands-on activities, and games, as well as an interactive website with science adventures, all organized to support particular standards drawn from the Benchmarks for Science Literacy (American Association for the Advancement of Science, 1993). The evaluation included a pre- and post-test based on the program's learning goals, which included concepts pertaining to animal biology (e.g., classification and adaptation). Students also completed a creative writing activity that incorporated their understanding of the scientific concepts covered in the program. Mean scores for both components of the evaluation (pre/post-tests and the writing task) increased after completion of the program, suggesting that students acquire content knowledge through participation. Johnson also compared the effects on program participants who had access to an additional computer-based component of the program (the Kinetic City website) with those who did not. She found that the inclusion of the website component led to significantly greater positive impact on students' science knowledge.

Three other programs—Gateway; Mathematics, Engineering, Science Achievement (MESA); and the Gervirtz Summer Academy—have shown positive effects using academic outcome measures. Gateway is an out-of-school-time mathematics and science program for high school students from nondominant groups. It includes an academic summer program and separate mathematics and science classes during the school day that involve only Gateway students (Campbell et al., 1998). The study of the impact of Gateway included a matched comparison group of students who were not in the program. It found that participants had better high school graduation rates, better SAT scores, and were more likely to complete high school mathematics and science classes than students in the control group. And 92 percent of students who completed the Gateway program attended college, and the colleges they attended had mean SAT scores higher than the students' own scores. Although the Gateway results show that programs supporting science and math can have significant effects on important school-based measures, it is important to note that, because Gateway consisted of many different forms of support (e.g., summer and in-school), it is unclear whether to attribute impact to one or another component or to a synergy among the program components.

The MESA Schools Program is designed to improve middle and high school students' success in mathematics and science and increase the numbers of students from nondominant cultural backgrounds who pursue careers in science, technology, engineering, and mathematics. The program includes academic tutoring and counseling, peer supports (e.g., study groups, scheduling cohorts of participants in common courses), field trips, summer internships, and campus-based summer programs. The results of a study conducted in 1982 showed that MESA students had higher grade point averages than non-MESA students and, by senior year, the MESA students had taken more mathematics and science courses (Building Engineering and Science Talent, 2004).

The Gevirtz Summer Academy is an experimental five-week academic enrichment program. The curriculum reflects the local district curricular standards and takes an experiential and integrated instructional approach. The academy uses science as a unifying theme to teach language arts, mathematics, and science. A pre- and post-test evaluation examined the program's effect on student attitudes as well as on standardized test scores (Brenner et al., 2001). A total of 94 students participated in the evaluation the first year, and 120 students participated in the second and third years. A matched comparison group was recruited from the same schools as the study participants. Comparing pre- and post-measures, evaluators found significant increases in students' interest in science and in science careers and in their confidence and motivation in science. There were also improvements in students' science test scores, but not in their mathematics test scores.

The Gervitz evaluators (Brenner et al., 2001) also pointed to the limitations of using standardized tests as a measure of the learning that took place in the program. They explain: "It was mandated by the school district and the funding agencies that we had to use standardized test scores as documentation of the benefits of the program. It is somewhat unrealistic that a five-week program would be able to greatly influence the scores on a test that is designed to measure a school year of learning." They also point to the fact that the SAT 9 tests that they used, particularly the mathematics test, focused on basic skills, whereas the program curriculum was geared toward conceptual learning and the integration of mathematics, science, and language arts.

The problem of using standardized test scores as a measure of out-of-school-time learning is also noted by Kane (2004). He discusses the question of what are reasonable expectations of test impact for out-of-school-time programs. He points out that an entire year of classroom instruction is estimated to raise achievement test scores a quarter of a standard deviation. By this measure, an out-of-school-time program providing students with an hour of instruction five days per week could be expected to raise test scores 0.05 standard deviation (assuming there is 100 percent attendance every day). The Gervitz program chose to focus on a limited number of curricular standards, given the short amount of time that they had (five weeks), and as a result only a few questions on the standardized test pertained to the material that was covered. In the third year of the program, the teachers decided to design a mathematics test based on their own curriculum and found positive gains in the students' scores.

It is also important to note that the Gateway, MESA, and Gervitz programs all use elements beyond those typically used in after-school programs (e.g., extended day, integrated school subject matter). Similarly, the Kinetic City follow-up study found that an added media environment improved outcomes (for more on the impact of media, see Chapter 8). There is no conclusive finding here about how environments should be integrated or about the optimal relationship between out-of-school and school curricula, however, the positive outcomes for learners of integration is important to note. At the very least, these results support the assertion that helping learners extend their experiences across settings through multiple representations of concepts, practices, and phenomena is a promising design.

### Strand 3: Engaging in Scientific Reasoning

We identified no clear emphasis on Strand 3 in out-of-school programs, nor studies that evaluated the effectiveness of program emphasis on Strand 3 skills. However, in some instances, Strand 3 skills are clearly a part of programs. For example, the Service at Salado Program, described under Strand 5, is an environmental education and remediation program that introduces students to writing up scientific protocols, which typically includes testing and prediction, key elements of scientific reasoning.

### Strand 4: Reflecting on Science

Although we turned up little research that focused on reflection on science as an outcome of out-of-school programs, there is clearly some program emphasis in this area. For example, the Kinetic City evaluation (Johnson, 2005) described how participants were asked to write an essay requiring them to recall certain aspects of the program from the perspective of a creature in the rain forest and to integrate information acquired over the course of performing project-designed learning activities. In this case, participants were reflecting on the experience and what they learned, though with no clear emphasis on science.

Among the venues for science learning, out-of-school programs may be the most logical place to seriously pursue learning related to Strand 4. As discussed in Chapter 4, there is strong evidence that many children and adults struggle to understand science as a dynamic process in which knowledge is developed, vetted, and shared through sophisticated social processes. As settings in which participants can develop knowledge over longer periods of time with a common group of peers, out-of-school programs seem well suited to exploration of this important aspect of science learning.

### Strand 5: Engaging in Scientific Practices

Participation in science is a broad construct, which includes doing science, using specialized ways of talking about science, and using scientific tools. In a very general sense, one can point to the vast, expanding scale of out-of-school science programming as a crude estimate of participation in science. Participation in a more nuanced sense—for example, groups that work in an interdependent fashion to make intellectual progress on a complex problem—can be facilitated through specialized social structures that decentralize authority and create multiple ways in which even novice learners can participate. Box 6-3 describes one such program, The Fifth Dimension, which has had tremendous success in supporting learner participation though it is not science-specific.

A few science-specific efforts in out-of-school programs have also focused on participation. For example in the program Critical Science, students developed and implemented a plan to turn an empty lot in New York City into a community garden (Fusco, 2001). Students engaged in activities related to a variety of middle school science performance standards defined by the school system, such as science connections, scientific thinking, scientific tools and technology, and scientific communication. A product-oriented model of assessment similar to portfolio assessment, in which descriptions and artifacts reflecting students' participation in the program was used as evidence of learning.

In Service at Salado, an after-school science program combining service and learning, middle school students, undergraduate student mentors, and university-based scientists work together to learn about an urban riverbed habitat through classroom lessons and service and learning activities (Saltz, Crocker, and Banks, 2004). This program includes many of the components evident in the successful Fifth Dimension Program described in Box 6-3. In the classroom, students were taught about ecology in an urban setting and also learned about teamwork and leadership. Groups visited a local urban riverbed habitat two or three times during the semester to explore and apply what they learned in the classroom. Toward the end of the program, the interns worked with the students on producing products that will benefit the urban riverbed habitat.

When evaluating Service at Salado, the evaluators used the participatory goals of the program—the specific things participants would *do* as opposed to what they would *know*—to frame their evaluation (Saltz et al., 2004). Outcomes included students being able to implement a scientific protocol and write up, present, and defend their results, as well as showing awareness of urban ecology issues. The program evaluators used an observation inventory and focus group responses to measure student use of scientific protocols and technology, and they used an observation protocol to assess teamwork and leadership. A short-answer pre-post survey was used to gauge student interest in postsecondary education and careers, and a Likert scale pre-post survey was used to assess civic responsibility. On the basis of these measures, participation was associated with increases in participants use of scientific protocols and technologies, improved teamwork and leadership, greater interest in careers in which they could help people or animals, and a better understanding of civic responsibilities. However, no comparison group was used, so causal claims about the impact of the experience are not supportable.

### Strand 6: Identifying with the Scientific Enterprise

We came across little use of the construct "identity" in research and evaluation of out-of school science programs. However, a number of studies examine a suite of outcome measures that collectively may point to identity development. For example, several studies show that science programs, when deeply embedded in community issues and attuned to students' cultural backgrounds, can support development of strong science interest that is sustained long after participation, particularly among minority and low-income students or students living in disadvantaged communities (Au, 1980; Davidson, 1999; Erickson and Mohatt, 1982; Zacharia and Calabrese Barton, 2003).

There is also attention to creating spaces that are conducive to interweaving science with one's own identity. Moje, Collazo, Carillo, and Marx (2001), documenting the clash between competing school and community discourses in a science classroom, argue for the necessity of constructing a "third space" for science learning that bridges the classroom and the community (see discussion of third spaces in Chapters 2 and 4). "In many ways, the construction of congruent third spaces in classrooms requires the deconstruction of boundaries between classroom and community, especially for students who are often at the margins of mainstream classroom life" (p. 492). Moje and colleagues recommend bringing together students' home lives and school lives by creating spaces in which students' everyday discourses are intentionally brought into the classroom to enhance scientific learning, instead

## BOX 6-3  The Fifth Dimension After-School Program

The Fifth Dimension Program aims to teach students technological literacy skills as well as a range of basic literacy, mathematical, and problem-solving skills in after-school settings (Cole and Distributed Literacy Consortium, 2006). This program is designed specifically to function under the real constraints of an after-school program. Specifically, it seeks to balance fun with academic goals, assumes a very modest budget, and strategically leverages human resources to make up for low levels of trained staff.

The Fifth Dimension uses a unique social structure to engage learners and facilitate earnest engagement with complex tasks. In its ideal form, the program deploys university faculty and undergraduate students as instructional resources. Undergraduates enrolled in courses that cover the program's developmental principles serve as co-participants in children's play and learning. Following each session, the undergraduates write up field notes that are used to plan subsequent sessions and to communicate with faculty. A web-based Wizard distributes tasks to participants, which they solve collectively over the course of one or several sessions.

Participants work in groups of two or three on a computer and spend much of their time playing games, such as Boggle and chess, or engaging in activities such as origami. The Wizard periodically communicates with participants to assign tasks and adventures, which require participants to learn new games or skills, or to test particular problem-solving strategies in games in their groups. The Wizard stays in touch with the group through its webpage, e-mail, or chats

of trying to compete with it. Out-of-school programs are well positioned to be such a third space, navigating among schools, families, and communities (Noam, 2001; Noam et al., 2003).

As we have observed, numerous studies show that out-of-school-time science programs are associated with interest in science and science careers among children and adolescents. Studies also provide evidence that some programs have documented associations with graduation rates, grades, and test scores. Evaluations show that through participation in out-of-school science programs, students may increase their science content knowledge, learn scientific skills, and develop their ability to think scientifically. Repeated studies, increasingly rigorous designs, and careful definition of science-specific learning measures could help fortify these promising findings.

to facilitate and support efforts to fulfill tasks.

A series of controlled evaluations across three Fifth Dimension programs, including pre-post assessments and a controlled quasi-experimental design, showed significant gains across a range of outcomes. Studying students after 10-20 sessions, Mayer and colleagues (1997) found positive outcomes for computer skills (knowledge of new terms and facts, operating a computer), reading comprehension, and problem-solving skills. Although there were no science-specific learning outcomes, the broad range of positive results suggests that the program design is promising.

One other feature of the program deserves mention here. The Fifth Dimension Program succeeded in creating an after-school environment in which heterogeneous groups regularly engage in joint, dialogic problem solving. Building these cognitively rich activities into a program organized on a drop-in basis is extremely rare and can be quite difficult to create, even in classrooms in which training levels and other resources are more abundant. Brown and Cole (1997) attribute this success to the social structure—children, undergraduates as helpers and co-participants, and the computer-based Wizard—which decentralizes authority and invites and supports participants' curiosity and sustained engagement.

Future research and development could examine which elements of this approach could be emulated in science-specific programming and specifically test ways of structuring learner and facilitator roles to build productive, engaged scientific inquiry.

While still relatively new, the study of out-of-school science programs holds great potential. To realize this potential, it will be necessary not only to greatly expand the body of literature regarding out-of-school science programs, but also to define the hoped-for outcomes. Basing these outcomes on the specific science learning that takes place in each individual program, rather than defining outcomes using standardized test criteria or interest in science careers, is perhaps a more effective strategy.

## PROGRAMS FOR ADULT SCIENCE LEARNING

Adults' interest in science tends to be more pragmatic than children's interest. Adult science learning experiences are often self-motivated and closely connected to individual interest or life circumstances. Adults tend not to be generalists in their pursuit of science learning; instead, they tend to become experts in specific domains of interest in relation to the problems of everyday life (as discussed in Chapter 4). Thus, they become knowledgeable and conversant about concepts and explanations in specific domains (Sachatello-Sawyer, 2006).

Many of the venues in which adults engage with science, both as facilitators and as learners, cross somewhat artificial boundaries between "everyday life" and out-of-school programs. For example, as chaperones of family or school groups visiting informal institutions, adults support the science learning of others by answering questions, leading group discussions, and using various other strategies. In their everyday experiences, they build their own understanding of science through observations of the natural world, attending to media-based science programming, and through conversations with other adults. Of particular interest for this chapter is that adults may then choose to engage in more program-based learning to pursue topics of interest. However, adult learners have repeatedly been found to identify informal institutions as essentially geared toward children, not their own adult learner interests (Sachatello-Sawyer, 2006). This point is critical to understanding science programming for adult learners, because the adult perception is often mistaken. In fact, informal institutions host and organize many adult programs, and they could potentially engage more adults if they were perceived as interested in adult learners.

This section describes a variety of programs for adult science learning in informal settings. It includes programs associated with cultural institutions (e.g., museums, universities, science centers, labs, clinics) and ones developed and sustained by self-organized science enthusiasts and activists. It also includes health-related programs and programs designed for K-12 teachers and science educators in informal settings. We also examine the unique considerations of, and programs designed for, older adults. Most of the studies the committee reviewed are descriptive and did not focus on learning outcomes, so a strand by strand synthesis of the literature was not plausible. Instead, relevant strands of learning are highlighted in the description of each program type.

## Characteristics of Adult Programs

Sachatello-Sawyer and colleagues (2002) surveyed over 100 institutions that offer science learning experiences across the country to assess the number and type of adult programs. The studies surveyed staff and participants from informal institutions of varied sizes and types (e.g., art institutes, natural and cultural history museums, science centers, botanical gardens) and across types of programs (e.g., credited and noncredited classes, teacher training classes, guided tours, lectures). They found that nearly all institutions offered some sort of adult museum program (94 percent), but the majority of the programs (63 percent) were designed for families or children.

They also found that institutions reported offering more adult programs than ever before (Sachatello-Sawyer et al., 2002). However, interviews with 508 museum program participants, 75 instructors, and 143 museum planners indicated that many of the programs were struggling to connect with and attract an audience. Lectures were the most commonly cited program offered, and they were viewed as dull from the adult learner's perspective. Adults wanted to learn more from museum programs and wanted exposure to unique people, places, and objects. They had positive impressions of programs that exposed them to new perspectives, attitudes, insights, and appreciations. It is also important to bear in mind a limitation to the findings: the programs reported attracting a highly homogenous population that was more white and more highly educated than the communities in which the institutions were located.

The study found that no single teaching or facilitation methodology worked best across situations. It was most important that the facilitator or instructor related to the needs and interests of the learners and helped them discuss, integrate, reflect on, and apply new insights. In fact, many participants indicated that it was their relationship with the facilitator that was the most important aspect of the program. From the data collected, the authors argue that science centers and museums have great potential to develop exhibits and programs to reach adult learners. This can be achieved for older adults—and in preparation for the movement of the baby boom age group into their retirement years—by incorporating what is known about this group into the instructional framework and addressing issues of diversity, including cultural issues and age-related disabilities.

A wide range of impacts were reported by the program participants. Sachatello and colleagues depict the effects as a pyramid, with the most common and basic effect—acquiring new knowledge—at the bottom and the less common life-changing experiences at the apex. Between these extremes are four levels of participant-reported impact: expanded or new relationships, increased appreciation, changed attitudes, and transformed perspectives. These findings suggest that adult programs can impact each of the strands of learning. More detailed analysis of these impacts is found in the next sections, which look at three categories of adult programs (those on which we found the most relevant literature): citizen science, health, and teacher professional development programs.

## Citizen Science and Volunteer Monitoring Programs

Citizen science and volunteer monitoring programs encourage networks of volunteers, including both adults and children, to engage in scientific practice (Strand 5) through the collection of data for scientific investigations, providing adults with opportunities to gain scientific knowledge (Strand 2), test and explore the physical world (Strand 3), understand science as a way of knowing (Strand 4), and develop positive attitudes toward science (Strand 6). They are often organized and administered through scientific organizations, such as university-based labs and local environmental groups. The broad goals of citizen science include enabling scientists to conduct research in more feasible ways than they could without the participation of volunteers and promoting the public understanding of science. As Krasny and Bonney (2004) have noted, citizen science may also engage nonscientists in decision making about policy issues that have technical or scientific components and engage scientists in the democratic and policy processes. The specific focus of a given volunteer science program may include basic scientific goals, such as tracking migratory species, gathering climate data, or documenting species behavior (Cornell Lab of Ornithology, 2008). Or programs may focus on changing behavior (e.g., preservation/environmental goals) or be closely linked to informing particular policy concerns.

A project funded by the National Science Foundation (NSF) designed to enhance the ability of citizen science projects to achieve success had identified, by November 2007, more than 50 published scientific articles based on citizen science data, along with other articles assessing the data quality, educational processes, and impact of citizen science projects (http://www.citizenscience.org). For example, the Community Collaborative Rain, Hail and Snow Network is primarily concerned with a basic science issue—gathering data on weather patterns in the central Great Plains region of the United States (Albright, 2006). Trained volunteers, including adults and children, used inexpensive instruments to measure precipitation across the region, which typically was highly variable. Data collected through the network provided scientists with detailed local precipitation data and supported more sophisticated weather modeling.

Other programs focus on basic scientific questions that have clear policy implications. For example, Lee, Duke, and Quinn (2006) reported on Road Watch in the Pass, which engaged citizens in reporting wildlife sightings along a stretch of highway. The dataset generated new insights into the location of automobile-wildlife collisions that were not evident in models previously developed, providing important, empirically established guidelines for policy makers as they planned road maintenance and construction.

The number and scale of citizen science programs is increasing (Cohen, 1997; http://pathfinderscience.net/; http://www.citizenscience.org). The effects of these projects on participants' knowledge and attitudes toward science have rarely been documented (Brossard, Lewenstein, and Bonney, 2005), although efforts to increase assessment are actively in progress (http://www.citizenscience.org). The current evidence base sheds some light on participant learning; however, it is limited and in some ways contradictory, as illustrated in the literature reviewed below. To show the promising nature of some of these programs and describe areas that need further analysis, we consider findings from two studies closely.

Brossard et al. (2005) studied participants in The Birdhouse Network (TBN) to explore the hypotheses that participation in this program resulted in new knowledge of bird biology and behavior (Strand 2), a richer sense of science as practiced by scientists (Strand 4), and more positive attitudes toward science (Strand 6). Participants were asked to place "one or more nest boxes in their yards or neighborhoods, then to observe and report data on the nest boxes and their inhabitants while following one or more of four different protocols focusing on the clutch size of each nest; the calcium intake by the birds; the feathers used in the nests; and the nest site selection. Participants receive detailed explanations of the scientific protocols to be followed, biological information about cavity-nesting species, and practical information concerning nest box design, construction, and monitoring. Interaction with TBN staff by phone, email, or through an electronic mailing list is strongly encouraged" (p. 1103).

Using a quasi-experimental design Brossard and colleagues administered pre- and post-surveys to a nonrandom sample of TBN participants (300 pre-, 200 post-, and 400 science-interested new TBN member nonparticipants in the control group). The response rates for the treatment group were 57 percent (at pretest) and 63 percent (at posttest), and for the control group, 29 and 53 percent, respectively. To measure knowledge, attitude, and interest in science, the researchers used several instruments that are commonly used repeatedly in science education research (such as the Attitudes Towards Organized Science Scale, ATOSS). In addition, a team of scientists, science communicators, and science educators developed an instrument to test participants' knowledge of 10 specific concepts and facts pertaining to bird behavior and biology, reflected in such statements as "Most songbirds lay one egg per day during the breeding season," "Clutch size refers to the number of eggs a female bird can fit in her nest," and "All birds line their nest with feathers."

There was found significant improvement in the treatment group's specific knowledge of bird biology and behavior, while the control group showed no significant change. There were no significant changes in the treatment or comparison group's understanding of the scientific process or in attitudes toward science and the environment. This may have been due to a ceiling effect. Both the control and the comparison groups were part of TBN and thus may have been interested in and knowledgeable about science and the environment prior to the study. The authors theorized that the program had the potential to influence participants' understanding of the nature of science, but that this particular goal would need to be made explicit to them.

Results from a study of participants in a different program revealed a different pattern of outcomes. Overdevest and colleagues (2004) studied participants in the Water Action Volunteer (WAV) Program, a volunteer stream monitoring program, to discern impact on individuals' knowledge of stream-related topics (Strand 2), their levels of participation in resource-related management issues (Strand 5), and the degree of community networking regarding resource-related management issues. Like TBN, WAV is an ongoing program in which individual volunteers track a scientific issue—in this case, water quality—over time, using scientific protocols and under the auspices of a scientific organization.

Overdevest and colleagues used a nonequivalent group, quasi-experimental design with two groups: 155 experienced participants, who had been involved in the group for at least a year, and 105 inexperienced participants, who had expressed interest in the group at the beginning of the study but who had not yet participated in WAV activities.

In contrast to the findings of Brossard and colleagues, Overdevest and colleagues found that experienced participants exhibited greater participation in political issues related to water quality, enhanced their personal networks, and built community connections among the group of volunteers. But compared to inexperienced participants, experienced participants did not demonstrate greater knowledge of streams as a result of their participation.

Looking at these two studies side-by-side suggests that adult participants can develop various capabilities as a result of these kinds of programs, but does not clearly answer questions such as: Which capabilities are best developed in particular types of programs? What specific program features are associated with learning outcomes? What kinds of programs or program features support the learning of concepts and facts (Strand 2)? And what aspects of these programs are associated with participation in the activities of science (Strand 5)? What would programs look like that support the other strands? While looking at just a pair of studies about two programs is far from a sufficient basis for conclusive observations, we use this pair of studies to explore questions that the field may wish to take up. We also do so with full knowledge that the programs in question may support additional learning outcomes that were not reported. We urge readers to keep this in mind.

One obvious difference between the two programs is that one is explicitly linked to environmental stewardship, and the other is more closely associated with a basic scientific mission of documenting animal behavior. These differences in primary goal may impact who chooses to participate in the programs, as well as the particular skills and knowledge they develop through participation. However, understanding how the nature of the task relates to participation and how participation relates to specific learning outcomes will require considerably more research.

## Health Education

Another group of studies examines adult programs that relate specifically to managing human health. These programs typically focus on improving patients' understanding of human health (Strand 2) and can influence their attitudes toward science (Strand 6). A handful of studies examine informal support networks for individuals diagnosed with medical conditions, such as multiple sclerosis (Pfohl, 1997) and diabetes (Gillard et al., 2004) and others related to such practices as breastfeeding (Abbott, Renfrew, and McFadden, 2006; Lottero-Perdue, 2008).[2]

Informal health programs are typically organized and administered by health care agencies and serve as a way to extend the impact of medical professionals through discussion groups, lectures, and distribution of relevant literature. For example, Pfohl (1997) reports on a program that prepares multiple sclerosis patients for treatment. The program is designed to be relaxed and enjoyable but also technically valuable for the administration of medication and management of side effects. Patients whose treatment requires regular injections, for example, are given anatomy lessons (Strand 2). To bolster learning and ease anxiety that may be associated with feelings of isolation, groups of patients convene to share stories about their illness and treatment.

The few studies assessing informal health programs that the committee was able to identify focused on issues other than participant learning. The studies we identified have focused on measuring levels of participation of health care professionals, identifying sustainability factors (Abbott et al., 2006), building cases of innovative practices in health care (Pfohl, 1997), and examining how broader social phenomena, such as biases and attitudes toward science (Strand 4), mediate the impact of programs for informal health education (Lottero-Perdue, 2008).

Gillard and colleagues (2004), however, did examine participants' behavioral outcomes in a pilot study of screening clinics designed to detect and treat diabetes-related eye disease. During three annual visits to the clinic, patients received a physical examination and diabetes screening, engaged in unstructured discussions about diabetes and treatments with other patients and a diabetes expert, had access to pamphlets on diabetes and eye care, and were able to discuss the results of their examination with a health professional. Patients received a letter that included their test results and the implications of their results in the mail 30 days after their visit. Self-reported management of glucose levels from the first to the third clinic visit were used to determine whether changes in self-management behavior resulted from participation. The researchers found significant and desirable changes in self-management in terms of insulin use and self-monitoring of glucose. In this pilot study, the researchers reached no specific conclusions about the qualities of the experience that led to behavioral change, nor did they track participants' ideas or attitudes about particular concepts or practices.

---

[2]A related body of work examining the social marketing of health practices in international development is discussed in Chapter 8.

Given the positive health behavior outcomes, however, they did urge health practitioners to view diabetes patients' learning as "part of every diabetes care encounter" (p. 42).

## Programs for Science Teachers

As with adult learners facing a health issue, science teachers constitute a particular adult group with a great need to learn many aspects of science (National Research Council, 2007). Teacher professional development has been an area of significant growth over the past several decades. Program activity, interest among education leaders, and research on teacher professional development have grown in concert with the standards-based reform movement. Science has received considerable attention as several major school reform initiatives funded through the NSF, including the state, local, and urban and rural systemic initiatives, have emphasized teacher professional development in order to address teachers' knowledge of and comfort with science and appropriate pedagogy. Institutions that support science learning in informal settings have been identified as critical participants in this effort, premised on the notion that their emphasis on phenomena-rich, learner-driven interactions with science resonates with the notion of inquiry underlying K-12 science education reform.

Although many institutions have long-standing professional development programs for science teachers, until recently their role in teacher professional development has been relatively undocumented. Just a decade ago, a well-known national study described these institutions as an "invisible infrastructure" of science education supporting K-12, yet it did not include data on teacher professional development (Inverness Research Associates, 1996). However, researchers from the Center for Informal Learning and Schools recently attempted to document teacher professional development efforts in these institutions in a study describing the scale and qualities of these programs (Phillips, Finkelstein, and Wever-Frerichs, 2007). The study was designed to answer two questions:

1. What are the design features of teacher professional development based in informal science institutions?
2. To what extent do teacher professional development programs based in informal science institutions integrate particular aspects that are known from research to produce measurable effects on teacher practice?

Phillips and colleagues mailed a survey to specific individuals in 305 institutions, including 279 who had previously responded to a survey indicating that they provided teacher professional development programs, as well as to individuals from an additional 26 institutions known to offer programs. The survey asked respondents to characterize their programs and the educational credentials of the staff and describe the unique features of their program or what their programs "provide for teachers that other programs were unable or unlikely to provide."

With a relatively low response rate (28 percent) for the 305 mailed surveys, the study reports that these institutions are devoting considerable energy to teacher professional development and that their programs are focused on supporting teachers to learn activities they can use in their classrooms, as well as how to integrate their institution's resources into their curriculum. How these offerings influence teacher knowledge and practice is yet unknown.

The committee also reviewed two case studies of in-service teacher preparation programs that integrate informal experiences (Anderson, Lawson, and Mayer-Smith, 2006; Zinicola and Delvin-Scherer, 2001). In these programs, teachers may learn content and how to teach it, as well as how to identify and create curriculum materials, and organize and manage students and instruction in their particular subject. They may explore new epistemologies and different ways of personally connecting with science. Through relationships built during the programs they also begin to build a network to nurture their own ongoing professional education.

Anderson and colleagues describe an aquarium-based preservice teacher program designed to wrap around a school-based teaching practicum. Preservice teachers participated in a three-day orientation to the educational programs of the aquarium, its student-centered, hands-on pedagogy, and the institution's educational goals, described as "developing inspiration, curiosity and marine stewardship . . . the importance of ecosystems; promoting awareness of the historical and economic aspects of the fishing industry . . . knowledge of (local) marine invertebrates" (Anderson, Lawson, and Mayer-Smith, 2006, p. 344). Following the orientation, they spent 10 weeks in a school-based assignment, after which they returned to the aquarium to work in the educational programs under the tutelage of aquarium staff for three weeks.

Anderson and colleagues conducted two focus groups with the teachers, analyzed reflective essays they wrote during the semester, and conducted ethnographic observations at the aquarium. They drew conclusions about areas of impact primarily based on teachers' reflections on their experiences. These included broadening teachers' sense of education and enhancing their understanding of educational theory, improving their classroom skills, enhancing their sense of autonomy and self-efficacy, strengthening their commitment to collaborative work, and helping them recognize the power of hands-on experiences in learning science (Anderson, Lawson, and Mayer-Smith, 2006, p. 350). While based primarily on self-report data from a single case, the results suggest this is a promising approach to integrating teacher education and informal educational institutions; clearly, further research and development are needed.

## PROGRAMS FOR OLDER ADULTS

Older adults are a unique population to which informal institutions are increasingly attending. Their abilities, needs, and interests—like those of other learners—require special attention in order to create programs that serve them. Although there have been few studies of older adult science learners in informal settings, a review of the general literature on learning in older adults is useful for understanding what issues in science learning might be best explored.

Like other populations and groups (discussed in Chapter 7), older adults are often misunderstood. One aspect of older learners that gets little attention, but which is especially important for thinking about educational programming in informal environments, is their extensive experience base and knowledge. Older adults have a long history of family life, occupational experiences, and leisurely pursuits. In contrast to children, who are "universal novices" (Brown and DeLoache, 1978), older adults draw on decades of experience. They have rich histories and knowledge that they can elaborate on and from which they can draw analogies to access new concepts and insights.

Older adults can also be stereotyped as suffering from memory decline and other aspects of mental slowing, and this tends to lead to an erroneous assumption that they lack ability. Such stereotyped views are often conveyed and upheld broadly, including by older adults themselves (Parr and Siegert, 1993; Ryan, 1992). Craik and Salthouse (2000) have reviewed the literature and report that older adults do face a steady loss in what is called fluid intelligence or processing capacity. This decline can adversely affect the performance of everyday tasks and learning through a weakened capacity for attention (Salthouse, 1996), processing speed (Madden, 2001), and various types of memory performance (Bäckman, Small, and Wahlin, 2001). Because older adults often also face declines in hearing, vision, and motor control, these deficits in fluid intelligence can appear exaggerated. Studies by McCoy et al. (2005) concluded that the extra effort expended by a hearing-impaired listener in order to successfully perform a task comes at the cost of processing resources that would otherwise be directed at memory encoding.

Studies of declines in fluid intelligence on computer use in older adults indicate that older adults make more errors and perform at a lower level than younger people on a variety of common tasks (Charness, Schumann, and Boritz, 1992; Czaja, 2001; Czaja and Sharit, 1993; Echt, Morrell, and Park, 1998). In addition, they demonstrate a relative difficulty with editing out unnecessary information (Rogers and Fisk, 2001). As the baby boom generation ages, its familiarity with computers and the web will increase, and the majority of boomers in the United States will use the web on a regular basis (Czaja et al., 2006). Website designers and web-assisted programmers who serve these aging populations should strongly consider these findings and make adjustments.

Not all human functions decline with age. The discovery that humans continue to generate new neurons throughout life in the hippocampal region and that new neuronal connections are constantly being formed in response to life experience should help reshape thinking on lifelong learning (McKhann and Albert, 2002). Knowledge of general facts and information about the world (crystallized intelligence) does not change with age, and experience and life skills lead to a more comprehensive understanding of the world (Baltes, 1987; Beier and Ackerman, 2005; Heckhausen, 2005; Schaie, 2005). Self-worth, autonomy, and control over emotions increase or remain stable with age (Brandstadter, Rothermund, and Schmitz, 1998; Sheldon, Houser-Marko, and Kasser, 2006). Studies indicate stabilized limbic and autonomic nervous system activity in older adults (Lawton, Kleban, and Dean, 1993; Lawton, Kleban, Rajagopal, and Dean, 1992; Levenson, Carstensen, Friesen, and Ekman, 1991).

There is evidence to suggest that older adults regulate negative emotions better than young adults and experience positive emotions with similar intensity and frequency (Carstensen, Pasupathi, Mayr, and Nesselroade, 2000). Mather and colleagues (2004) showed that older people's memory for positive imagery was strikingly better than for neutral or negative images. Functional MRI data indicate that amygdala activation increased only in response to positive stimuli (Lindberg, Carstensen, and Carstensen, 2007). Carstensen and her colleagues have developed a socioemotional selectivity theory that suggests that older adults experience an improved sense of well-being by pursuing experiences that are meaningful and are tied to emotional information.

Benbow (2002) produced a useful list of implications for teaching to support effective learning by older adults:

- Instruction must respond to the experience, skills, and understanding of the big picture that adults bring to the learning environment. It may also require time spent correcting preexisting misunderstandings.
- Instruction should include how older learners can encode information and new processes to stimulate recall.
- Because stereotypes about memory loss can impact the ability to learn, instruction should be directed to reinforce the belief that people can remember and should be strengthened by practice opportunities.
- As people age, there is an increased interest in connecting learning to an impact on society. Instruction should therefore be designed to relate to both simple and complex situations in real life.
- Instruction should build on the strong emotional bonds toward people, objects, and beliefs that develop as people age.

Jolly (2002) reminds the informal science education community that it must make a bid to educate the large group of older adults who will begin to avail themselves of opportunities in museums and science centers between

2010 and 2030 by developing programs that result in "sustainable diversity." This will require a deep integration of policies and practices that incorporate diversity into institutional frameworks. He enumerates some important goals for consideration by the community:

- Building boards of trustees and hiring staff that can represent the appropriate perspectives of the aging community.
- Addressing issues of age-related disabilities in all program design (i.e., vision, hearing, mobility, and fine motor coordination). Programs resulting from this process will end up appealing to learners of every age.
- Producing more on-the-go and virtual programming that can travel to populations that cannot come to museums and science centers.
- Increasing collaboration between the informal science community and the local network of aging services. This will foster the development of programming tailored to the culture of the older adults in the area and result in incorporation of experiences that increase trust and respect from them.
- Incorporating assistive technology and equal access to all possible venues, including field sites, to increase participation by adults with age-related disabilities.

Although there is scant empirical analysis of programs designed for older science learners, several driving propositions derived from practice, basic human development research, and several current programs instantiate (to varying degrees) these principles. Although untested, the following practices are worthy of further development and empirical scrutiny:

- Develop and foster dialogue and partnerships with local and area networks of aging services.
- Incorporate representation from this community in program and exhibit design.
- Use the principles of universal instructional design in exhibit and program materials.
- Incorporate findings about the adult learner in program design.
- Seek funding for assistive technology to support learning.
- Design and structure outcomes and evaluations that will provide data to inform the informal science community.

Some programs designed for older adults are taking first steps toward addressing these concerns. We mention two of these here.

## Explora

Explora, a museum in Albuquerque, New Mexico, runs a science club for 30 members, ages 54 to 101, at the Laguna Pueblo. The club is part of an outreach exploration program offered at several senior, assisted-living, and nursing care centers. Explora has produced a guide containing 44 science, technology, and art programs for middle-aged and older adults. They have hosted adults-only nights for people 18 and older and altered space to include ample seating, wheelchair access, assistive technology, and modified materials. Older adults from all local communities are included in program design, and Explora hires seniors as educators in the programs and on the museum floor (Leigh, 2007).

## Meadowlands Environment Center

Project SEE (Senior Environmental Experiences), at the Ramapo College of New Jersey, is supported by a grant from NSF and represents a partnership between the Meadowlands Environment Center, Ramapo College, and regional aging community services, including the Bergen County Division of Senior Services. The project is using interactive videoconferencing technology to educate and enhance science learning among senior citizens in assisted living facilities, nursing homes, and senior community centers in the Meadowlands District of New Jersey and in facilities in the northern area of the state. Participants gather information and take part in an ongoing dialogue with environmental scientists. SEE provides videoconferencing equipment, staff to set up and take down the technology and conduct all program activities, and pre- and post-conference materials.

## CONCLUSION

The potential of programs for science learning is great, given the broader population patterns in society. Two demographic issues are relevant to science learning programs. One is a vast demand and infrastructure for quality programs for children and youth in out-of-school time. The other is the aging of the baby boom generation. These demographic issues warrant careful consideration.

To understand the full potential of out-of-school programs to function as a large-scale delivery system for science learning outside schools, tools and resources are needed to that end, and their development would benefit from empirical research. As we have observed one of the limitations in this area of inquiry is that the literature is primarily made up of evaluations, which are not necessarily built upon a peer reviewed body of evidence and linked to other inquiries. It would be constructive to integrate findings from across studies of science learning and perhaps with the broader evidence base on non-science-specific out-of-school and adult programs.

There is also a specific need for examination of the type of science learning occurring in programs for older adults. These learners will require special accommodations to serve their science interests and needs, and it will be necessary to plan learning experiences that are accessible to them. Developing and improving programs for older learners will require substantial growth in research. Currently the knowledge base consists of general cognitive and developmental research and descriptions of programs designed for older learners. In a broad sense, adults of all ages need to understand that the science learning resources are intended to serve them, not just children.

There is evidence that programs can result in scientific learning and understanding across the strands. For the types of programs we reviewed, we found science-specific learning outcomes for school-age participants and a few studies on adults. However, there is no clear, organized and synthesized body of knowledge on science-specific program effects or on qualities of effective programs for science learning. In this chapter we have begun to organize some of the relevant studies. There may be more evaluation reports that examine science-specific outcomes than we reviewed in this chapter. It would be helpful to further integrate the literature in future research.

In the long run, identifying a set of best practices that can be applied across programs would also be beneficial. This task would involve a complex set of issues: curricular choices, staff training, management issues, space, and many others. Given the potential to vastly increase the participation of children, youth, and adults in these programs, it seems a worthwhile investment.

Finally, we urge the field to attend carefully to the goals and measures used in program development and evaluation, drawing and building on the strands as an important resource. Identification of goals can make it possible for staff, participants, and evaluators to approach their experiences and work with greater focus and can facilitate efforts to build strong empirical bases for theory and practice.

## REFERENCES

Abbott, S., Renfrew, M.J., and McFadden A. (2006). "Informal" learning to support breastfeeding: Mapping local problems and opportunities. MIRU No: 2006.28. *Maternal and Child Nutrition, 2*(4), 232-238.

Afterschool Alliance. (2008). *21st century learning centers providing supports to communities nationwide.* Available: http://www.afterschoolalliance.org/researchFact Sheets.cfm [accessed October 2008].

Albright, L. (2006). *Summative evaluation: Bringing CoCoRaHS to the central great plains.* Colorado State University. Available: http://www.informalscience.org/evaluation/show/89 [accessed October 2008].

American Association for the Advancement of Science. (1993). *Benchmarks for science literacy*. New York: Oxford University Press.

Anderson, D., Lawson, B., and Mayer-Smith, J. (2006). Investigating the impact of a practicum experience in an aquarium on pre-service teachers. *Teaching Education, 17*(4), 341-353.

Archer, E., Fanesali, C., Froschl, M., and Sprung, B. (2003). *Science, gender, and afterschool: A research action agenda*. New York: Educational Equity Concepts and Academy for Educational Development.

Au, K.H. (1980). Participant structures in a reading lesson with Hawaiian children: Analysis of a culturally appropriate instructional event. *Anthropology and Education Quarterly, 11*(2), 91-115.

Bäckman, L., Small, B.J., and Wahlin, A. (2001). Aging and memory: Cognitive and biological perspectives. In J.E. Birren, and K.W. Schaie (Eds.), *Handbook of the psychology of aging* (5th ed., pp. 288-312). San Diego: Academic Press.

Baltes, P.B. (1987). Theoretical propositions of life-span developmental psychology: On the dynamics between growth and decline. *Developmental Psychology, 23*(5), 611-626.

Beier, M.E., and Ackerman, P.L. (2005). Age, ability, and the role of prior knowledge on the acquisition of new domain knowledge. *Psychology and Aging, 20*(2), 341-355.

Benbow, A.E. (2002). *Communicating with older adults: A guide for health care and senior service professionals and staff*. Seattle, WA: SPRY Foundation/Caresource Healthcare Communications.

Bissell, J.S., Cross, C.T., Mapp, K., Reisner, E., Vandell, D.L., Warren, C., and Weissbourd, R. (2003). Statement released, May 10, by members of the Scientific Advisory Board for the 21st Century Community Learning Center evaluation. Available: http://childcare.wceruw.org/pdf/publication/statement.pdf [accessed February 2009].

Brandtstadter, J., Rothermund, K., and Schmitz, U. (1998). Maintaining self-integrity and efficacy through adulthood and later life: The adaptive functions of assimilative persistence and accommodative flexibility. In J. Heckhausen and C.S. Dweck (Eds.), *Motivation and self-regulation across the life span* (pp. 365-388). New York: Cambridge University Press.

Brenner, M., Hudley, C., Jimerson, S., and Okamoto, Y. (2001). *Three-year evaluation of the Gevirtz Summer Academy, 1998-2000*. Santa Barbara: University of California, Gevirtz Research Center, Gevirtz Graduate School of Education.

Brossard, D., Lewenstein, B., and Bonney, R. (2005). Scientific knowledge and attitude change: The impact of a citizen science project. *International Journal of Science Education, 27*(9), 1099-1121.

Brown, A., and DeLoache, J.S. (1978). Skills, plans and self-regulation. In R. Siegler (Ed.), *Children's thinking: What develops?* (pp. 3-35). Mahwah, NJ: Lawrence Erlbaum Associates.

Brown, K., and Cole, M. (1997). Fifth dimension and 4H: Complementary goals and strategies. *Youth Development Focus: A Monograph of the 4-H Center for Youth Development University of California, Davis. 3*(4). Available: http://cyd.ucdavis.edu/publications/pubs/focus/pdf/FO97V3N4.pdf [accessed April 2009].

Building Engineering and Science Talent. (2004). *What it takes: Pre-K-12 design principles to broaden participation in science, technology, engineering, and mathematics.* Available: http://www.bestworkforce.org/publications.htm [accessed October 2008].

Campbell, P.B., Wahl, E., Slater, M., Iler, E., Moeller, B., Ba, H., and Light, D. (1998). Paths to success: An evaluation of the gateway to higher education program. *Journal of Women and Minorities in Science and Engineering, 4*(2-3), 297-308.

Carstensen, L.L., Pasupathi, M., Mayr, U., and Nesselroade, J. (2000). Emotional experience in everyday life across the adult life span. *Journal of Personality and Social Psychology, 79*(4), 644-655.

Charness, N., Schumann, C.E., and Boritz, G.M. (1992). Training older adults in word processing: Effects of age, training technique, and computer anxiety. *International Journal of Technology and Aging, 5*(1), 79-106.

Cohen, K.C. (Ed.). (1997). *Internet links for science education: Student-science partnerships.* New York: Plenum Press.

Cole, M., and the Distributed Literacy Consortium. (2006). *The fifth dimension: An after-school program built on diversity.* New York: Russell Sage.

Cornell Lab of Ornithology. (2008). *Citizen science central.* Available: http://www.birds.cornell.edu/citscitoolkit/ [accessed October 2008].

Craik, F.I.M., and Salthouse, T.A. (Eds.). (2000). *The handbook of aging and cognition.* Mahwah, NJ: Lawrence Erlbaum Associates.

Czaja, S.J. (2001). Technological change and the older worker. In J.E. Birren and K.W. Schaie (Eds.), *Handbook of the psychology of aging* (5th ed., pp. 547-568). San Diego: Academic Press.

Czaja, S.J., and Sharit, J. (1993). Age differences in the performance of computer-based work. *Psychology and Aging, 8*(1), 59-67.

Czaja, S.J., Charness, N., Fisk, A.D., Hertzog, C., Nair, S.N., Rogers, W.A., and Sharit, J. (2006). Factors predicting the use of technology: Findings from the Center for Research and Education on Aging and Technology Enhancement (CREATE). *Psychology and Aging, 21*(2), 333-352.

Davidson, A.L. (1999). Negotiating social differences: Youths' assessments of educators' strategies. *Urban Education, 34*(3), 338-369.

DeHaven, M., and Weist, L. (2003). Impact of a girls' mathematics and technology program on middle school girls' attitudes towards mathematics. *Mathematics Educator, 13*(2), 32-37.

Dorph, R., Goldstein, D., Lee, S., Lepori, K., Schneider, S., and Venkatesan, S. (2007). *The status of science education in the Bay area: Research study e-report.* Lawrence Hall of Science, University of California, Berkeley.

Dynarski, M., James-Burdumy, S., Moore, M., Rosenberg, L., Deke, J., and Mansfield, W. (2004). *When schools stay open late: The national evaluation of the 21st century community learning centers program: New findings.* National Center for Education Evaluation and Regional Assistance. Washington, DC: U.S. Department of Education.

Echt, K.V., Morrell, R.W., and Park, D.C. (1998). Effects of age and training formats on basic computer skill acquisition in older adults. *Educational Gerontology, 24*(1), 3-25.

Erikson, F.D., and Mohatt, G. (1982). Cultural organization in two classrooms of Indian students. In G.D. Spindler (Ed.), *Doing the ethnography of schooling: Educational anthropology in action* (pp. 132-175). New York: Holt, Rinehart and Winston.

Fancsali, C. (2002). *What we know about girls, STEM, and afterschool programs.* Prepared for Education Equity Concepts. New York: Academy for Educational Development.

Ferreira, M. (2001). *The effect of an after-school program addressing the gender and minority achievement gaps in science, mathematics, and engineering.* Arlington, VA: Educational Research Spectrum, Educational Research Services.

Friedman, L.N. (2005). *Where is after-school headed and how do science learning opportunities fit into the after-school landscape.* Available: http://www.afterschool resources.org/kernel/images/tascsci.pdf [accessed January 2009].

Fusco, D. (2001). Creating relevant science through urban planning and gardening. *Journal of Research in Science Teaching, 38*(8), 860-877.

Gibson, H., and Chase, C. (2002). Longitudinal impact of an inquiry-based science program on middle school students' attitudes towards science. *Science Education, 86*(5), 693-705.

Gillard, M., Nwankwo, R., Fitzgerald, J.T., Oh, M., Musch, D.C., Johnson, M.W., and Anderson, R. (2004). Informal diabetes education: Impact on self-management and blood glucose control. *Diabetes Educator, 30*, 136-142.

Halpern, R. (2002). A different kind of child development institution: The history of after-school programs for low-income children. *Teachers College Record, 104*(2), 178-211.

Harvard Family Research Project. (2003). *A profile of the evaluation of SECME raising interest in science and engineering.* Cambridge, MA: Author. Available: http://www.hfrp.org/out-of-school-time/ost-database-bibliography/database/secme-rise-raising-interest-in-science-engineering [accessed October 2008].

Heckhausen, J. (2005). Competence and motivation in adulthood and old age: Making the most of changing capacities and resources. In A.J. Elliot and C.S. Dweck (Eds.), *Handbook of competence and motivation* (pp. 240-256). New York: Guilford.

Institute of Medicine. (2002). *Community Programs to Promote Youth Development.* Committee on Community-Level Programs for Youth. J. Eccles and J.A. Gootman (Eds.). Washington, DC: National Academy Press.

Inverness Research Associates. (1996). *An invisible infrastructure: Institutions of informal science education, executive summary.* Washington, DC: Association of Science-Technology Centers.

Jarman, R. (2005). Science learning through scouting: An understudied context for informal science education. *International Journal of Science Education, 27*(4), 427-450.

Jarvis, C. (2002). *SECME RISE (Raising Interest in Science & Engineering): Final evaluation report, September 1, 1998-August 31, 2001.* Miami: Miami Museum of Science.

Jeffers, L. (2003). *Evaluation of NYC first!* New York: Education Development Center, Center for Children and Technology. Available: http://cct.edc.org/report_summary.asp?numPublicationId=141 [accessed October 2008].

Johnson, A. (2005). *Summative evaluation of Kinetic City afterschool.* Report for the American Association for the Advancement of Science. Available: http://www.kcmtv.com/juneevaluation.pdf [accessed October 2008].

Jolly, E. (2002, February). *Confronting demographic denial: Retaining relevancy in the new millennium.* Washington, DC: Association of Science and Technology Centers. (Reprinted 2002 in *Journal of Museum Education, 27*(2-3).)

Kane, T. (2004). *The impact of after-school programs: Interpreting the results of four recent evaluations.* New York: William T. Grant Foundation.

Krasny, M., and Bonney, R. (2004). Environmental education through citizen science and participatory action research. In E.A. Johnson and M.J. Mappin (Eds.), *Environmental education or advocacy: Perspectives of ecology and education in environmental education.* New York: Cambridge University Press.

Lawton, M., Kleban, M., and Dean, J. (1993). Affect and age: Cross-sectional comparisons of structure and prevalence. *Psychology and Aging, 8*(2), 165-175.

Lawton, M., Kleban, M., Rajagopal, D., and Dean, J. (1992). Dimensions of affective experience in three age groups. *Psychology and Aging, 7*(2), 171-184.

Lee, T., Duke, D., and Quinn, M. (2006). Road watch in the pass: Using citizen science to identify wildlife crossing locations along Highway 3 in the Crows Nest Pass of Southwestern Alberta. Poster presentation in C.L. Irwin, P. Garrett, and K.P. McDermott (Eds.), *Proceedings of the 2005 International Conference on Ecology and Transportation* (p. 638). Raleigh: North Carolina State University, Center for Transportation and the Environment, North.

Leigh, K. (2007, Jan/Feb). *In their own right: Adult learning at Explora.* Washington, DC: Association of Science-Technology Centers.

Levenson, R., Carstensen, L., Friesen, W., and Ekman, P. (1991). Emotion, physiology, and expression in old age. *Psychology and Aging, 6,* 28-35.

Lindberg, C.M., Carstensen, E.L., and Carstensen, L.L. (2007). *Lifelong learning and technology.* Background paper for the Committee on Learning Science in Informal Environments. Available: http://www7.nationalacademies.org/bose/Lindberg_et%20al_Commissioned_Paper.pdf [accessed October 2008].

Lottero-Perdue, P. (2008). *Critical analysis of science-related texts in a breastfeeding information, support, and advocacy community of practice.* Unpublished manuscript, Towson University, MD.

Madden, D.J. (2001). Speed and timing of behavioral processes. In J.E. Birren and K.W. Schaie (Eds.), *Handbook of the psychology of aging* (5th ed., pp. 288-312). San Diego: Academic Press.

Mather, M., Canli, T., English, T., Whitfield, S., Wais, P., Ochsner, K., Gabrieli, J.D.E., and Carstensen, L.L. (2004). Amygdala responses to emotionally valenced stimuli in older and younger adults. *Psychological Science, 15,* 259-263.

Mayer, R.E., Quilici, J., Moreno, R., Duran, R., Woodbridge, S., Simon, R., Sanchez, D., and Lavezzo, A. (1997). Cognitive consequences of participation in a fifth dimension after-school computer club. *Journal of Educational Computing Research, 16*(4), 353-369.

McCoy, S.L., Tun, P.A., Cox, L.C., Colangelo, M., Stewart, R.A., and Wingfield, A. (2005). Hearing loss and perceptual effort: Downstream effects on older adults' memory for speech. *Quarterly Journal of Experimental Psychology, 58*(1), 22-33.

McKhann, G., and Albert, M. (2002). *Keep your brain young.* New York: Wiley.

McLaughlin, M. (2000). *Community counts: How youth organizations matter for youth development.* Washington, DC: Public Education Fund Network.

McMurrer, J. (2007). *Choices, changes, and challenges: Curriculum and instruction in the NCLB era.* Washington, DC: Center for Education Policy.

Miller, B.M. (2003). *Critical hours: After-school programs and educational success.* Quincy, MA: Nellie Mae Education Foundation.

Moje, E., Collazo, T., Carillo, R., and Marx, R. (2001). "Maestro, what is quality?": Language, literacy, and discourse in project-based science. *Journal of Research in Science Teaching, 38*(4), 469-498.

National Center for Education Statistics. (2006). *Digest of education statistics: 2006 digest tables.* Available: http://nces.ed.gov/programs/digest/2006menu_tables. asp [accessed October 2008].

National Research Council. (2007). *Taking science to school: Learning and teaching science in grades K-8.* Committee on Science Learning, Kindergarten Through Eighth Grade. R.A. Duschl, H.A. Schweingruber, and A.W. Shouse (Eds.). Washington, DC: The National Academies Press.

Noam, G. (2001). *After-school time: Toward a theory of collaboration.* Paper presented at the Urban Seminar Series on Children's Mental Health and Safety, Out of School Time, Cambridge, MA.

Noam, G., Biancarosa, G., and Dechausay, N. (2003). *After-school education: Approaches to an emerging field.* Cambridge, MA: Harvard Education Press.

Overdevest, C., Huyck Orr, C., and Stepenuck, K. (2004). Volunteer stream monitoring and local participation in natural resource issues. *Research in Human Ecology, 11*(2), 177-185.

Parr, W.V., and Siegert, R. (1993). Adults' conceptions of everyday memory failures in others: Factors that mediate the effects of target age. *Psychology and Aging, 8*(4), 599-605.

Pfohl, D.C. (1997). A multiple sclerosis (MS) center injection training program. *Axone (Dartmouth, N.S.), 19*(2), 29-33.

Phillips, M., Finkelstein, D., and Wever-Frerichs, S. (2007). School site to museum floor: How informal science institutions work with schools. *International Journal of Science Education, 29*(12), 1489-1507.

Project Exploration. (2006). *Project exploration youth programs evaluation.* Available: http://www.projectexploration.org [accessed October 2008].

Rogers, W.A., and Fisk, A.D. (2001). Understanding the role of attention in cognitive aging research. In J.E. Birren and K.W. Schaie (Eds.), *Handbook of the psychology of aging* (5th ed., pp. 267-287). San Diego, CA: Academic Press.

Ryan, E.B. (1992). Beliefs about memory changes across the adult life span. *Journal of Gerontology: Psychological Sciences, 47*(1), 41-46.

Sachatello-Sawyer, B. (2006). *Adults and informal science learning.* Presentation to the Committee on Learning Science in Informal Environments, December 13-14, Keck Center, National Research Council, Washington, DC.

Sachatello-Sawyer, B., Fellenz, R.A., Burton, H., Gittings-Carlson, L., Lewis-Mahony, J., and Woolbaugh, W. (2002). *Adult museum programs: Designing meaningful experiences.* American Association for State and Local History Book Series. Blue Ridge Summit, PA: AltaMira Press.

Salthouse, T.A. (1996). Constraints on theories of cognitive aging. *Psychonomic Bulletin and Review, 3*, 287-299.

Saltz, C., Crocker, N., and Banks, B. (2004). *Evaluation of service at the Salado for fall 2004.* Tempe: Arizona State University International Institute for Sustainability. Available: http://caplter.asu.edu/explorers/riosalado/pdf/fall04_report. pdf [accessed October 2008].

Schaie, K.W. (2005). *Developmental influences on adult intelligence: The Seattle longitudinal study*. New York: Oxford University Press.

Sheldon, K.M., Houser-Marko, L., and Kasser, T. (2006). Does autonomy increase with age? Comparing the goal motivations of college students and their parents. *Journal of Research in Personality, 40*(2), 168-178.

U.S. Department of Education, Office of the Under Secretary. (2003). *When schools stay open late: The national evaluation of the 21st century community learning centers program: First-year findings*. Washington, DC: U.S. Government Printing Office.

Zacharia, Z., and Calabrese Barton, A. (2003). Urban middle-school students' attitudes toward a defined science. *Science Education, 88*(2), 1-27.

Zinicola, D., and Devlin-Scherer, R., (2001). A university-museum partnership for teacher education field experiences in science. *Clearing House, 74*(5), 248-250.

# Part III

# Cross-Cutting Features

# 7

# Diversity and Equity

An important value of informal environments for learning science is being accessible to all. Socioeconomic, cultural, ethnic, historical, and systemic factors, however, all influence the types of access and opportunities these environments afford to learners (Heath, 2007). "Being born into a racial majority group with high levels of economic and social resources—or into a group that has historically been marginalized with low levels of economic and social resources—results in very different lived experiences that include unequal learning opportunities, challenges, and potential risks for learning and development" (Banks, 2007, p. 15).

The challenges in engaging nondominant groups in the sciences are reflected in studies showing

1. inadequate science instruction exists in most elementary schools, especially those serving children from low-income and rural areas;
2. girls often do not identify strongly with science or science careers;
3. students from nondominant groups perform lower on standardized measures of science achievement than their peers;
4. although the number of individuals with disabilities pursuing post-secondary education has increased, few pursue academic careers in science or engineering; and
5. learning science can be especially challenging for all learners because of the specialized language involved (Banks, 2007; Allen and Seumptewa, 1993; Cajete, 1993; MacIvor, 1995; Malcom and Matyas, 1991; Snively, 1995).

These findings suggest the barriers that exist to engaging those from nondominant groups in science. It is critical to consider diversity issues and the science learning of nondominant groups for several reasons: to ensure equitable treatment of all individuals; to continue to develop a well-trained workforce; to develop a well-informed, scientifically literate citizenry; and to increase diversity in the pool of scientists and science educators who can bring new perspectives to science and the understanding of science.

Scientific discourse, teaching, and learning are not culturally neutral, although people tend to see and represent them as acultural or neutral or, in the case of science, as representing a unique culture unto itself. An important perspective on science learning in informal environments emphasizes that, although treating the construct of culture as a homogeneous categorical variable is problematic, people nonetheless do "live culturally" (Nasir, Rosebery, Warren, and Lee, 2006; Gutiérrez and Rogoff, 2003). From this perspective, a key object of study is the wide, varied repertoire of sense-making practices that people participate in, especially in everyday contexts.

Gutiérrez and Rogoff (2003) point out that "individual development and disposition must be understood in (not separate from) cultural and historical context" (p. 22). All people engage in sophisticated learning shaped by the cultural and contextual conditions in which they live. In this sense, all people learn, but a given group may learn different knowledge and practices and may organize its learning differently. This chapter addresses diversity issues related to learning science in informal environments. Among the many dimensions of diversity, here we take a cultural-historical perspective on learning and illustrate the implications for science learning and the structuring of informal environments where science learning takes place.

Before we review the research literature on the experiences of diverse populations with science and their access to it, we first define culture and equity. We then focus on science learning in four nondominant groups for which a research tradition has developed: girls and women, American Indians, individuals from rural communities, and individuals with disabilities. In reviewing the research involving these groups, we explore such issues as engagement, identity, self-efficacy, and border crossing, which are related to diversity and science learning. We end with a set of guiding principles to develop culturally responsive and effective informal environments for science learning.

## CULTURE AND EQUITY

Culture is a complex concept that is difficult to define succinctly. Most scholars agree, however, that culture includes the symbols, stories, rituals, tools, shared values, and norms of participation that people use to act, consider, communicate, assess, and understand both their daily lives and their images of the future (Brumann, 1999). Disagreements arise concerning the costs and benefits of treating culture as a noun, in which case it may lend itself to stereotyping, versus treating culture as a modifier—as in "people live culturally." A closely related issue is how culture and cultural processes should be studied (Medin and Atran, 2004).

If the study of culture is conceptualized as identifying shared norms and values, it is natural to assume that individuals become part of a culture through a process of socialization—that is, they acquire culture. If culture is instead seen as dynamic, contested, and variably distributed within and across groups, it is natural to see cultural learning as involving a reciprocal relationship between individuals' goals, perspectives, abilities, and values and their environment (Hirschfeld, 2002). In this view, for example, in the earliest years of life, one's socialization partially depends on agents or others who are caregivers as well as an individual's interpretation of and reaction to their environment. Furthermore, as one grows older, associates, friends, organizations, and institutions become part of varying socialization processes, but the influence of each is dependent on an individual's characteristics, and vice versa. Thus, socialization depends on access and opportunities, as well as the perspectives and attitudes that an individual brings to these opportunities. From this perspective, in fact, one can see that while culture is often used in reference to ethnic or racial background, any group with some shared affiliation (e.g., people with disabilities, women), might be seen as having some shared cultural values and resources.

Research on cultural variations in learning has tended to describe ethnic or racial cultural groups in a manner that is static. Although there are historically rooted continuities that connect individuals across generations (Lee, 2003), describing culture in categorical terms to distinguish groups of people often leads to statements that attempt to describe the "essence" of groups. This can lead to stereotypes, such as the idea that Asian children are good at math or that girls struggle in science. Such statements treat culture as a fixed configuration of traits and assume that all group members share the same set of experiences, skills, and interests (Gutiérrez and Rogoff, 2003). Thus, they tend to obscure the heterogeneity of nondominant (and dominant) cultures. In addition, even when stereotypes are framed in an effort to illustrate the strength of a nondominant group or to compare groups, this reductive tendency can have negative impacts on members of a group (Steele, 1997). For example, there may be greater pressure placed on Asian children by their teachers and parents to excel in mathematics. Such statements can impact the self-esteem of children who do not excel in the manner that the statement claims.

A cultural-historical perspective on how individuals and groups learn offers a way to move beyond the assumption that characteristics of cultural groups are homogeneous and solely located within individuals. This perspective stresses that culture is not a static set of traits but is something more dynamic and develops through an individual's history of engagement in various practices. From this perspective, culture becomes a question of situating the social practices and histories of groups and less about attributing certain styles to groups. In other words, culture is "the constellations of practices historically developed and dynamically shaped by communities in order to accomplish the purposes they value" (Nasir et al., 2006).

## Diversity and Equity

Over the past several decades, concerns about equitable access to science for nondominant groups (as well as underutilization of the nation's human resources) have been strong motivators in the issue of science equity. To that end, equity in science education has primarily focused on defining and identifying science content standards—that is, what students are expected to learn and achieve in science classrooms (Lee, 1999). Within these standards science has typically been represented as objective, universal knowledge—and culturally neutral. Moreover, some educators have stressed science as a set of practices that define a singular "culture of science" that would-be scientists must acquire. This view assumes, implicitly or explicitly, that the culture of science does not reflect the cultural values that people bring to science. We question this assumption, which is analogous to assuming that learners of a second language naturally speak without accent, without any trace of their first language. This assumption has resulted in an approach to equity that does not adequately address systemic factors that might restrict access or hinder individuals from nondominant groups from engaging and identifying with science (Secada, 1989, 1994).

Thus, science equity has often resulted in attempts to provide equal access to opportunities already available to dominant groups, without consideration of cultural or contextual issues. Science instruction and learning experiences in informal environments often privilege the science-related practices of middle-class whites and may fail to recognize the science-related practices associated with individuals from other groups. In informal venues for learning science, for example in museums, some initiatives are aimed at introducing new audiences to existing museum science content, such as outreach initiatives offering reduced-cost admission or bringing existing science programming that is, already offered to mainstream groups, to nondominant communities. The goal of such initiatives is to enable students to become members of the science community without changing existing science systems (Good, 1993, 1995; Matthews, 1994; Williams, 1994). This view of science equity has been called the assimilationist view of science equity (Lee, 1999). The logic of this view is that particular groups have not had sufficient access to science learning experiences. So to remedy that situation, educators deliver to nondominant groups the same kinds of learning experiences that have served dominant groups.

Participation and achievement in science, however, are mediated by a complex set of sociocultural and systemic factors not often recognized in such science equity efforts. Principal among these is the idea that one's social world and context shape values, skill sets, and expectations (Nasir et al., 2006). Thus, the act of exposing all individuals to the same learning environments does not result in science equity, because the environments themselves are designed in a manner that supports the cultural repertoire of the dominant culture.

Alternatively, a group of theories portrays equity in science learning as a political process (Lee, 1999, 2005). This view assumes that as students from underrepresented populations gain access to science, they learn to appropriate the language and discourse of science and use it to address local or personal concerns. This perspective assumes that engagement in science by underrepresented populations will lead to a politically driven shift in the nature of science to better reflect the cultural practices and concerns of those underrepresented populations, which may result in more equitable power structures (Calabrese Barton, 1997, 1998a, 1998b; Calabrese Barton and Osborne, 1998; Eisenhart, Finkel, and Marion, 1996; Howes, 1998; Keller, 1982; Mayberry, 1998; Rodriguez, 1997). Thus, this orientation is a major departure from the assimilationist view, which sees science as the central goal to be reached by students who are at the margins and assumes the practices of science will remain unchanged by their participation (Calabrese Barton, 1998a, 1998b).

A third perspective on science equity stems from the cultural anthropological perspective. From this perspective, equity in science learning occurs when individuals from diverse backgrounds participate in science through opportunities that account for and value alternative views and ways of knowing in their everyday worlds (Aikenhead, 1996; Cobern and Aikenhead, 1998; Costa, 1995; Gallard et al., 1998; Maddock, 1981; Pomeroy, 1994), while also providing access to science as practiced in the established scientific community. This approach centers on making science accessible, meaningful, and relevant for diverse students by connecting their home and community cultures to science. Lee (1999) likens this perspective to biliteracy or biculturalism, whereby an individual can successfully bridge the culture of science.

Carol Lee (1993, 1995, 2001) has used this approach to design learning environments that leverage knowledge associated with everyday experiences to support subject matter learning (in her case, literacy practices). Lee's approach, termed cultural modeling, works on the assumption that students who are speakers of African American vernacular English (AAVE) already tacitly engage in complex reasoning and interpretation of literary concepts, such as tropes and genres. She engages students in metacognitive conversations in which students make explicit the evidence and reasoning they are using in their discussions. The conversations might focus, for example, on how students know that rap lyrics are not intended to be taken literally and the strategies they use to interpret and reconstruct the intended meaning. These conversations reflect AAVE norms, such as multiparty talk and signifying.

From this framework, cultural practices are seen as providing different perspectives. In other words, there is no cultureless or neutral perspective, no more than a photograph or painting could be without perspective. Everything is cultured (Rogoff, 2003), including the layout of designed experiences, such as museums (Bitgood, 1993; Duensing, 2006), and the practices associated with teaching science in school (Warren et al., 2001). For example, in a study of a collaborative of nine museums, Garibay, Gilmartin, and Schaefer (2002) found that participants who previously did not regularly visit museums initially needed more staff facilitation to help them better understand the learning and experiential goals of exhibits. Thus, the more one understands the role of culture and context in learning, particularly in science learning, the more effectively one can ensure that science is available to all children and adults.

## Learning Is a Cultural Process

Working from the perspective that learning is a fundamentally cultural process (Nasir et al., 2006; Rogoff, 2003) in which conceptions of learning are historically and locally situated, science learning is viewed as a socio-cultural activity. Its practices and assumptions reflect the culture, cultural practices, and cultural values of scientists. In this section, we first describe the cultural nature of learning generally and then focus in on the specific aspects of science learning that make it a cultural activity (see Chapter 2 for related discussion).

Focusing on the strengths of parents in working-class households, González, Moll, and Amanti (2005) have shown that children develop "funds of knowledge"—historically developed and accumulated strategies (skills, abilities, ideas) or bodies of knowledge that prove useful in a household, group, or community. This represents a fundamental shift in analysis and discussion of learning for nondominant groups. The traditional viewpoint often implies or even explicitly states that the cultural values and knowledge that circulate in nondominant cultural groups are deficient, not useful, or even counterproductive (Lareau, 1989, 2003; Rogoff and Chavajay, 1995). However, close analysis of parenting and childrearing practices shed new light on the productive exchanges and values in nondominant cultural groups and illustrate for researchers and educators how those can be leveraged in educational practice.

Children all over the world explore their world and have conversations about causes and consequences, and the particular topics they discuss and the ways they learn to explore the world are likely to vary, depending on the cultural practices with which they grow up (Heath, 2007; Rogoff, 2003). People live in different environments across their life span, with varied exposure to activities relating to different science domains (e.g., fishing, farming, computer technology). What counts as learning and what types of knowledge are seen as "important" are closely tied to a community's values and what is useful in that community context (Bruner, 1996; McDermott and Varenne, 2006).

Everyday contexts and situations that are meaningful and important in children's lives not only influence their repertoires of practice, but also are likely to afford the development of complex cognitive skills. This is evident in the studies of meaningful activities for individuals from various American cultures (Nasir, 2000, 2002; Nasir and Saxe, 2003; Rose, 2004). Nasir (2002) illustrated that playing basketball can be linked to an improved understanding of statistics and other mathematical concepts and that complex cognitive strategies are developed playing the game of dominoes. These studies illustrate that deep participation in such hobbies is linked to cognitive gains associated with knowledge valued by these cultures. For example, Nasir studied African American elementary school, high school, and adult dominoes tournament participants. Her findings show that players developed important general cognitive abilities, including perspective taking, numerical competence, and the ability to weigh multiple factors and goals at once. The development of these skills is intertwined with changes in the sociocultural setting of dominoes. The analysis of these data depicts the cognitive shifts that occurred among players of different age groups, the manner in which the sociocultural setting became intertwined with the cognitive shifts, and the shifting nature of the social setting.

Rose's (2004) depiction of the cognitive and physical skills developed by various blue-collar workers is a further illustration of the sociocultural nature of learning. In the workplace, groups and organizations develop specialized language, rituals, shared values, and norms of participation. Through their experiences and interactions with others in these settings, adults learn the various cognitive and physical skills needed to be successful at their jobs. The work lives of waiters, hair stylists, plumbers, welders, carpenters, and electricians are not usually associated with learning or learning science. However, Rose's case studies illustrate how learning and even science learning occurs in the informal context of their work.

The cognitive and physical skills of blue-collar work are learned in a manner that reflects the defining characteristics of learning in informal environments, such as direct access to phenomena and learning with others (such as through apprenticeship relationships) (Rose, 2004). For example, in his observations of a carpentry class, Rose shows that high school students learned by planning and building objects in class and as volunteers at Habitat for Humanity sites. While working in small groups to build cabinets, tables, and homes, students learned many of the physical skills (e.g., measuring, sanding, sawing) required of carpenters. In these groups, students learned from "guided participation." The more experienced students coached or facilitated more novice students' use of tools or their understanding of how all the pieces come together.

Students also learned important lessons just by being around others doing work. For example, one student said "You see work going on all around you. You see people making small, small mistakes, and you learn from that" (Rose, 2004, p. 76). The teacher also played an important role in the classroom. His assistance often came in the form of sharing tricks of the trade that he developed from years of experience. For example, when he noticed a student who was struggling to hammer a nail into a board, he explained that if the student moved his hand down on the tool he would produce more force. When the student made the adjustment, he was surprised at the different feel of swinging the hammer and that the hammer now seemed more powerful. Rose explains that such interactions not only lead to learning a physical skill, but also lead to an awareness of the connection between the work and such scientific principles as force, friction, and balance. There is, of course, a substantial difference between knowing where to hold a hammer to exert the most force on a nail head and mastering a scientific explanation of the same. However, as diSessa (1993) has argued, learners may quickly develop embodied knowledge or "phenomenological principles" through such experiences. Later the learner may relate these phenomenological principles to more abstract concepts (e.g., force, momentum, leverage).

The cultural and historical nature of learning relates not only to the accumulation of facts and concepts, but also to identity development. As Lave and Wenger (1991) explain, "Learning involves the construction of identities. . . . [It is] an evolving form of membership" (p. 53). "Our identities are rich and complex because they are produced within the rich and complex set of relations of practice" (Wenger, 1998, p. 162). When speaking about identity, people often first consider such demographic characteristics as age, gender, socioeconomic status, race, and ethnicity. Although these factors no doubt have the potential to influence people's attitudes and behavior, as well as the ways in which others may treat them in society, Fienberg and Leinhardt (2002) suggest: "Another conception of identity is that it includes the kinds of knowledge and patterns of experience people have that are relevant to a particular activity. This second view treats identity as part of a social context, where prominence of any given feature varies, depending on which aspects of the social context are most salient at a given time" (p. 168).

This discussion of learning as a cultural process illustrates that how learning occurs and what is learned are influenced by personal and contextual factors from early childhood through adulthood. Applying a sociocultural perspective to the different modes of learning and valued knowledge across and within cultures can move the discussion from one based on a deficit model to one that recognizes and values the contributions of a wide variety of cultural groups.

## Science Learning Is Cultural

Too often cultural diversity in science learning is studied by comparing the skills and knowledge of children from nondominant groups with those from the dominant group (Chavajay and Rogoff, 1995). In these comparisons, mainstream skills and upbringing are considered "normal" and variations observed in nondominant groups are taken as aberrations that produce deficits, lending support to a deficit model of diversity. Such studies do not appropriately account for the cultural nature of education environments or the diverse practices of science.

Science has been described by some as a social construct, "heavily dependent on cultural contexts, power relationships, value systems, ideological dogma and human emotional needs" (Harding, 1998, p. 3). Although this view of science is a contested one, seeing science as "a culturally-mediated way of thinking and knowing suggests that learning can be defined as engagement with scientific practices" (Brickhouse, Lowery, and Schultz, 2000, p. 441). This, in turn, can lead to expectations and limitations that greatly impact who engages in science and how science is conducted. When people enter into the practices of science, they do not shed their cultural world views at the door. Calabrese Barton (1998b) argues for allowing science and science understanding to grow out of lived experiences and that, in doing so, people "remove the binary distinction from doing science or not doing science and being in science or being out of science . . . allow[ing] connections between [learners'] life worlds and science to be made more easily . . . [and] providing space for multiple voices to be heard and explored" (p. 389). This view is a very powerful one when one considers the goals of informal environments for learning science.

It has also been argued that the field of science itself is quite diverse in the methods it employs. Nobel laureate physicist P.W. Bridgeman argued that "there is no scientific method as such" (Dalton, 1967, cited in Bogdan and Biklen, 2007). He continued by stating that "many eminent physicists, chemists, and mathematicians question whether there is a reproducible method that all investigators could or should follow, and they have been shown in their research to take diverse, and often unascertainable steps in discovering and solving problems" (Dalton, 1967, p. 41). This conception of science illustrates the need to cultivate various ways of knowing, learning, and evaluating evidence.

Ways of knowing, learning, and evaluating evidence are connected to the language and discourse styles accepted in science and science learning. Traditional classroom practices have been found to be successful for students whose discourse practices at home resemble those of school science—mainly students from middle-class and upper-middle class European American homes (Kurth, Anderson, and Palincsar, 2002). Such practices create an exclusionary aspect to science in which the discourse of science functions as a gatekeeper barring individuals from nondominant groups, because their science-related practices may not be acknowledged (Lee and Fradd, 1998; Lemke, 1990; Moje, Collazo, Carillo, and Marx, 2001; Brown, 2006).

Recognizing that language use and discourse patterns may vary across culturally diverse groups, researchers point to the importance of recognizing the use of informal and native language, as well as culturally developed communication and interaction patterns in science education (e.g., Lee and Fradd, 1996; Warren et al., 2001; Moschkovich, 2002). Lee and Fradd (1996) noted distinct patterns of discourse (e.g., use of simultaneous or sequential speech) around science topics in groups of students from different backgrounds. As mentioned earlier, Rosebery, Warren, and Conant (1992) identified connections between Haitian Creole students' skills in story-telling and argumentation and science inquiry, using those connections to support their learning of both the content and the practices of science. Hudicourt-Barnes (2001) demonstrated how *bay odyans*—the Haitian argumentative discussion style—can be a great resource for students as they practice science and scientific discourse.

Children's experience with scientific thinking also varies a great deal, depending on a range of issues, such as culture, gender, and parents' educational, financial, and occupational background. For example, Valle (2007) found that parents with college majors in engineering were more likely to discuss scientific evidence with their children in the context of conflicting claims (e.g., the relative advantages and disadvantages of food additives) than were parents with a background in the humanities.

The cultural nature of science described in this section illustrates the need to expand the perspective on what counts as scientific thinking and competence. Science education often tends to privilege certain ways of demonstrating understanding of a phenomenon or topic (Ballenger, 1997). Therefore it is often difficult for students of diverse backgrounds to reconcile their own discursive norms with the norms of scientific discourse typically presented in both formal and informal environments for learning. A potential consequence of this narrow view of science practices is that students may dis-identify with science, perceiving it as incompatible with their own cultural values (Lederman, Abd-El-Khalick, Bett, and Schwartz, 2002).

## CULTURE AND SCIENTIFIC KNOWLEDGE

Research exploring the access to and participation in science of specific groups is generally limited. However, there is an emergent research base related to science learning in informal environments for a small set of underrepresented cultures. Here, we synthesize research on four groups and their experiences with learning science in informal environments. In this synthesis we illustrate common themes that underlie the experiences of individuals with varied cultural and historical backgrounds.

# Gender

The largest body of research with regard to access and equity in science learning focuses on gender with specific attention to underrepresentation of women. Gender can be viewed, and ultimately studied, from a range of perspectives. The prevailing view of gender in the field is that it is not a fixed attribute, but it is constructed in social interactions (Murphy and Whitelegg, 2006). Gender is only one component of diversity, and, despite the overlapping similarities among women, issues of ethnicity, class, culture, and the like all contribute to socialization and play a role in learning.

Statistically, a case can be made that gender impacts career success and pursuits in ways that are inconsistent with women's level of achievement. Although there is convincing evidence that gender does not define capability, its impact on skill and capacity building is unclear.

## Statistical Evidence of Gender Disparities

Statistics suggest continued areas of inequity, but overall, there are great improvements in science participation by gender. Recent statistics suggest that, since 2000, women have earned more science and engineering bachelors degrees than men (National Science Foundation, 2007). However, the numbers are less favorable when separated by area of science. For example, the gap in male and female degree earners in computer sciences has widened over the past few years (National Science Foundation, 2007). In their review of research on gender differences in mathematics and science learning, Halpern and colleagues (2007) found small mean differences between male and female science achievement and ability in comparison to the large variance within male and within female scores. The variance in male scores is consistently greater than that found in female scores, leading to more men than women scoring in the highest and lowest quartiles in tests of science achievement and ability.

In general, the differences between male and female participation in science have been decreasing over the past 20 years (National Science Foundation, 2002). Women constituted a greater percentage of science graduate students in 2004 than in 1994, growing from 37 to 42 percent. This varied by field of science. In 2004, women made up 74 percent of the graduate students in psychology, 56 percent in biology, and 53 percent in social sciences. However, women accounted for only 22 percent of graduate students in engineering and 27 percent in computer sciences, with a 30-45 percent representation in most other science fields (National Science Foundation, 2007). Disparity in participation in science increases further along the educational continuum (Lawler, 2002; Mervis, 1999; Sax, 2001). Seymour and Hewitt (1997) found that undergraduate women were more likely to leave the sciences than similarly achieving men.

Changes in the science workforce have been slower to emerge. In fact, there are some indications that the percentage of women in the science workforce actually decreased from 1999 (46 percent) to 2002 (24 percent; National Science Foundation, 2002). Recent data also illustrate that women are less likely to obtain tenure (29 percent of women compared with 58 percent of men at four-year colleges) or to achieve the rank of full professor in science and engineering fields (23 percent of women compared with 50 percent of men; Ginther and Kahn, 2006). Male doctoral science and engineering faculty outnumber female ones by more than 2 to 1 (National Science Foundation, 2007, p. 20). Eisenhart (2001) suggests that the structure and expectations of physical science programs are more rigid and thus alienating to women with additional agendas, such as families, hobbies, and the like.

These differences are not occurring only in the United States. Results from the *Trends in International Mathematics and Science Study* (National Center for Education Statistics, 2003) revealed no significant difference between fourth grade male and female students' science scores. However, in eighth grade, on average, across all countries, boys scored significantly higher than girls. In 28 of the 46 participating countries, boys scored significantly higher than girls, while girls scored significantly higher than boys in seven countries. In countries in which achievement gaps have narrowed and even closed—Uganda, the Philippines, Ghana, Finland, and Japan—overall engagement in science remains unequal. The reasons for the gender differences in science achievement and engagement in the United States and other countries remain unclear. A European Commission publication on gender equality in science calls for "sociocultural understanding of gender and multidisciplinary gender research" (European Commission, 2008). This is reiterated by Calabrese Barton and Brickhouse (2006): "It seems important . . . to understand why it is that achievement does not necessarily lead to access to high-status science. If one wants to understand why it is that access to many areas of science continues to be a struggle, one must look beyond achievement and examine more broadly how gendered identities are constructed and how they interact with an educational system that serves an important gate keeping function" (p. 227).

### *Sociocultural Influences: Experiences Vary by Gender*

The Institute of Education Sciences (2007) identified three areas in which consistent gender differences emerged and could be influenced: (1) beliefs about abilities, (2) perceptions of the importance of careers, and (3) the importance of sparking an interest and then cultivating it throughout the school year. Girls may be succeeding on measures of standard success, however they are not necessarily identifying with science (Calabrese Barton and Brickhouse, 2006). Growing areas of research center on related questions. How are beliefs and identities linked to future choices? What does it mean to identify with science, and how can identity development be enhanced? How and why do achievement and actual engagement in science differ? What is the timing of developmental differences, if they exist, or of sociocultural influences that have positive or negative impacts?

***Identity.*** Lips (2004) and Packard and Nguyen (2003) have begun to examine a framework to consider how girls' images of themselves as possible scientists can influence future choices. For example, self-efficacy beliefs have been linked to mathematics and science-related choices (Simpkins, Davis-Keans, and Eccles, 2006). Focusing on physical science, researchers looked for longitudinal association between students' mathematics- and science-related activities, beliefs, and course-taking practices from fifth through twelfth grade. The participation of youth in out of school mathematics and science activities during fifth grade predicted self-concepts about the fields and level of interest and perceived importance in subsequent years. Related to self-conceptions is the study of stereotype threat (Steele and Aronson, 1995). McGlone and Aronson (2006) compared male and female performance when primed with positive achieved identities and negative stereotypes. They saw corresponding variation in performance, suggesting that social context and mind sets may be important.

The point at which gendered identities arise with regard to science is unknown. Substantial evidence documents the many ways in which girls and boys are exposed to gendered messages, experiences, and stereotypical perspectives from their earliest days, beginning at home, and continuing throughout their school years and in out-of-school programs and contexts. Parents' differential socialization of girls and boys has frequently been suggested as a possible influence on the gender differences in perceptions of and participation in science. In fact, some studies have shown differences in parental and adult encouragement in science depending on the gender of their child. Differences in the ways parents engage children of different genders is evident in conversations, questioning, access to resources, expectations, and perceptions of capabilities with regard to science learning, interest, and achievement (Crowley et al., 2001a). Specifically, parents are more likely to believe that science is less interesting and more difficult for daughters than sons (Tenenbaum and Leaper, 2003). Mothers underestimate the mathematical abilities of daughters and overestimate those of sons (Frome and Eccles, 1998). Fathers tend to use more cognitively demanding speech with sons than with daughters while engaged in science tasks (Tenenbaum and Leaper, 2003); and, when playing games with their children, mothers are more likely to talk about related scientific process when interacting with boys than with girls (Tenenbaum, Snow, Roach, and Kurland, 2005). Girls' interest in mathematics was observed to decrease as father's gender stereotypes increased, while boys' interest increased (Jacobs et al., 2005). Exposure to science toys, computers, and science-related experiences overall has been shown to differ for children depending on their gender (Kahle, 1998; Kahle and Meece, 1994; Sadker and Sadker, 1992, 1994).

Teachers, like parents, have been observed to question children of different genders in different ways and to encourage science-related skills (question-asking, use of tools) variably according to gender. For example, in science classes, teachers are more likely to encourage boys to ask questions and to explain concepts (American Association of University Women, 1995; Jones and Wheatley, 1990). This calls attention to the critical role adults can play in supporting science learning and the importance of adults' roles as facilitators across multiple contexts (Crowley et al., 2001a; Falk and Dierking, 1992, 2000; McCreedy, 2005).

Many efforts outside of home and school exist and have been developed specifically to address concerns about gendered science trajectories (Seeing Gender, 2006). However, while many programs see immediate impacts (albeit often self-reported), few programs have the benefit of funding and opportunity to look longitudinally at their impact (Gender Equity Expert Panel, 2000). As discussed in Chapter 5 a handful of studies have specifically looked at gender relations and the interactions of families in museum contexts which have documented, among other things, variable participation and interaction structures for boys and girls (Borun et al., 1998; Crowley et al., 2001b; Diamond, 1986; Dierking, 1987; Ellenbogen, 2002; Laetsch, Diamond, Gottfried, and Rosenfeld, 1980).

A review of research on girls' participation in physics in the United Kingdom (Murphy and Whitelegg, 2006) reinforces how differences in perceptions may influence strategies for engagement. Girls and boys differed in what they considered relevant when solving problems. These differences have the potential to lead to differing perceptions of competence. Differences between what girls and boys have learned is relevant and has a valuable effect on the problems they perceive. Girls are more likely to give value to the social context in which tasks are posed in defining a problem; boys are more likely than girls not to "notice" the context (Murphy and Whitelegg, 2006). What learners pay attention to—or learn to value as useful information—may influence what they learn and may also result in negative perceptions of their competence among educators and parents.

***Career choices.*** With regard to career choices, some have focused on early intervention due to concerns about decreases in girls' perception of their science ability over years of schooling (Jovanovic and King, 1998). In looking at career patterns of youth first questioned in middle school and then followed into their adult lives, Tai, Liu, Maltese, and Fan (2006) document the importance of career expectations for young adolescents and suggest that early elementary experiences (before eighth grade) may be critical. Fadigan and Hammrich's (2005) longitudinal study of high school girls who participated in an after-school, summer, and weekend program offered by the Academy of Natural Sciences documented the impact of these experiences on career choices. In particular, they found that of 152 women from urban, low-income, single-parent families who participated in the program, 109 enrolled in college, and the majority reported that their educational and career decisions were influenced by the opportunity to talk to staff and develop job skills and in having the museum as a safe place to go.

Many adults are involved in children's daily lives, including immediate family members or guardians, teachers, and adults with whom children spend out-of-school time (such as youth group leaders, after-school facilitators, and child care providers). The influence that early experiences and role models can have in supporting women's engagement in science is further reflected in the retrospective studies of what launched female scientists down their career paths. These women often cite particular individuals or contexts outside schools as significant influences on their pursuit of science careers (Baker, 1992; Fort, Bird, and Didion, 1993). In a study of barriers and strategies for success among female scientists (Hathaway, Sharp, and Davis, 2001), women reflected on the importance of finding informal networks and supporters through family as well as outside routes. In addition, as Eisenhart and Finkel (1998) found, "once outside the confines of conventional school science and engaged in more meaningful activities, women seemed to lack neither an interest in science nor the ability to learn it" (p. 239). Thus, it seems imperative to understand more about the nontraditional contexts and individuals instrumental in influencing young women in science, as well as the ways in which opportunities offered in nontraditional and intergenerational contexts available in informal environments can challenge the ways gendered messages about science are reproduced.

Overall, inequities persist in science participation by gender; however, there has been a positive trend toward reducing these gender inequities in science participation and achievement. The disparities continue to be more apparent in each successive level of education and career. Contextual and personal factors are related to these issues of inequity. Self-efficacy and gender stereotypes have been associated with girls' participation in science, and connected to the different types of encouragement provided to boys and girls by their parents, teachers, and other adults. Engagement with scientists and with science outside the context of formal environments for learning shows promise in mediating the impacts of self-efficacy and gender stereotype issues for young women.

## Native Americans

For people from nondominant groups, negotiating between various systems and communities can be stressful and problematic. Aikenhead (1996) described this process in relation to science education as one in which students must engage in "border crossings" from their own everyday culture into the subculture of science. These border crossings often involve code switching (different discourse practices and forms of argumentation) and therefore require students to be proficient in more than one linguistic tradition (McCarthy, 1980). To illustrate these points, we draw on the contrast between Native American science and Western science (Brayboy and Castagno, 2007).

It is important to keep in mind that there is not one native culture and to resist essentializing tribal cultures. There are more than 500 federally recognized tribes, and as many or more languages from more than 50 language families. There are some similarities in the epistemologies and ontologies of different tribal peoples, but this does not imply that a single or unified native science or native epistemology characterizes all tribal nations or all indigenous people.

It is evident from history that indigenous peoples have long been scientists and inventors of scientific ideas. Indigenous peoples in the Americas created toboggans to carry the heavy carcasses of deer and caribou; built seaworthy kayaks and canoes; constructed snowshoes and snow goggles; domesticated a wide range of plants, including corn, potatoes, squash, beans, and peanuts; built architectural masterpieces in which they lived and ovens in which they cooked; used petroleum to create rubber and stars to successfully navigate the continent; and found ways to dry meat for storage and future use.

Awareness of the need to improve science education for indigenous students is not new. Thirty years ago, the American Association for the Advancement of Science (AAAS) noted that one primary obstacle to indigenous participation in science was the lack of relevance of science to their lives. Based on this observation, the AAAS issued a number of recommendations for improving science teaching and learning for native youth. These recommendations included using an ethnoscientific approach to teaching science and a bilingual approach in particular contexts. In response, scholars have called for science education that directly relates to the lives of indigenous students and tribal communities. Most scholars agree that, to be most effective, learning environments must be connected and relevant to the local community, rather than some perceived unitary indigenous community (Aikenhead, 2001; Allen and Seumptewa, 1993; Cajete, 1988, 1999; Davison and Miller, 1998).

The goal of science education through a multicultural or culturally responsive lens is not only to connect science to indigenous students' lives, but also to create better scientists and students with stronger critical thinking skills. These goals are shared by scholars and tribal community members alike. Kawagley (1999) and Martin (1995) have found that tribal elders from Yup'ik and Iroquoian communities want their youth to learn multiple world views and be able to operate in both the dominant and tribal communities. A further goal of science education ought to be to foster more positive attitudes toward science among indigenous communities. Indeed, researchers have found that incorporating culturally responsive approaches into science education results in a more positive attitude toward science, which in turn impact academic achievement (Matthews and Smith, 1994; Ritchie and Butler, 1990). Indeed, if the primary goal is more effective science education for indigenous students, epistemological and sociocultural issues should be recognized.

The issue of world views or indigenous epistemologies is especially relevant to culturally based science education as Nelson-Barber and Estrin (1995) note:

> In considering what would constitute a curriculum and an approach to instruction that is valid for a given cultural group, we must first consider the customary ways of knowing and acquiring knowledge of that group. We are faced with essential epistemological questions such as, "What counts as important knowledge or knowing?," "What counts as evidence for claiming something to be true?," and "How and when should knowledge or understanding be expressed or shared?"... A blanket approach to students that fails to take socio-cultural factors into consideration is not likely to succeed in reaching all students (p. 22).

The concept of an indigenous science recognizes the role of culture, subjectivity, and perspective in making sense of the world and draws attention to the notion that people interpret reality through a particular cultural lens. Epistemological concerns and sociocultural factors must be central to the discussion of native or indigenous science and to efforts to provide a more culturally responsive science education to indigenous students.

Haukoos and LeBeau (1992) further elaborate this point:

> Science is also problematic because it fails to consider the socio-cultural environments in which students and communities live, it presents scientific knowledge as objective and universal, and thus fails to recognize that scientific knowledge is itself socially constructed. . . . This presumed objectivity and universalism of Western Science rationalizes our failure to acknowledge other ways of knowing. And, as Snively and Corsiglia (2001) have pointed out, "many scientists and science educators continue to view the contributions of Indigenous science as 'useful,' but outside the realm of 'real science'" (p. 15).

As discussed throughout this chapter, science is itself a subculture of Western culture, thus engaging in science education is already a cross-cultural event for many students (Aikenhead, 1998; Cobern and Aikenhead, 1998). Many indigenous students attempting to learn Western science must cross cultural borders and acquire facility in another culture. They must be able to use the linguistic traditions of both their own and the majority culture. Delpit (1988, 1995) argues that teachers must explicitly teach their students the norms and codes of the "culture of power" so that students who are not members of that culture obtain the necessary skills to negotiate the culture when they choose to do so. A similar effort needs to be made to make these norms and codes explicit for learning science in informal environments.

Finally, more detailed studies of native world views and understandings of nature have implications for designed environments. For example, the common Western view that nature is something external, something to be preserved, and something that is at its best when humans are visitors, not residents (e.g., national parks), may lead to depictions of ecosystems that do not include human beings, even though people are likely to play a dominant role in the viability of these same ecosystems. American Indians, who see themselves as a part of nature, may be puzzled by this omission (Bang, Medin, and Atran, 2007).

In summary, students from nondominant cultures, such as American Indians, must engage in border crossings from their everyday culture to the subculture of science when participating in science. To develop more effective science education in informal environments for nondominant cultures, epistemological and sociocultural issues must be recognized and taken into account. For example, the differing world views of the natural world in American Indian cultures are often not valued and can make engaging and participating in science especially difficult and confusing.

## People with Disabilities

Variation in cognitive, physical, and sensory abilities is another aspect of diversity to be considered and mediated in informal environments for science learning. Among school-age children, some 6.7 million are categorized as disabled under the Individuals with Disabilities in Education Act (IDEA) (National Center for Education Statistics, 2006). Among adults ages 25-64, about 24.4 million are categorized as disabled under the Americans with Disabilities Act, about 16.1 million of whom are categorized as having a severe disability (Steinmetz, 2006). According to the 2002 census, the rates of disabilities are higher among older people than younger people. For example, 8.4 percent of children under age 15 were categorized as disabled, 11 percent of people ages 25-44, 19.4 percent of people ages 45-54, and 72 percent of people over the age of 80. People with disabilities make up a sizable population (about 18 percent of the U.S. population), and they can be well served by science learning experiences in informal environments.

There are many constraints on access to science for people with disabilities, including navigation of physical spaces and access to and processing of language. Constraints on access are often multiple and act in concert, resulting in limitations on opportunity to learn science for those who experience a disability. For example, people with hearing impairment may feel cut off from science across multiple settings that are typically available to others. While others may passively consume science news stories as "background noise" on television or radio during the workday or at home, a hearing impairment prevents this. Given limitations to their access to spoken language, hearing-impaired students' may have less access to specialized forms of science language (e.g., Lemke, 1990; Lehrer and Schauble, 2006; National Research Council, 2007). This compounding of limitations on learning science for people with disabilities presents both a serious challenge and an exciting opportunity for learning science in informal venues.

The literature on science learning in informal environments for people with disabilities is extremely thin, yet it offers some useful analyses of the factors to be considered and the practices that may enable or enhance participation. There are two prominent ways of framing the issue. On one hand, educators and researchers explore the specific challenges associated with accessing science learning experiences in informal environments as those experiences are currently construed. This includes analysis of the gaps between the skills and practices required to participate in informal venues and the ability profile of learners in order to develop interventions and technologies that will enable participation. On the other hand, disability can be thought of as situated and culturally determined (McDermott and Varenne, 1995, 1996). From this perspective, the notion of ability is defined in light of a particular task and setting, and an individual's ability to complete it (or not) and conventional labels used to characterize "disability" are not valid. The label of disability is instead applied to the interaction between a particular individual and a particular task.

In addressing accessibility, educators and researchers attend to a variety of concerns, from simply getting participants through the door to how to make experiences relevant and accessible to people with physical and sensory disabilities. The cost of enrolling in science learning programs in informal settings and visiting informal institutions for learning science may be prohibitive for people with disabilities, as they experience higher rates of unemployment (National Center for Education Statistics, 2006). Physical access to place-based science learning and programs can be complicated or even impossible for many individuals.[1] People who are visually impaired may struggle to navigate designed spaces in order to find exhibits of interest. Programming interactive science experiences for a diverse public (e.g., participatory labs, field-based investigations) also requires analysis of the ways in which people with disabilities can and cannot engage. For example, there may be limitations in how a physically disabled person can participate

---

[1]The Americans with Disabilities Act requires any entity that receives federal funding to make reasonable accommodations to ensure that facilities are accessible to people with disabilities. See the accessibility guidelines at http://www.access-board.gov/adaag/html/adaag.htm.

in a species count in a local ecosystem. Similarly, hands-on demonstrations may require use of sight and sound.

Adaptive practices and technologies can facilitate some of these access constraints. For example, several interesting innovations facilitate navigation of exhibits. The New York Hall of Science has experimented with cell phones that allow visitors to call exhibits that are equipped with bells that activate when calls come in. People then follow the sound to locate the exhibit. Reich, Chin, and Kunz (2006) report on the use of virtual personal digital assistants that use American Sign Language in the context of a science/science-fiction exhibition at the Museum of Science, Boston. Study participants reported feeling they were freed from reliance on interpreters and other hearing participants. They reported greater freedom to pursue their own agenda.

People with learning disabilities also face unique challenges to learning science, and a limited body of research has characterized the barriers to their participation. Most of this work has examined children's experience in inquiry-oriented classrooms. The barriers identified include science being presented in highly abstract theoretical forms, overreliance on students' written forms of communication, reliance on individual (rather than group) scientific tasks, and peer group exclusion (Morocco, 2001; Palincsar, Collins, Marano, and Magnusson, 2000). Although approaches to mediating science for people with learning disabilities have not been studied thoroughly and almost no work has taken place in informal settings for science learning, several promising ideas have emerged. These include linking real-world scenarios to scientific abstractions, using peer conversation, providing support with writing tasks, and allowing children to try out their thinking with a teacher or aid before presenting it to the class (Palincsar et al., 2000; Rivard, 2004). Researchers have also observed specific research practices to be used with children with learning disabilities. They call for assessment tasks that model appropriate language for them (rather than requiring them to generate language) and using multiple measures of student thinking (Carlisle, 1999).

While adaptive technologies and practices may ease access to informal environments for science learning, there are also more fundamental cultural issues to address that entail holistic reassessment of the practices of informal venues for science education, as well as research and development frameworks. Understanding and engaging the disability community may lie beyond the scope of adaptive technologies. As suggested by McDermott and Varenne (1995, 1996), it may be more accurate to think about disability as cultural, where participation is an intersection of the cultures of science and science learning institutions with the communities of people with disabilities.

In this sense, the barriers to participation are culturally produced and culturally overcome. Like other underrepresented groups, people with disabilities may tend to dis-identify with science, face language barriers, and experience political and ideological tension between the norms of science and host institutions and those of their cultural group. For example, Molander,

Pedersen, and Norell (2001) studied and observed that a core group of deaf students rejected science. Two sets of interviews with deaf students were carried out to assess if there were differences between how deaf and hearing students reason about science. A survey from the National Evaluation of Compulsory Schools was used so that the results of the interviews with deaf students could be compared with the responses by hearing students that were previously reported. The first group interview, with three 15-year-old eighth grade students, was carried out to study how likely they were to use scientific concepts or models to answer the interview questions related to scientific phenomena. In the second set of interviews, seven 17-year-old tenth grade students were given the same questions and were also shown a scientific experiment described by one of the students in the first interviews to explain the process of recycling matter. Unlike other students, who, in the context of the interviews, freely mixed their personal experiences with scientific observations, a significant portion of deaf students did not. These students also made negative statements about their abilities in science. The researchers interpreted this as cultural resistance, speculating that students felt that joining a scientific culture would mean rejecting deaf culture. Similarly, in a study of deaf students, ages 7 to 17, about their understanding of cosmology, Roald and Oyvind (2001) observed that young deaf students performed as well as their hearing peers, whereas older deaf students did not.

Universal design for learning is a philosophy and educational practice based on a cultural conception of ability and learning that aims to create learning environments that are better for everyone. Tenets of universal design, according to the Center for Applied and Specialized Technologies (CAST), include representing information in multiple formats and media, providing multiple pathways to engage students' action and expression, and providing multiple ways to engage students' interest and motivation (Rose and Meyer, 2002). As a framework for research and development, universal design is in its infancy, but it may be a particularly useful framework for informal venues for science learning.

For example, Reich, Chin, and Kunz (2006) conducted a number of case studies on the accessibility of computer kiosks in a science museum. Her sample of 16 included learners ages 17 to 77 with a range of abilities and disabilities. She set out to understand the usefulness of three distinct interactive computer displays in the Museum of Science, Boston. Reich's study validated certain design elements common to the three exhibits (e.g., button interface design), which were used successfully by all participants, as well as specific aspects of exhibit design that inhibited participation. Reich's work also validated the idea that ability is situational. For example, she observed nondisabled computer users struggling with computer kiosks and a visually impaired noncomputer user who thrived in a computer environment.

In summary, this literature explores how adaptive technologies can ease access to science learning in informal environments. The general tenor of the research suggests that viewing disability as cultural leads to greater understanding and engagement of disabled learners. Seeing barriers to participation as cultural will require informal venues to make holistic reassessments of their practices. Emerging frameworks for research and development, such as universal design, illustrate the potential impacts of making such holistic reassessments.

## Urban and Rural Environments

The nature of the environments to which individuals are exposed influences their conceptions of scientific principles and ways of knowing. There is evidence that outdoor experiences foster social development and academic success (Hattie, Marsh, Neill, and Richards, 1997) and that being in nature is a stress reducer for children (Wells and Evans, 2003). Given this emphasis, it is surprising that few have directed attention toward science learning within different outdoor (nondesigned) environments. In this section, we describe a handful of studies that suggest that some aspects of children's culture are influenced by whether they grow up in either urban or rural environments, and that these differences in culture impact people's understanding of biology.

Most research studies on children's biology have been carried out with urban, middle-class children. One claim growing out of this traditional research pattern is that young children are strongly anthropocentric—that is, that they tend to interpret entities in the biological world by comparing them to a single (human) standard.

The predominant evidence supporting this claim comes from young children's performance on a category-based induction task. In this task, children are introduced to a single base (e.g., a dog, a bee, a human), hear a novel property attributed to that base (e.g., "dogs have *andro* inside of them"), and are then asked whether this property holds for other bases, both biological and nonbiological (e.g., birds, raccoons, fish, trees, bicycles).

Using this procedure with young children, Carey (1985) reported several striking results. First, children made far more inductive generalizations to other animals when introduced to a human rather than a nonhuman animal base (either a dog or a bee). The resulting pattern violated generalizations based on biological similarity. For example, 4- to 5-year-olds generalized more from a human to a bug than from a bee to a bug. In addition, strong asymmetries existed; children were more likely, for example, to generalize from a human base to a dog than from a dog base to a human.

Carey argues that this asymmetry reveals the central status of humans in biological reasoning. Going further, she argues that this early anthropocentrism must be overturned if children are to embrace the Western scientific view in which humans are not the most central exemplar or prototype, but rather are one among many biological entities.

Why might young children be especially anthropocentric? One factor may be that they are presumably exposed more to humans than to other biological kinds. Another idea is that children are reluctant to generalize from any base without extensive knowledge about that biological kind. In support of this notion and as discussed in Chapter 4, Inagaki (1990) examined generalization of biological properties by children in Tokyo, some of whom had extensive experience raising goldfish. She found that children who had no pets showed a familiar anthropocentric pattern of generalization, whereas children raising goldfish showed two generalization gradients—one around humans and one around goldfish (e.g., they generalized from goldfish to turtles).

If intimate experience with biological kinds governs patterns of generalization, then rural children may not show anthropocentrism at all. Ross, Medin, Coley, and Atran (2003) examined inductive generalizations from different bases among urban children, rural European American children, and rural American Indian children using a procedure similar but not identical to that employed by Carey (1985).

For both groups of rural children, human was not a better base for generalization than a nonhuman mammal. Young urban children showed broad and relatively undifferentiated generalization. Ross et al. (2003) also found evidence for an alternative strategy for generalization. Older rural European American and American Indian children of all ages sometimes generalized in terms of ecological or causal relations. For example, when they were told that bees have "andro" inside them, they might reply that bears also have andro inside them, justifying their judgments by saying that andro might be transmitted to bears when bees sting them, or that andro might also be in honey (which bears eat).

Other results suggest that evidence for anthropocentrism in young children depends on the details of tasks and procedures (Waxman and Medin, 2007) but that it is seen only in young urban children. Although anthropocentrism may reflect a lack of intimate experience with the biological world, it may also reflect an anthropocentric cultural model, as seen, for example, in Disney movies and in the way urban pets are often treated (e.g., dogs are typically seen as part of the family).

Related work reinforces the idea that urban (as differentiated from rural) environments influence the development of children's biology. For example, Coley and associates (Coley, Vitkin, Seaton, and Yopchick, 2005) have examined taxonomic and ecological generalization as a function of age and experience. Rather than dichotomizing children as urban versus rural, Coley used the continuous measure of population density. He found that taxonomic generalization shows little, if any, variation as a function of age or population density (see Waxman and Medin, 2007, for similar results using a different paradigm) but that ecological generalization increased systematically with age and decreased systematically with population density. In addition, the distinction between properties that may be distributed by ecological agents versus intrinsic biological properties also increases with age and decreases with population density. In short, sensitivity to ecological relations appears to vary as a function of culture and geography.

Studies conducted in Poland also suggest that environment matters. Using a category-based induction task patterned after Ross et al. (2003), Tarlowski (2006) found that urban 4- to 5-year-olds generalized in a broad, relatively undifferentiated manner from a human, nonhuman mammal, and insect base, whereas rural children generalized as a function of biological (taxonomic) similarity and showed no evidence for anthropocentrism. Tarlowski added an interesting twist to his studies with the variable of whether the child had a parent who was a biological expert. The findings associated having an expert parent with greater differentiation of generalization to biological versus nonbiological kinds. In general, the effects of "rural versus urban" and "expert versus layperson" parents appeared to be additive.

Overall, these studies tend to associate children's exposure to a rural rather than an urban environment with reduced anthropocentrism and greater sensitivity to ecological relations. Having a parent with expertise in biology also apparently helps young children display a more mature understanding of biology. This research also calls into question the current practice of treating urban, middle-class children as the gold standard for claims about cognitive development in science learning in general—and science learning in informal environments in particular.

## SCIENCE LEARNING IN INFORMAL SETTINGS FOR DIVERSE POPULATIONS

### Ownership and Outreach

As we have argued, informal settings for science learning are themselves embedded in cultural assumptions that may tend to privilege the world view, discourse practices, and contextualizing elements of the dominant culture. People from nondominant cultural groups may tend to see these institutions as being owned and operated by this same group. Garibay (2006a, 2006b, 2007) identified a number of factors—particularly the lack of diverse staff, perceptions that content was not culturally relevant, and the unavailability of bilingual or multilingual resources—that resulted in second-generation Latinos feeling unwelcome in museums.

When museum staffs conceptualize efforts to broaden participation as "outreach," they implicitly endorse this view of ownership. The term "outreach" implies that some communities are external to the institution. Collaboration, partnership, and diversity in power and "ownership" may provide greater opportunity for nondominant groups to see their own ways of sense-making reflected in informal settings, designed environments, and practices.

Attendance patterns appear to reflect this disconnect. For example, several studies have noted that informal institutions for science learning (e.g., museums, nature centers, zoos, etc.) face challenges in reaching and serving nondominant groups. In a study of adult programming in museums, participants had very high levels of education (just over 70 percent had college or postgraduate degrees) compared with the general U.S. adult population (Sachatello-Sawyer, 1996). Similarly, Rockman Et Al (2007) noted that the audience for science media tends to be a predominantly white, older, wealthier, and more educated segment of society. A study in Chicago about cultural participation, which also included several science-focused informal institutions, found that participation was highest in predominantly white, high-income sections of the metropolitan area (LaLonde et al., 2006) despite the fact that many museums are located in areas that are populated with large proportions of families from nondominant cultural groups.

The informal science learning community and many related institutions are making efforts to address inequity. These efforts typically aim to introduce new audiences to existing science programming, through outreach initiatives, reduced-cost admission, or other methods. They do not often take into account the contexts, perspectives, and needs of diverse populations.

## Design for Diverse Populations

Although research on how to structure science learning opportunities to better serve nondominant groups is sparse, it does include several promising insights and practices. These practices should serve as the basis for an ongoing research and development agenda.

Environments should be developed in ways that expressly draw on participants' cultural practices, including everyday language, linguistic practices, and local cultural experiences. Designers of informal programs and spaces for science learning have long recognized the importance of prior knowledge that participants and visitors bring to schools and other learning environments. This knowledge is typically considered culturally neutral (Heath, 2007; McDermott and Varenne, 2006). Much more attention should be paid to the ways in which culture shapes knowledge, orientations, and perspectives.

These and other findings undermine the view that typical scientific practices are largely abstract logical derivations not associated with everyday experience of the natural world. This observation also underlines the opportunity of educators working in designed environments (Cobb et al., 2003; Bell et al., 2006; Bricker and Bell, 2008) to take better advantage of the cultural practices that a diverse set of learners might bring to the environment.

In designed environments, such as museums, bilingual or multilingual labels cannot only provide access to the specific content, but also can facilitate conversations and sense-making among groups. Bilingual interpretation, for example, can enhance social interaction and learning in intergenerational groups with varying language abilities (Garibay and Gilmartin, 2003; Garibay, 2004a). Garibay observed that, in such groups, bilingual interpretive labels (English and Spanish) allowed adult members who were less proficient in English to read the labels and then discuss the content with their children, directly increasing the attention of these groups to the exhibition and learning outcomes.

The work of Ash (2004) with Spanish-speaking families in museums showed that the science themes of interest were similar across families with different backgrounds, but that the emergence of scientific dialogue was made possible by providing additional support, such as a Spanish-speaking mediator. Ash discusses the importance of distributed expertise, joint productive activity, and progressive sense-making in promoting dialogic inquiry. The dynamic changes, however, for non-English-speaking families who cannot use signs or read English. Wheaton and Ash's research (2008) on science education in informal programming found that participating girls welcomed and enjoyed the bilingual program because they learned science terminology and concepts in both languages and thus could better communicate with their parents (who were predominantly Spanish speaking) about what they were doing and learning in camp. This increased their confidence and helped bridge camp and home environments.

Having community-based contacts that are familiar and safe can also be critical in engaging families in science exploration and conversations and even, at a more basic level, in helping diverse groups see museums as less enigmatic places and as viable destinations for their families (Garibay, 2004b). Members of diverse cultural groups can play a critical role in the development and implementation of programs, serving as designers, advisers, front-line educators, and evaluators of such efforts.

Lee (2001) emphasizes the need to acknowledge and use a learner's linguistic resources, pointing to the importance of a balanced orientation, which values a learner's cultural identity. The Native Waters project, for example, strives to deliver culturally sensitive water education that includes programmatic components grounded in American Indian world views. The Algebra project, aimed explicitly at serving low-income and minority children, uses students' lived experiences and local environments as the starting point to help them build an understanding of mathematical concepts. For example, drawing on urban students' experiences riding the subway, participants might take a train ride and then reconstruct their trip using a map to represent a number line where they explore algebraic concepts, such as equivalent and positive and negative numbers ("how many" and "which direction"). Both Native Waters and the Algebra project consider community involvement central to their work and include community members (e.g., elders, college-age tutors) in their design process.

The cultural variability of social structures (e.g., family structure, norms governing gender relations, patterns of underrepresentation) should be reflected in design of informal environments for science learning as well. In a study of a nine-museum collaborative, for example, one of the initial problems was that one component of the project was designed for nuclear families where a parent was to bring their 9- to 10-year-old child, and did not account for the fact that local communities expected to participate in extended family groups. The expectation seemed to stem equally from families' desire to spend time together and real limitations that parents faced regarding childcare arrangements (Garibay, Gilmartin, and Schaefer, 2002). Designed spaces that serve families should include consideration of visits by extended families. In another study, Basu and Calabrese Barton (2007) noted that educational environments must value the relationships that learners themselves value. Barron (2006) found that there is typically an adult somewhere in a child's social network who has relevant knowledge and works with learners. This person may or may not be a parent. Developing peer networks may also be particularly important to foster sustained participation of nondominant groups in informal environments for science learning.

In sum, an informal environment designed to serve particular cultural groups and communities should be developed and implemented with the interests and concerns of these groups in mind. Project goals should be mutually determined by educators and the communities and cultural groups they serve.

It is also important to develop strategies that help learners identify with science in personally meaningful ways. Wong (2002) promotes "helping students experience the stories of individual scientists 'as if' they are the scientist and not the outside observers" (p. 396). DeBoer (1991) suggests that science education, "as all education, should lead to independent self-activity. It should empower individuals to think and to act. It should give individuals new ideas, and investigative skills that contribute to self-regulation, personal satisfaction, and social responsibility" (p. 240). Calabrese Barton (1998b) frames her analysis through pedagogical questions of representation in science (what science is made to be) and identity in doing science (who I think I must be to engage in that science). She comments: "Pedagogy involves the production of values and beliefs about how scientific knowledge is created and validated, as well as who we must be to engage in that process. . . . The way teachers choose to represent science to students leaves room for particular kinds of engagements, particular kinds of activities, and particular kinds of identities" (Calabrese-Barton, 1988b, p. 380).

Ultimately construction of science in educational settings results from a teacher (or other adult) and learner collaboration in the process, a "joint act [that] is influenced by the kinds of connections [they] can make between their lives, experiences, values, beliefs, and science" (Calabrese-Barton, 1988b, p. 380). Studies in schools, out of schools, and in family contexts are beginning to examine how personal frameworks and identities in science can be influenced.

Gallagher and Hogan (2000, p. 108) suggest that "examining science-learning experiences that expand the boundaries of typical schooling gives new meaning to the term systemic educational reform when 'the system' embraces the community at large . . . [and] encourage[s] others to create, implement and systematically study models of intergenerational and community-based science education. . . . Such inquiries have potential to provoke new thinking that could expand our field's basic conceptions of what it means to learn and practice science."

## CONCLUSION

There is no cultureless or neutral perspective on learning or on science—no more than a photograph or painting could be without perspective. Science is a sociocultural activity; its practices and epistemological assumptions reflect the culture, cultural practices, and cultural values of its scientists. Diversity in the pool of scientists and science educators is critical. It will benefit science by providing new perspectives in research, and it will benefit science education by providing a better understanding of science. Informal environments for science learning are themselves embedded in cultural assumptions. People from nondominant cultural groups may tend to see these institutions as being owned and operated by the dominant cultural group. Furthermore, science may be broadly construed as an enterprise of the elite.

Informal institutions concerned with science learning are making efforts to address inequity and encourage the participation of diverse communities. However, these efforts typically stop short of more fundamental and necessary changes to the organization of content and experiences to better serve diverse communities. Much more attention needs to be paid to the ways in which culture shapes knowledge, orientations, and perspectives. A deeper understanding is needed of the relations among cultural practices in families, practices preferred in informal settings for learning, and the cultural practices associated with science. The conceptions of what counts as science need to be examined and broadened in order to identify the strengths that those from nondominant groups bring to the field.

We highlight two promising insights into how to better support science learning among people from nondominant backgrounds. First, informal environments for learning should be developed and implemented with the interests and concerns of community and cultural groups in mind: Project goals should be mutually determined by educators and the communities and cultural groups they serve. Second, the cultural variability of social structures should be reflected in educational design. For example, developing peer networks may be particularly important to foster sustained participation of nondominant groups. Designed spaces that serve families should include consideration of visits by extended families.

More generally, environments should be developed in ways that expressly draw upon participants' cultural practices, including everyday language, linguistic practices, and common cultural experiences. Members of diverse cultural groups can play a critical role in the development and implementation of programs, serving as designers, advisers, front-line educators, and evaluators of such efforts.

# REFERENCES

Aikenhead, G. (1996). Science education: Border crossing into the subculture of science. *Studies in Science Education, 27*, 1-52.

Aikenhead, G. (1998). Many students cross cultural borders to learn science: Implications for teaching. *Australian Science Teachers' Journal, 44*(4), 9-12.

Aikenhead, G. (2001). *Cross-cultural science teaching: Praxis.* A paper presented at the annual meeting of the National Association for Research in Science Teaching, St. Louis, March 26-28.

Allen, G., and Seumptewa, O. (1993). The need for strengthening Native American science and mathematics education. In S. Carey (Ed.), *Science for all cultures: A collection of articles from NSTA's journals* (pp. 38-43). Arlington, VA: National Science Teachers Association.

American Association for the Advancement of Science. (1976). *A report on the barriers obstructing entry of Native Americans into the sciences.* Washington, DC: Author.

American Association of University Women. (1995). *Growing smart: What's working for girls in school.* Researched by S. Hansen, J. Walker, and B. Flom at the University of Minnesota's College of Education and Human Development.

Ash, D. (2004). Reflective scientific sense-making dialogue in two languages: The science in the dialogue and the dialogue in the science. *Science Education, 88,* 855-884.

Baker, D. (1992). I am what you tell me to be: Girls in science and mathematics. *Association of Science-Technology Centers Newsletter, 20*(4), 5, 6, 14.

Ballenger, C. (1997). Social identities, moral narratives, scientific argumentation: Science talk in a bilingual classroom. *Language and Education, 11*(1), 1-14.

Bang, M., Medin, D.L., and Atran, S. (2007, August). Cultural mosaics and mental models of nature. *Proceedings of the National Academy of Sciences, 104*(35), 13868-13874.

Banks, J.A. (2007). *Educating citizens in multicultural society* (2nd ed.). New York: Teachers College Press.

Barron, B. (2006). Interest and self-sustained learning as catalysts of development: A learning ecology perspective. *Human Development, 49*(4), 193-224.

Basu, S.J., and Calabrese Barton, A. (2007). How do urban minority youth develop a sustained interest in science? *Journal of Research in Science Teaching, 44*(3), 466-489.

Bell, P., Bricker, L.A., Lee, T.R., Reeve, S., and Zimmerman, H.T. (2006). Understanding the cultural foundations of children's biological knowledge: Insights from everyday cognition research. In S.A. Barab, K.E. Hay, and D. Hickey (Eds.), *Proceedings of the seventh international conference of the learning sciences* (pp. 1029-1035). Mahwah, NJ: Lawrence Erlbaum Associates.

Bitgood, S. (1993). Social influences on the visitor museum experience. *Visitor Behavior, 8*(3), 4-5.

Bogdan, R., and Biklen, S.K. (2007). *Qualitative research for education: An introduction to theory and methods* (5th ed.). Boston: Pearson/Allyn and Bacon.

Borun, M., Dritsas, J., Johnson, J.I., Peter, N.E., Wagner, K.F., Fadigan, K., Jangaard, A., Stroup, E., and Wenger, A. (1998). *Family learning in museums: The PISEC perspective*. Philadelphia: Franklin Institute.

Brayboy, B.M.J., and Castagno, A.E. (2007). *How might native science inform "informal science learning"?* Background paper for the Committee on Learning Science in Informal Environments. Available: http://www7.nationalacademies.org/bose/Brayboy_and%20Castagno_Commissioned_Paper.pdf [accessed October 2008].

Bricker, L.A., and Bell, P. (2008). *Evidentiality and evidence use in children's talk across everyday contexts*. Everyday Science and Technology Group, University of Washington.

Brickhouse, N.W., Lowery, P., and Schultz, K. (2000). What kind of a girl does science? The construction of school science identities. *Journal of Research in Science Teaching, 37*(5), 441-458.

Brown, B.A. (2006). "It isn't no slang that can be said about this stuff": Language, identity, and appropriating science discourse. *Journal of Research in Science Teaching, 43*(1), 96-126.

Brumann, C. (1999). Writing for culture: Why there is no need to discard a successful concept. *Current Anthropology, 40*(1), S1-S27.

Bruner, J. (1996). *The culture of education*. Cambridge, MA: Harvard University Press.

Cajete, G. (1988). *Motivating American Indian students in science and math*. Las Cruces, NM: ERIC Clearinghouse on Rural Education and Small Schools.

Cajete, G. (1993). *Look top the mountain: An ecology of Indian education*. Skyland, NC: Kivaki Press.

Cajete, G. (1999). The Native American learner and bicultural science education. In K. Swisher and J. Tippeconnic (Eds.), *Next steps: Research and practice to advance Indian education* (pp. 135-160). Charles Town, WV: ERIC Clearinghouse on Rural Education and Small Schools.

Calabrese Barton, A. (1997). Liberatory science education: Weaving connections between feminist theory and science education. *Curriculum Inquiry, 27*, 141-163.

Calabrese Barton, A. (1998a). Teaching science with homeless children: Pedagogy, representation, and identity. *Journal of Research in Science Teaching, 35*(4), 379-394.

Calabrese Barton, A. (1998b). Reframing "science for all" through the politics of poverty. *Educational Policy, 12*, 525-541.

Calabrese Barton, A., and Brickhouse, N.W. (2006). Engaging girls in science. In C. Skelton, B. Francis, and L. Smulyan (Eds.), *Handbook of gender and education* (pp. 221-235). Thousand Oaks, CA: Sage.

Calabrese Barton, A., and Osborne, M.D. (Eds.). (1998). Marginalized discourses and pedagogies: Constructively confronting science for all. *Journal of Research in Science Teaching, 35*(4), 339-340.

Carey, S. (1985). *Conceptual change in childhood*. Cambridge, MA: MIT Press.

Carlisle, J.F. (1999). Free recall as a test of reading comprehension for students with learning disabilities. *Learning Disability Quarterly, 22*(1), 11-22.

Chavajay, P., and Rogoff, B. (1995). What's become of research on the cultural view of cognitive development? *American Psychologist, 50*(10), 859-877.

Cobb, P., Confrey, J., diSessa, A., Lehrer, R., and Schauble, L. (2003). Design experiments in education research. *Education Researcher, 32*(1), 9-13.

Cobern, W.W., and Aikenhead, G.S. (1998). Cultural aspects of learning science. In B. Fraser and K. Tobin (Eds.), *International handbook of science education* (Part One, pp. 39-52). Dordrecht, Netherlands: Kluwer Academic.

Coley, J.D., Vitkin, A.Z., Seaton, C.E., and Yopchick, J.E. (2005). Effects of experience on relational inferences in children: The case of folk biology. In B.G. Bara, L. Barsalou, and M. Bucciarelli (Eds.), *Proceedings of the 27th Annual conference of the cognitive science society* (pp. 471-475). Mahwah, NJ: Lawrence Erlbaum Associates.

Costa, V.B. (1995). When science is "another world": Relationships between worlds of family, friends, school, and science. *Science Education, 79*, 313-333.

Crowley, K., Callanan, M.A., Jipson, J., Galco, J., Topping, K., and Shrager, J. (2001a). Shared scientific thinking in everyday parent-child activity. *Science Education, 85*(6), 712-732.

Crowley, K., Callanan, M.A., Tenenbaum, H.R., and Allen, E. (2001b). Parents explain more often to boys than to girls during shared scientific thinking. *Psychological Science, 12*, 258-261.

Davison, D.M., and Miller, K.W. (1998). An ethnoscience approach to curriculum issues for American Indian students. *School Science and Mathematics, 98*(5), 260.

DeBoer, G. (1991). *A history of ideas in science education: Implications for practice.* New York: Teachers College Press.

Delpit, L. (1988). The silenced dialogue: Power and pedagogy in educating other people's children. *Harvard Educational Review, 58*(3).

Delpit, L. (1995). *Other people's children: Cultural conflict in the classroom.* New York: W.W. Norton.

Diamond, J. (1986). The behavior of family groups in science museums. *Curator, 29*(2), 139-154.

Dierking, L.D. (1987). *Parent-child interactions in a free choice learning setting: An examination of attention-directing behaviors.* Unpublished doctoral dissertation, University of Florida, Gainesville.

diSessa, A.A. (1993). Toward an epistemology of physics. *Cognition and Instruction, 10*(2-3), 105-225.

Duensing, S. (2006). Culture matters: Science centers and cultural contexts. In Z. Bekerman, N. Burbules, and D. Silberman-Keller (Eds.), *Learning in places: The informal education reader* (pp. 183-202). New York: Peter Lang.

Eisenhart, M. (2001). Moving women from school to work in science: Curriculum demands, adult identities, and life transitions. *Journal of Women and Minorities in Science and Engineering, 7*, 199-213.

Eisenhart, M., and Finkel, E. (1998). *Women's science: Learning and succeeding from the margins.* Chicago: University of Chicago Press.

Eisenhart, M., Finkel, E., and Marion, S.F. (1996). Creating the conditions for scientific literacy: A re-examination. *American Educational Research Journal, 33*, 261-295.

Ellenbogen, K.M. (2002). Museums in family life: An ethnographic case study. In G. Leinhardt, K. Crowley, and K. Knutson (Eds.), *Learning conversations in museums* (pp. 81-101). Mahwah, NJ: Lawrence Erlbaum Associates.

European Commission. (2008). *Gender Impact Assessment of the Specific Programmes of the Fifth Program, An Overview.* Brussels, Belgium.

Fadigan, K.A., and Hammrich, P.L. (2005). Informal science education for girls: Careers in science and effective program elements. *Science Education Review,* 4(3), 83-90.

Falk, J.H., and Dierking, L.D. (1992). *The museum experience.* Washington, DC: Howells House.

Falk, J., and Dierking, L.D. (2000). *Learning from museums.* Walnut Creek, CA: AltaMira Press.

Fienberg, J., and Leinhardt, G. (2002). Looking through the glass: Reflections of identity in conversations at a history museum. In G. Leinhardt, K. Crowley, and K. Knutson (Eds.), *Learning conversations in museums* (pp. 167-211). Mahwah, NJ: Lawrence Erlbaum Associates.

Fort, D.C., Bird, S.J., and Didion, C.J. (1993). *A hand up: Women mentoring women in science.* Washington, DC: Association for Women in Science.

Frome, P.M., and Eccles, J.S. (1998). Parents' influence on children's achievement-related perceptions. *Journal of Personality and Social Psychology,* 74(2), 435-452.

Gallagher, J., and Hogan, K. (2000). Intergenerational, community-based learning and science education. *Journal of Research in Science Teaching, 37*(2), 107-108.

Gallard, A., Viggiano, E., Graham, S., Stewart, G., and Vigliano, M. (1998). The learning of voluntary and involuntary minorities in science classrooms. In B. Fraser and K. Tobin (Eds.), *International handbook of science education* (Part Two, pp. 941-954). Dordrecht, Netherlands: Kluwer Academic.

Garibay, C. (2004a). *Animal secrets bilingual labels formative evaluation.* Unpublished manuscript, Garibay Group, Chicago.

Garibay, C. (2004b). *Museums and community outreach: A secondary analysis of a museum collaborative program.* Unpublished master's thesis, Saybrook Graduate School and Research Center, San Francisco.

Garibay, C. (2006a). *Washington metropolitan area Latino research study.* Unpublished manuscript, Garibay Group, Chicago.

Garibay, C. (2006b). *Chicago area Latino audience research for the Brookfield Zoo.* Unpublished manuscript, Garibay Group, Chicago.

Garibay, C. (2007). *Palm Springs Art Museum: Latino audience research.* Unpublished manuscript, Garibay Group, Chicago.

Garibay, C., and Gilmartin, J. (2003). *Chocolate summative evaluation at the Field Museum.* Unpublished manuscript, Garibay Group, Chicago.

Garibay, C., Gilmartin, J., and Schaefer, J. (2002). *Park Voyagers summative evaluation.* Unpublished manuscript, Garibay Group, Chicago.

Gender Equity Expert Panel. (2000). *Exemplary and promising gender equity programs 2000.* Washington, DC: U.S. Department of Education. Available: http://www.ed.gov/pubs/genderequity/gender_equity.pdf [accessed October 2008].

Ginther, D.K., and Kahn, S. (2006). *Does science promote women? Evidence from academia 1973-2001.* Cambridge, MA: National Bureau of Economic Research.

González, N., Moll, L.C., and Amanti, C. (Eds.). (2005). *Funds of knowledge: Theorizing practices in households, communities, and classrooms.* Mahwah, NJ: Lawrence Erlbaum Associates.

Good, R. (1993). Editorial: The slippery slopes of postmodernism. *Journal of Research in Science Teaching, 30*(5), 427.

Good, R.G. (1995). Comments on multicultural science education. *Science Education, 79*(3), 335-336.

Gutiérrez, K.D., and Rogoff, B. (2003). Cultural ways of knowing: Individual traits or repertoires of practice. *Educational Researcher, 32*(5), 19-25.

Halpern, D.F., Benbow, C.P., Geary, D.C., Gur, R.C., Hyde, J.S., and Gernsbacher, M.A. (2007). The science of sex differences in science and mathematics. *Psychological Science in the Public Interest, Supplement, 8*(1), 1-51.

Harding, S. (1998). *Is science multicultural? Postcolonialisms, feminisms, and epistemologies.* Bloomington: Indiana University Press.

Hathaway, R.S., Sharp, S., and Davis, C.S. (2001). Programmatic efforts affect retention of women in science and engineering. *Journal of Women and Minorities in Science and Engineering, 7*, 107-124.

Hattie, J.A., Marsh, H.W., Neill, J.T., and Richards, G.E. (1997). Adventure education and Outward Bound: Out-of-class experiences that make a lasting difference. *Review of Educational Research, 67*(1), 43-87.

Haukoos, G., and LeBeau, D. (1992). Inservice activity that emphasizes the importance of culture in teaching school science. *Journal of American Indian Education, 32*(1), 1-11.

Heath, S., (2007). *Diverse learning and learner diversity in "informal" science learning environments.* Background paper for the Committee on Science Education for Learning Science in Informal Environments. Available: http://www7.nationalacademies.org/bose/Learning_Science_in_Informal_Environments_Commissioned_Papers.html [accessed November 2008].

Hirshfeld, L.A. (2002). Why anthropologists don't like children. *American Anthropologist, 104*(2), 611-627.

Howes, E.V. (1998). Connecting girls and science: A feminist teacher research study of a high school prenatal testing unit. *Journal of Research in Science Teaching, 35*(8), 877-896.

Hudicourt-Barnes, J. (2001). Bay odyans: Argumentation in Haitian Creole classrooms. *Hands On!, 24*(2), 7-9.

Inagaki, K. (1990). The effects of raising animals on children's biological knowledge. *British Journal of Developmental Psychology, 8*, 119-129.

Institute of Education Sciences. (2007). *Encouraging girls in math and science: IES practice guide* (NCER 2007-2003). Washington, DC: U.S. Department of Education.

Jacobs, J.E., Davis-Kean, P., Bleeker, M., Eccles, J.S., and Malanchuk, O. (2005). "I can, but I don't want to": The impact of parents, interests, and activities on gender differences in math. In A.M. Gallagher and J.C. Kaufman (Eds.), *Gender differences in mathematics: An integrative psychological approach* (pp. 246-263). New York: Cambridge University Press.

Jones, G., and Wheatley, J. (1990). Gender differences in teacher-student interactions in science classrooms. *Journal of Research in Science Teaching, 27*(9), 861-874.

Jovanovic, J., and King, S.S. (1998). Boys and girls in the performance-based science classroom: Who's doing the performing? *American Educational Research Journal, 35*(3), 477-496.

Kahle, J.B. (1998). Equitable systemic reform in science and mathematics: Assessing progress. *Journal of Women and Minorities in Science and Engineering, 4,* 91-112.

Kahle, J.B., and Meece, J. (1994). Research on gender issues in the classroom. In D. Gabel (Ed.), *Handbook of research on science teaching and learning* (pp. 542-557). New York: Macmillan.

Kawagley, A.O. (1999). Alaska native education: History and adaptation in the new millennium. *Journal of American Indian Education, 39*(1), 31-35.

Keller, E. (1982). Feminism and science. *Signs: Journal of Women in Culture and Society, 7*(3), 589-602.

Kurth, L.A., Anderson, C., and Palincsar, A.S. (2002). The case of Carla: Dilemmas of helping all students to understand science. *Science Education, 86*(3), 287-313.

Laetsch, W.M., Diamond, J., Gottfried, J.L., and Rosenfeld, S. (1980). Children and family groups in science centres. *Science and Children, 17*(6), 14-17.

LaLonde, R.J., O'Muircheartaigh, C., Perkins, J., English, N., Grams, D., and Joynes, C. (2006). *Mapping cultural participation in Chicago.* Monograph. Chicago: University of Chicago, Cultural Policy Center.

Lareau, A. (1989). *Home advantage: Social class and parental intervention in elementary education* (2nd ed.). Lanham, MD: Rowan and Littlefield.

Lareau, A. (2003). *Unequal childhoods: Class, race, and family life.* Berkeley: University of California Press.

Lave, J., and Wenger, E. (1991). *Situated learning: Legitimate peripheral participation.* New York: Cambridge University Press.

Lawler, A. (2002). Engineers marginalized, MIT report concludes. *Science, 295*(5563), 2192.

Lederman, N.G., Abd-El-Khalick, F., Bell, R.L., and Schwartz, R.S. (2002). Views of the nature of science questionnaire: Towards valid and meaningful assessment of learners' conceptions of the nature of science. *Journal of Research in Science Teaching, 39*(6), 497-521.

Lee, C. (1993). *Signifying as a scaffold for literary interpretation: The pedagogical implications of an African American discourse genre.* Urbana, IL: National Council of Teachers of English.

Lee, C. (1995). A culturally based cognitive apprenticeship: Teaching African American high school students skills in literary interpretation. *Reading Research Quarterly, 30*(4), 608-630.

Lee, C. (2001). Is October Brown Chinese? A cultural modeling activity system for underachieving students. *American Educational Research Journal, 38*(1), 97-141.

Lee, C. (2003). Toward a framework for culturally responsive design in multimedia computer environments: Cultural modeling as a case. *Mind, Culture, and Activity, 10*(2), 42-61.

Lee, O. (1999). Equity implications based on the conceptions of science achievement in major reform documents. *Review of Educational Research, 69*(1), 83-115.

Lee, O. (2005). Science education and English language learners: Synthesis and research agenda. *Review of Educational Research, 75*(4), 491-530.

Lee, O., and Fradd, S.H. (1996). Interactional patterns of linguistically diverse students and teachers: Insights for promoting science learning. *Linguistics and Education: An International Research Journal, 8*(3), 269-297.

Lee, O., and Fradd, S.H. (1998). Science for all, including students from non-English-language backgrounds. *Educational Researcher, 27*(4), 12-21.

Lehrer, R., and Schauble, L. (2006). Scientific thinking and science literacy. In K.A. Renninger, W. Damon, I.E. Sigel, and R.M. Lerner (Eds.), *Handbook of child psychology: Child psychology in practice* (vol. 4, 6th ed.). Hoboken, NJ: Wiley.

Lemke, J.L. (1990). *Talking science: Language, learning and values.* Norwood, NJ: Ablex.

Lips, H.M. (2004). The gender gap in possible selves: Divergence of academic self-views among high school and university students. *Sex Roles, 50*(5/6), 357-371.

MacIvor, M. (1995). Redefining science education for aboriginal students. In M. Battiste and J. Barman (Eds.), *First nations education in Canada: The circle unfolds* (pp. 73-98). Vancouver: University of British Columbia Press.

Maddock, MN. (1981). Science education: An anthropological viewpoint. *Studies in Science Education, 8,* 1-26.

Malcom, S.M., and Matyas, M.L. (Eds.) (1991). *Investing in human potential: Science and engineering at the crossroads.* Washington, DC: American Association for the Advancement of Science.

Martin, K. (1995). The foundational values of cultural learning: The Akhkwasahsne science and math pilot project. *Winds of Change, 10*(4), 50-55.

Matthews, C., and Smith, W. (1994). Native American related materials in elementary science instruction. *Journal of Research in Science Teaching, 41*(3), 363-380.

Matthews, M.R. (1994). *Science teaching: The role of history and philosophy of science.* New York: Routledge.

Mayberry, M. (1998). Reproductive and resistant pedagogies: The comparative roles of collaborative learning and feminist pedagogy in science education. *Journal of Research in Science Teaching, 35*(4), 443-459.

McCarthy, T.L. (1980). Language use by Yavapai-Apache students with recommendations for curriculum design. *Journal of American Indian Education, 20*(1), 1-9.

McCreedy, D. (2005). Engaging adults as advocates. *Curator, 48*(2), 158-176.

McDermott, R., and Varenne, H. (1995). Culture as disability. *Anthropology and Education Quarterly, 26,* 324-348.

McDermott, R., and Varenne H. (1996). Culture, development, disability. In R. Jessor, A. Colby, and R.A. Shweder (Eds.), *Ethnography and human development: Context and meaning in social inquiry* (pp. 101-126). Chicago: University of Chicago Press.

McDermott, R., and Varenne, H. (2006). Reconstructing culture in educational research. In G. Spindler and L. Hammond (Eds.), *Innovations in educational ethnography* (p. 3-31). Mahwah, NJ: Lawrence Erlbaum Associates.

McGlone, M.S., and Aronson, J. (2006). Stereotype threat, identity salience, and spatial reasoning. *Journal of Applied Developmental Psychology, 27*(5), 486-493.

Medin, D.L., and Atran, S. (2004). The native mind: Biological categorization, reasoning and decision making in development across cultures. *Psychological Review, 111*(4), 960-983.

Mervis, J., (April, 1999). High level groups study barriers women face. *Science, 284*(5415), 727.

Moje, E., Collazo, T., Carillo, R., and Marx, R. (2001). "Maestro, what is quality?": Language, literacy and discourse in project-based science. *Journal of Research in Science Teaching, 38*(4), 469-498.

Molander, B., Pedersen, S., and Norell, K. (2001). Deaf pupils' reasoning about scientific phenomena: School science as a framework for understanding or as fragments of factual knowledge. *Journal of Deaf Studies and Deaf Education, 6*(3), 200-211.

Morocco, C.C. (2001). Teaching for understanding with students with disabilities: New directions for research on access to the general education curriculum. *Learning Disability Quarterly, 24*(1), 5-23.

Moschkovich, J.N. (2002). A situated and sociocultural perspective on bilingual mathematics learners. *Mathematical Thinking and Learning, 4*(2-3), 189-212.

Murphy, P., and Whitelegg, E. (2006). Girls and physics: Continuing barriers to belonging. *Curriculum Journal, 17*, 281-305.

Nasir, N.S. (2000). "Points ain't everything": Emergent goals and average and percent understandings in the play of basketball among African American students. *Anthropology and Education Quarterly, 31*(3), 283-305.

Nasir, N.S. (2002). Identity, goals, and learning: Mathematics in cultural practice. *Mathematics Thinking and Learning, 2*(2-3), 213-248.

Nasir, N., and Saxe, G. (2003). Ethnic and academic identities: A cultural practice perspective on emerging tensions and their management in the lives of minority students. *Educational Researcher, 32*(5), 14-18.

Nasir, N.S., Rosebery, A.S., Warren B., and Lee, C.D. (2006). Learning as a cultural process: Achieving equity through diversity. In R.K. Sawyer (Ed.), *The Cambridge handbook of the learning sciences* (pp. 489-504). New York: Cambridge University Press.

National Center for Education Statistics. (2003). *Trends in international mathematics and science study (TIMSS)*. Available: http://nces.ed.gov/timss/results03.asp [accessed October 2008].

National Center for Education Statistics. (2006). *Digest of education statistics: 2006 digest tables*. Available: http://nces.ed.gov/programs/digest/2006menu_tables.asp [accessed October 2008].

National Research Council. (2007). *Taking science to school: Learning and teaching science in grades K-8*. Committee on Science Learning, Kindergarten Through Eighth Grade. R.A. Duschl, H.A. Schweingruber, and A.W. Shouse (Eds.). Board on Science Education, Center for Education, Division of Behavioral and Social Sciences and Education. Washington, DC: The National Academies Press.

National Science Foundation. (2002). *Gender differences in the careers of academic scientists and engineers*. (NSF 04-323.) Arlington, VA: Author.

National Science Foundation. (2007). *Women, minorities, and persons with disabilities in science and engineering*. (NSF 07-315.) Arlington, VA: Author.

Nelson-Barber, S., and Estrin, E. (1995). *Culturally responsive mathematics and science education for Native students*. Washington, DC: Native Education Initiative of the Regional Educational Labs.

Packard, B.W.-L., and Nguyen, D. (2003). Science career-related possible selves of adolescent girls: A longitudinal study. *Journal of Career Development, 29*(4), 251-263.

Palincsar, A.S., Collins, K.M., Marano, N.L., and Magnusson, S.J. (2000). Investigating the engagement and learning of students with learning disabilities in guided inquiry science teaching. *Language, Speech, and Hearing Services in the Schools, 31*, 240-251.

Pomeroy, D. (1994). Science education and cultural diversity: Mapping the field. *Studies in Science Education, 24*, 49-73.

Reich, C., Chin, E., and Kunz, E. (2006). Museums as forum: Engaging science center visitors in dialogue with scientists and one another. *Informal Learning Review, 79*, 1-8.

Ritchie, S., and Butler, J. (1990). Aboriginal studies and the science curriculum: Affective outcomes from a curriculum intervention. *Research in Science Education, 20*, 249-254.

Rivard, L.P. (2004). Are language-based activities in science effective for all students, including low achievers? *Science Education, 88*(3), 420-442.

Roald, I., and Oyvind, M. (2001, April). Configuration and dynamics of the earth-sun-moon system: An investigation into conceptions of deaf and hearing pupils. *International Journal of Science Education Journal, 23*(4), 423-440.

Rockman Et Al. (2007). *Media-based learning science in informal environments.* Background paper for the Committee on Learning Science in Informal Environments. Available: http://www7.nationalacademies.org/bose/Rockman_et%20al_Commissioned_Paper.pdf [accessed October 2008].

Rodriguez, A.J. (1997). The dangerous discourse of invisibility: A critique of the National Research Council's national science education standards. *Journal of Research in Science Teaching, 34*(1), 19-37.

Rogoff, B. (2003). *The cultural nature of human development.* New York: Oxford University Press.

Rogoff, B., and Chavajay, P. (1995). What's become of research on the cultural basis of cognitive development? *American Psychologist, 50*, 859-877.

Rose, D., and Meyer, A. (2002). *Teaching every student in the digital age: Universal design for learning.* Alexandria, VA: Association for Supervision and Curriculum Development.

Rose, M. (2004). *Mind at work: Valuing the intelligence of the American worker.* New York: Penguin Group.

Rosebery, A.S., Warren, B., and Conant, F. (1992). Appropriating scientific discourse: Findings from language minority classrooms. *Journal of the Learning Sciences, 2*, 61-94.

Ross, N., Medin, D., Coley, J.D., and Atran, S. (2003). Cultural and experiential differences in the development of folk biological induction. *Cognitive Development, 18*(1), 25-47.

Sachatello-Sawyer, B. (1996). *Coming of age: An assessment of the status of adult education methodology in museums.* Doctoral dissertation, Montana State University, Bozeman.

Sadker, M., and Sadker, D. (1992). Ensuring equitable participation in college classes. *New Directions for Teaching and Learning, 49*, 49-56.

Sadker, M., and Sadker, D.M. (1994). *Failing at fairness: How America's schools cheat girls.* New York: Scribner.

Sax, L.J. (2001). Undergraduate science majors: Gender differences in who goes to graduate school. *Review of Higher Education, 24*(2), 153-172.

Secada, W.G. (1989). Educational equity versus equality of education: An alternative conception. In W.G. Secada (Ed.), *Equity in education* (pp. 68-88). Philadelphia: Falmer Press.

Secada, W.G. (1994). Equity and the teaching of mathematics. In M.M. Atwater, K. Radzik-Marsh, and M. Strutchens (Eds.), *Multicultural education: Inclusion of all* (pp. 19-38). Athens: University of Georgia.

Seeing Gender. (2006). *Seeing gender: Tools for change.* Available: http://www.meac. org/Resources/ed_services/SG_WEB/SeeingGender/Resources.html [accessed October 2008].

Seymour, E., and Hewitt, N.H. (1997). *Talking about leaving: Why undergraduates leave the sciences.* Boulder, CO: Westview Press.

Simpkins, S.D., Davis-Kean, P.E., and Eccles, J.S. (2006). Math and science motivation: A longitudinal examination of the links between choices and beliefs. *Developmental Psychology, 42*, 70-83.

Snively, G. (1995). Bridging traditional science and western science in the multicultural classroom. In G. Snively and A. MacKinnon (Eds.), *Thinking globally about mathematics and science education* (pp. 53-75). Vancouver: University of British Columbia, Centre for the Study of Curriculum and Instruction.

Snively, G., and Corsiglia, J. (2001). Discovering indigenous science: Implications for science education. *Science Education, 85*(1), 6-34.

Steele, C. (1997). A threat in the air: How stereotypes shape intellectual identify and performance. *American Psychologist, 52*(6), 613-619.

Steele, C.M., and Aronson, J. (1995). Stereotype threat and the intellectual test performance of African Americans. *Journal of Perspective Social Psychology, 69*(5), 797-811.

Steinmetz, E. (2006). Americans with disabilities: 2002. *Current Population Reports,* P70-P107.

Tai, R.H., Liu, C.Q., Maltese, A.V., and Fan, X. (2006). Planning early for careers in science. *Science, 312*(5777), 1143-1144.

Tarlowski, A. (2006). If it's an animal it has axons: Experience and culture in preschool children's reasoning about animates. *Cognitive Development, 21*(3), 249-265.

Tenenbaum, H.R., and Leaper, C. (2003). Parent-child conversations about science: Socialization of gender inequities. *Developmental Psychology, 39*(1), 34-47.

Tenenbaum, H., Snow, C., Roach, K., and Kurland, B. (2005). Talking and reading science: Longitudinal data on sex differences in mother-child conversations in low-income families. *Journal of Applied Developmental Psychology, 26*(1), 1-19.

Valle, A. (2007, April). *Developing habitual ways of reasoning: Epistemological beliefs and formal bias in parent-child conversations.* Poster presented at biennial meeting of the Society for Research in Child Development, Boston.

Warren, B., Ballenger, C., Ogonowski, M., Rosebery, A.S., and Hudicourt-Barnes, J. (2001). Rethinking diversity in learning science: The logic of everyday sensemaking. *Journal of Research in Science Teaching, 38*(5), 529-552.

Waxman, S., and Medin, D. (2007). Experience and cultural models matter: Placing firm limits on childhood anthropocentrism. *Human Development, 50*(1), 23-30.

Wells, N., and Evans, G. (2003). Nearby nature: A buffer of life stress among rural children. *Environment and Behaviour, 35*(3), 311-330.

Wenger, E. (1998). *Communities of practice: Learning, meaning and identity.* New York: Cambridge University Press.

Wheaton, M., and Ash, D. (2008). Exploring middle school girls' ideas about science at a bilingual marine science camp. *Journal of Museum Education, 33*(2), 131-143.

Williams, H. (1994). A critique of Hodson's "In search of a rationale for multicultural science education." *Science Education, 78*(5), 515-519.

Wong, E.D. (2002). To appreciate variation between scientists: A perspective for seeing science's vitality. *Science Education, 86*(3), 386-400.

# 8

# Media

It's 7:00 pm on a Sunday evening, and you have just returned home from a long day at the local aquarium. Your family saw many exotic fish and read about their behaviors on signs posted near their tanks. You also watched an IMAX® film that showed some of these fish in their natural habitats. Now that you are home and relaxing, your daughter wants to see more fish, so she asks to watch the Disney/Pixar film, *Finding Nemo*. Afterward, you decide to sit down and watch some television before going to bed. One channel is showing *The Life Aquatic with Steve Zissou*, a Hollywood film inspired by the career of Jacques-Yves Cousteau, the great science filmmaker. Meanwhile, upstairs, the long-running news program, *60 Minutes*, is on another channel showing a segment on vacationers diving into ocean waters to observe sharks up close and personal, as well as the consequences of invading their territories. This segment intrigues your son, so he goes to the *60 Minutes* website to see a long list of people posting their comments on the show's content in real time.

It is unlikely that a family would be able to find this many opportunities to learn about aquatic life on a single day, but that should not downplay the fact that science learning in informal environments is often connected with various forms of media. Television documentaries, entertaining portrayals of science and nature in film, Internet websites, printed news stories, and online communities provide opportunities to communicate science content to individuals. These materials are often accessed voluntarily, making them an important part of science education in informal settings.

# A CONTEXT AND TOOL FOR SCIENCE LEARNING

"Media" can mean many things and take many forms. It can refer to the content of a printed story or a broadcast image. It can refer to the technology used to convey a particular form of information (e.g., television, newspapers, museum signs). It can be modified to indicate the affordances of a particular medium: "interactive media" or "targeted media" or "mass media." The field of media studies ranges from critical analyses of the content of particular story forms, through quantitative correlations of content analyses and public opinion, to detailed analyses of eye movements while interacting with websites. Traditional scholarly distinctions between "mass media" and "interpersonal communication" have in recent years been challenged by the need to create new perspectives that account for the interactions among these approaches.

In the context of science learning, however, the existing literature remains largely tied to older forms of analysis, dividing reasonably well into the traditional categories of "mass media" and "interactive media." It is also important to acknowledge that media may be used differently across social contexts. For example, a television documentary created for home viewing may also be shown in classrooms, as part of a museum display, or in a computer-based learning environment. In order to assess the effects of media on science learning, one must consider the ways in which they are appropriated and used across different informal settings.

In this chapter, we begin with summaries based on the traditional categorization of mass media. We then move on to suggest ways in which newer modes of analysis might shed light on learning science in informal environments. What tools exist, and how can they be made available to the public? How can individuals and groups access and leverage the knowledge of others through media? How can individuals and groups make their own insights more broadly accessible? We limit our analysis to areas in which research attends to learning outcomes and to issues of emerging or pressing interest in the field (such as new technological tools employed for educational purposes and the pervasive influence of digital technologies in everyday life).

# PRINT MEDIA

Although print media has the longest history, few studies have explored the specific effects of print on science learning. Many studies have identified the content of science books (both popular books and textbooks), magazines, including science specific magazines (such as *Popular Science* or *Scientific American*), and newspapers, making claims about the scientific quality and promotional or ideological effects of the content (Bauer, Durant, Ragnarsdottir, and Rudolfsdottir, 1995; Bauer, Petkova, Boyadjieva, and Gornev, 2006; Broks, 2006; Burnham, 1987; Dornan, 1989; Hansen and Dickinson, 1992;

Haynes, 1994; LaFollette, 1990). Few of these claims, however, have been subjected to empirical testing. Other studies have explored the production of printed media, focusing on the opportunities and constraints that shape their media content (Burnham, 1987; LaFollette, in press; Lewenstein, in press; Nelkin, 1987).

Particularly in the area of risk communication (the study and development of communicating the health implications of particular behaviors) some studies have examined the effects of particular print presentations of scientific information on individual perceptions of risk (Singer and Endreny, 1993; Walters, Wilkins, and Walters, 1989; Weiss and Singer, 1988; Wilkins and Patterson, 1990) In general, these studies have found that media do influence participants' perception of risk related to events (hazards, natural disasters) that may have immediate consequences for them. However, individuals' long-term considerations about these issues remain unaffected. This literature has also demonstrated that the social context in which stories are presented (e.g., the overall patterns of news coverage, the degree of trust that exists between readers and governmental or corporate institutions involved in the risk story) are typically more influential on participants' perceptions of risk than the genre of individual stories (e.g., whether they are sensational or measured and analytic).

In recent years, political scientists and other scholars concerned about political communication have tried to correlate public opinion about scientific and technological issues with media coverage of such controversies as nuclear power, biotechnology, and nanotechnology (Bauer and Gaskell, 2002; Brossard, Scheufele, Kim, and Lewenstein, 2008; Brossard and Shanahan, 2003; Ten Eyck, 1999, 2005; Ten Eyck and Williment, 2003; Gamson and Modigliani, 1989; Gaskell and Bauer, 2001; Gaskell, Bauer, Durant, and Allum, 1999; Nisbet, Brossard, and Kroepsch, 2003; Priest, 2001; Scheufele and Lewenstein, 2005). Although there is evidence that both demographic and psychological characteristics can influence opinion, and claims have been made about the link between those characteristics and exposure to particular media frames (Nisbet and Goidel, 2007; Nisbet and Huge, 2006), the evidence is not yet sufficiently strong to draw conclusions about the effect of particular print media on either broad public opinion or individuals' particular knowledge (Strand 2) and attitudes (Strand 6).

Particular science books are sometimes said to have had influence on the interests and career choices of later scientists, particularly Paul de Kruif's 1929 *Microbe Hunters* and James Watson's 1968 *Double Helix* (Lewenstein, in press), but little empirical evidence exists to show the direct effect of books on any of the strands of learning.

# EDUCATIONAL BROADCAST MEDIA

Perhaps the most studied area of learning science through media is the role of broadcasting, particularly television, in education. This literature explores the effects both of ubiquitous broadcast media and of broadcast media specifically intended for educational purposes.

Television and radio both offer science-themed programming that is broadcast widely and accessible to almost anyone in the developed world and a majority of people in the developing world. Television is present in over 98 percent of households in the United States, Europe, and developing nations (Clifford, Gunter, and McAleer, 1995; Dowmunt, 1993). It has an enormous influence on many aspects of everyday life and is arguably the single most influential means of communication of modern time (Huston et al., 1992; Kubey and Csikszentmihalyi, 1990). Science radio takes the form of weekly 1-2-hour programs and weekly or brief (90-second) shorts on both pubic and commercial radio, most of which are targeted to an adult audience. Program format ranges from hosted call-in talk shows to documentaries and interviews with scientists. Contemporary radio plays an important role in disseminating science news, addressing health policy objectives (e.g., family planning, disease prevention), and, to a limited extent, conveying science through more purely entertainment-oriented programming.

While many educators have concerns about the value of broadcast media, especially television (Gunter and McAleer, 1997; Hartley, 1999), it is clearly one of the most accessible sources of information for literate and illiterate populations. Broadcast media are particularly easy to use for children, youth, and adults. Not surprisingly, television is the primary source in the United States for general information about science and technology (National Science Board, 2008).

> Science- and math-based television and radio programs reach some 100 million children and adults each year. Educational science programming on television, once primarily the domain of the Public Broadcasting System (PBS), can now also be found on several Discovery Channels, the National Geographic Channel, The Learning Channel (TLC), NASA TV, and others. Top-rated educational programming currently includes *Zoom* (WGBH, ages 5 to 11), *Cyberchase* (WNET, ages 8 to 12), *Dragonfly TV* (TPT, ages 9 to 12), and *PEEP and the Big Wild World* (WGBH/TLC and Discovery Kids, pre-K). ... Each of these programs also offers ancillary activities on the web, making pbs.org one of the most popular .org sites and informal resources for learning worldwide (Ucko and Ellenbogen, 2008, p. 253).

Since the early days of television broadcast science programming such as *Watch Mr. Wizard* (Ucko and Ellenbogen, 2008), science programming has increased with the U.S. Children's Television Act of 1990, which required networks to broadcast educational television programming for children (U.S. Congress, 1990). In 1996, the Federal Communications Commission created new rules to enforce the congressional mandate on children's television (Federal Communications Commission, 1996). These include requiring television stations to air at least three hours per week of core educational programming between the hours of 7:00 am and 10:00 pm, with those programs being regularly scheduled and at least 30 minutes in length. Broadcasters are also required to explicitly signal when core educational programming is on the air through announcements or graphics displayed on the screen.

Historically the evidence of impact of the television shows was largely anecdotal (Newsom, 1952). However, there has been a recent increase in evaluative and scholarly studies of science-related television (e.g., Fisch, 2004; Rockman Et Al, 1996). These studies characterize the impact of science television on children, youth, and adults. While the quality and quantity of research have increased, these studies are extremely hard to locate, as they often exist only in sponsors' and evaluators' file cabinets. It is also very difficult to determine overarching findings, as most report on individual programs. However, a few careful syntheses have brought together these studies. Rockman Et Al's synthesis of research on broadcast media, which was prepared to inform this report, observes (2007, p. 16):

> Much of this material is fugitive literature, and requests to producers and distributors—and even to some researchers—did not always yield a response. For many of our queries, respondents (both producers and researchers) were unsure as to whether their reports were public documents and therefore able to be shared without permission. Almost all of the reports we obtained were funded by the National Science Foundation. We were not able to obtain research reports on science programming found on commercial radio and television.

Programming and approaches to research vary somewhat by the age of the intended audience for a given program. Research on programming for children and youth has typically considered the effects of watching 10 to 40 episodes of a given program, asking participants to respond to questions about the specific science content presented in the program (Strand 2). Evaluations of adult science programs are less extensive, and their designs reflect a basic difference in the structure of the programs. Unlike children's programming, which typically establishes a conceptual or topical theme across multiple episodes (e.g., problem-solving strategies, the principle of mechanical advantage), adult science education programming generally presents single topics in a given episode of television or radio that are not referenced in subsequent episodes. Studies of adult learning typically use surveys and questionnaires that prompt learners to self-assess knowledge gains (Strand 2) related to particular programs or to recall specific information from a program itself (Rockman Et Al, 2007).

There is some evidence that participants develop knowledge of science through television and radio programming; however, it is focused primarily on children. Several popular programs for children and youth, including *3-2-1 Contact*, *Bill Nye the Science Guy*, *The Magic School Bus*, and *Cro*, have been shown to positively influence viewers' knowledge of science (Strand 2) (Rockman Et Al, 1996; Fisch, 2007). Evaluations of adult programs have documented participants' self-reported knowledge gains and self-reported influence on subsequent behavior. For example, a series of evaluations were conducted by Flagg (2000, 2005b) on two National Public Radio science programs: *Science Friday*, a call-in show, and *Earth & Sky*, a series of 90-second shorts. Listeners reported that they learned about science and scientific methods (Strands 3 and 4), sought out more information, and also spoke with peers about what they heard on the program. The studies of adults hint at science learning outcomes. However, as we have observed in other areas, there is no clear documentation or measurement of what participants learned, nor have the self-reports been triangulated with other measures.

Considerably less attention has been devoted to practices or the ways in which learners act in the world to advance their understanding of science. Studies of the *Magic School Bus*, for example, have examined children's recall of how characters in the program learn. Evaluations of *Bill Nye the Science Guy* and *Square One TV* have looked at how viewers themselves use science and mathematical processes. A quasi-experimental study of the impact of *Bill Nye the Science Guy* found that viewers made more observations and more sophisticated classifications than nonviewers (Rockman Et Al, 1996). In this study, assessment materials (pre and post) were collected from a total of 1,350 children in schools, approximately 800 among the viewing group and 550 in comparison classrooms. The participants were recruited from three urban regions: Sacramento, Philadelphia, and Indianapolis. Results from the pre- and post-assessments showed that students who viewed the show were able to provide more complete and more complex explanations of scientific concepts than they were before viewing. Furthermore, in hands-on assessments, students who viewed the program regularly were better able to generate explanations and extensions of scientific ideas (Strand 2).

Several evaluations have examined the impact of radio programs on behavior, in which radio has been the mechanism for communicating public health messages in rural and developing areas. Public health-oriented radio programming typically takes the form of "entertainment-education," integrating desired health messages (e.g., about water quality, safe sex) into ongoing soap opera-like dramas, shorts, or songs about family planning and safe sex. A group of studies show the wide reach of health radio programming, as well as a connection between the programs and family planning and other health behaviors (Kane et al., 1998; Piotrow et al., 1990; Piotrow, Kincaid, Rimon, and Rinehart, 1997; Singhal and Rogers, 1989, 1999; Valente et al., 1994, 1997; Valente, Poppe, and Merritt, 1996; Valente and Saba, 1998).

However, Sherry (1997) urges caution in interpreting these results. Sherry's review of 17 entertainment-education studies from 8 developing countries found that the evidence supporting the impact of these programs was problematic. The research was based exclusively on high inferential self-reports of impact. It is also hampered by study design issues, including self-selected samples. More recent studies implementing quasi-experimental designs clearly show that health policy-oriented radio programming has a wide reach and supports the impact of programs on family planning behaviors (Kane et al., 1998; Karlyn, 2001). However, the programs did not always reach the target audience. Furthermore, there is no indication that behavior changes are linked to increased knowledge of scientific concepts (Strand 2) or scientific reasoning (Strand 3). They seem to be linked to knowledge of the practical and social implication of contraceptive use and attitudes about the health, financial, and social impacts of unplanned pregnancy.

Broadcast science education programs have also shown mixed results in promoting interest in science (Strand 1). Fisch (2004) observed that studies of science television's influence on children's interest in science indicate a moderate-sized effect and the Rockman Et Al (1996) study of *Bill Nye the Science Guy* corroborates this finding. However, Rockman Et Al also suggests that this is a likely underestimate and that a ceiling effect may be to blame for lower than expected posttest scores. Rockman and colleagues observed that their participants, children ages 8-10, already expressed an extremely high level of interest in science, so a pre-post study design may have made it difficult to detect significant changes.

Similarly, studies focusing on the effects of educational science programming on gender stereotypes have demonstrated some effect on attitudes (Steinke, 1997, 1999, 2005; Steinke and Long, 1996). However, the study designs precluded identifying long-term effects.

Several studies have found correlations between television and radio viewers' and listeners' interest in science (Strand 1) and frequency of listening to or viewing science programs. For example, evaluation findings for the short-format science radio series *Earth & Sky* reported that program appeal and engagement were highest among regular listeners. Similar results were found in the evaluation for *Science Friday;* frequency of listening was found to be higher among those with higher interest in science and also with enjoyment of the program. However, these are correlations and do not suggest an impact of the program. Whether more frequent listening is the by-product of engagement and enjoyment or vice versa is not explored in any of the studies reviewed.

Fisch and colleagues (Fisch, 2004; Fisch et al., 1997) have looked deeper at the organization of programs to discern how presentation of content varies across programs. Fisch (2004) describes differences between educational content (the underlying concepts and messages that a program conveys) and the story line (the interactions between events, characters, and their goals) in telling a coherent story. The interplay of these aspects of educational television may have implications for what viewers learn. Take, for example, an episode of *Bill Nye the Science Guy* focused on environmental issues related to plants and trees. The science educational content includes how to estimate the age of a tree and concerns related to logging. The story line used to present these topics varied from Bill Nye illustrating a tree trunk and counting its rings to stories from loggers explaining their work.

Storyline content, that is the story that presents educational concepts, methods, and messages, can be decomposed into two broad categories—documentary and narrative formats. Fisch et al. (1997) compared the narrative style of *Cro* with the documentary style of *3-2-1 Contact*. They found that, in the narrative format, scientific explanations were broken up and spread among multiple characters in contrast to the more didactic approach of the documentary format. In the narrative format, content was also constrained by the need to fit the setting (e.g., the Ice Age). There are probably learning trade-offs associated with organizing science programming in either a documentary or a narrative fashion. While a documentary format allows for direct explanation of scientific phenomena, a narrative format allows the freedom to break from historical or journalistic commitments. Fisch makes this point by comparing *3-2-1 Contact*, an educational program for young adolescents that typically employs a documentary approach, with *Cro* (pp. 108-109):

> Where fairly straightforward demonstrations and explanations could be fit into *3-2-1 Contact* simply by having characters address the audience or host/interviewers directly, these had to be fit into a fictional narrative in *Cro*, and the fit had to seem natural. Characters in *Cro* could not suddenly break the "fourth wall" and interrupt the ongoing story to give a lengthy explanation to viewers; rather, such explanations needed to occur in the course of conversation among characters. To seem natural, this often meant that explanations had to be broken up and spread over the course of the story, rather than taking place in a single, lengthy speech.

For example, the topic of light and refraction was approached in *3-2-1 Contact* through demonstrations of the effects of different-shaped lenses (with a teenage host speaking directly to camera) and a visit to a lighthouse to learn how beams of light are focused to be visible at greater distances. By contrast, *Cro* approached light and reflection through a story in which the prehistoric characters discovered some shiny, reflective rocks that they dubbed "see-myselfers" (i.e., natural mirrors).

Another pocket of research attends to the effects of coparticipation in broadcast media (e.g., watching or listening to programming with others). A series of studies examined the influence of children coviewing educational television with parents and peers and compared their outcomes with those of children who viewed programming alone. These studies suggest that the participation of others in consumption of broadcast media may enhance learning (e.g., Fisch, 2004; Haefner and Wartella, 1987; Reiser, Tessmer, and Phelps, 1984; Reiser, Williamson, and Suzuki, 1988; Salomon, 1977).

Reiser and colleagues (1984) conducted a randomized experimental study of adult-facilitated viewing sessions of Sesame Street with 23 white, middle-class children ages 3 and 4. In the experimental group, adults intervened to ask children to name the letters and numbers depicted on the screen. Three days after viewing the program, these children were better able to name the letters and numbers. These findings—although the outcomes are neither science-specific nor particularly complex—suggest that lightly facilitated adult coviewing can support learning.

Haefner and Wartella (1987) conducted a randomized experiment to examine the influence of siblings on 42 first- and second-grade children viewing educational programming. In this study, older siblings were 0-6 years older than their siblings. The older siblings were asked to actively explain important plot elements to their younger siblings while coviewing. The researchers found that coviewing did result in some older sibling "teaching." However, the teaching rarely focused on critical events and did not facilitate children's interpretation of either child-oriented or adult-oriented programs. In part, this is explained by the kinds of questions that younger siblings asked—typically requests for simple clarifications or elaborations of nonessential events. They also observed that many of the older siblings' comments were not efforts to promote learning, thus limiting the potential effects. However, the researchers observed that nonexplicit teaching by large-interval older siblings was conducive to understanding. Through these actions, which included laughter and comments, "older children did influence the younger children's general evaluations of the program characters" (Haefner and Wartella, 1987, p. 165).

Findings on coviewing resonate with research reported earlier on family learning in science centers (Callanan, Jipson, and Soennichsen, 2002; Callanan and Jipson, 2001; Crowley and Callanan, 1998; Gleason and Schauble, 2000). Children can access science media programming alone—whether television programming or an interactive science center exhibit—and adult interaction and possibly sibling and peer interactions can enrich and extend their experience and learning.

In summary, the literature on science learning from broadcast media is limited but converges on several important insights. First, when children watch science-themed educational television programs regularly, they can make important gains in conceptual understanding (Strand 2) and in their understanding of science processes (Strand 4) (Fisch, 2006; Rockman Et Al, 1996). We should also note, however, that the research relies heavily on conscripted participation. How children choose to navigate science television programming and whether their naturalistic forms of participation result in similar gains are not yet understood. The committee found little inquiry into adult learning outcomes.

The evidence of the impact of interaction with other people on learning gains is promising (e.g., Fisch, 2004; Haefner and Wartella, 1987; Reiser et al., 1984, 1988; Salomon, 1977), but it seems to have had little influence on subsequent research. Additional analysis of the watching and listening practices of groups and social networks may offer useful insights into programming features.

## POPULAR FILM AND TELEVISION

Most of the broadcast media discussed thus far are deliberately designed for science education. However, science and scientists also appear in popular television programs, films, and other entertainment media. Representations of science in the popular media have rarely been studied in the context of learning, yet it seems obvious that most Americans are more familiar with fictional scientists like Dr. Frankenstein or the medical staff of *ER* than recent Nobel laureates (Gerbner, 1987; Weingart and Pansegrau, 2003). As in the case of print media, most studies have focused on the production and content of entertainment films and television (Kirby, 2003a, 2003b). In general, these studies have found no single dominant image of scientists ranging from bumbling buffoons and nerdy social misfits to evil geniuses and high-minded saviors of humanity (Hendershot, 1997; Jones, 1997, 2001; Kirshner, 2001; Sobchak, 2004; Vieth, 2001).

Popular films are occasionally used in formal educational settings to illustrate scientific and mathematical concepts (Strand 2). In these cases, educators rely on familiar movies to provide context and motivation for problem solving (Strand 1). For example, the Cognition and Technology Group at Vanderbilt (CTGV) used the opening 12 minutes of *Raiders of the Lost Ark* to engage students in mathematics learning (Bransford, Franks, Vye, and Sherwood, 1989). In that scene, the main character, Indiana Jones, is in a jungle trying to retrieve a valuable statue. Students watched the scene and were asked to plan a return trip to the jungle to look for artifacts that Indiana had left behind. They used approximate measurements from the film (e.g., Indiana Jones' height) to make calculations (e.g., the relative width of a pit that needed to be crossed) about the return trip. Although the film lacks explicit instructional sequences, mathematical data could be drawn from it to provide students with problem-solving opportunities.

Popular films have also been used to complement science education and support student understanding of scientific concepts (Strand 2). The University of Central Florida's *Physics in Film* course is designed to give nonscience undergraduate students an engaging introduction to the physical sciences (Efthimiou and Llewellyn, 2006, 2007). For example, one scene from the film *Armageddon* involves using a nuclear bomb to split an asteroid into two pieces, hence saving the planet from destruction. The scene is used to introduce such concepts as mass, conservation of momentum, energy, and deflection. In the end, students work through the physics to discover that the film's outcome, two smaller asteroids being deflected away from Earth, is physically impossible. Instead, they learn that the two smaller pieces would strike the planet's surface a few city blocks apart (Efthimiou and Llewellyn, 2006). An important part of the *Physics in Film* curricula is helping learners see that science on the big screen does not necessarily correspond to the laws of physics. The same approach has been used in biology and in other fields (Rose, 2003).

Many television dramas are also based on scientific concepts, especially medicine (Turow, 1989; Turow and Gans, 2002). Criminal programs like *Numb3rs* and *Crime Scene Investigation* (*CSI*) have received recent attention due to their influence on public perceptions of science. In fact, the term "*CSI* effect" has been used to describe two different phenomena that result from viewing popular science programming.

In one case, the forensic science aspects of shows like *CSI* are believed to result in jurors increasing their demand for physical evidence in court trials, since this is what they see in fictional television labs (Houck, 2006). For example, district attorneys suggest that jurors now expect advanced technology to be involved in all court proceedings and that DNA testing is required as evidence. There are alarming examples of court cases being dismissed because jurors lack DNA and other physical evidence that appears prominently on *CSI* and related programs. In one case, jurors fought for DNA evidence despite the defendant's admission of being at the crime scene (Houck, 2006).

This version of the *CSI* effect demonstrates how viewers may not understand differences between fictional accounts of science and the realities of practice. It also demonstrates the power of entertainment media to teach viewers what it means to do science, as these programs seem to increase expectations of what occurs in court trials. While *CSI* may occasionally lead to misconceptions about real science, it has also led to positive outcomes in terms of viewers' awareness of and interest (Strand 1) in forensics (Podlas, 2006).

The second interpretation of the *CSI* effect focuses on representations of scientists. Jones and Bangert (2006) asked a convenience sample of 388 ethnically diverse middle school students to participate in a version of the Draw-A-Scientist Test (DAST, Chambers, 1983) to understand children's beliefs about scientists. Their results showed seventh grade girls drawing a larger percentage of female scientists than their ninth and eleventh grade female counterparts. Additional interviews with a sample of female and male students found seventh grade girls mentioning *CSI, Killer Instinct*, and other programs that made forensics look "fun" while including male and female characters as scientific contributors. Although the research design precludes a conclusive finding, the authors propose that middle school girls may have different mental images of scientists than their older counterparts due to their exposure to new programming, like *CSI*. Unlike many television programs in the past, these shows do not characterize scientists as odd, eccentric people wearing lab coats (e.g., Gerbner, 1987), and they portray women in key scientific roles—portrayals that students, especially young women, are more likely to identify with (Strand 6).

Both versions of the *CSI* effect may be behind the large growth in forensic science programs in higher education (Houck, 2006; Jones and Bangert, 2006). It appears that exposure to these programs may help middle and high school students become interested in science as a career (Strand 6). In some cases, student interests have driven universities to create forensic science majors to meet growing demands (Houck, 2006). While *CSI* and related shows deal with forensics, other programs with science-related content (e.g., hospital dramas like *ER* and *Grey's Anatomy*) may also influence attitudes toward and perceptions of science and scientific practice (Strand 4). The research base around popular media as a tool for science learning in informal environments is limited; further studies are needed to understand the role of television and film on viewers' knowledge and attitudes.

## GIANT SCREEN FILM AND OTHER IMMERSIVE MEDIA

One particular type of film has been studied for its contributions to learning science in informal environments: giant screen theaters (primarily IMAX®, but other vendors as well). These theaters are located in approximately one-third of science museums as well as other venues and show science-based documentaries along with other films. While in some basic sense large-format film is similar to television and cinema—they all employ a screen and typically engage learners in observing a production in silence—there are important differences. The scale and setting of giant screen film may result in a uniquely immersive experience compared with other screen experiences. Because of the large frame size and extremely high resolution of the film, this technology immerses viewers into the projected image, whether photographed with special cameras or computer-generated.

Other types of immersive media include planetariums and laser-projection systems. Planetariums employ optical or digital projection systems to create shows that incorporate images of the sky, space, and occasionally other scientific subjects. Studies of planetarium experiences (e.g., Fisher, 1997) have focused on programming characteristics, such as humor, that have the potential to impact learning or appeal to specific audiences, such as school groups (e.g., Storksdieck, 2005). Laser projection systems, including 3-D versions, have been used in both planetarium and theater settings. These systems can yield spectacular scientific imagery that is simply not available to most people through any other means. Subject matter may include the natural phenomena scientists are inquiring about or representations of scientific inquiry (e.g., depictions of deep sea exploration).

With the exception of giant screen cinema evaluations, few studies have examined the learning potential of these immersive media. A recent article makes the case for digital, full-dome systems as a powerful tool for learning astronomy, calling for research studies on the best ways to use this technology (Yu, 2005). The most comprehensive study to date is a review of summative evaluations on 10 giant screen projects and associated supporting materials (Flagg, 2005a). The evaluators typically conducted pre-post studies to measure changes in scientific knowledge and perceptions of scientists when scientists were characters in the film. All 10 of the studies showed a positive impact on viewers' knowledge of scientific concepts (Strand 2).

Attitudes and interest have not been measured as frequently in these studies. In 5 of the 10 studies, pre-post measures of interest level were used, and 2 of the 5 found a significant positive impact. Viewers of these two films (*Stormchasers* and *Dolphins*) were found to have greater interest in learning more about related topics after viewing the films (Strand 1). A study of the film *Tropical Rainforests* measured attitude and found that adult, youth, and child viewers had a more positive attitude toward rain forests after viewing the film. In the three studies that measured perceptions of scientists or researchers half or more of viewers felt they learned something new about the lives and work of scientists and researchers (Strand 4). Given the continuing commercial success of giant screen films—since the mid-1990s, the format has moved from almost exclusively educational venues and products to largely commercial venues—and these positive evaluation results after only a single viewing, the immersive format appears to have value for viewers. But the there is a need for further research and perhaps a broader set of science learning outcome measures.

## DIGITAL ENVIRONMENTS

The final area of concentrated literature addresses the Internet and associated technologies that have grown rapidly since the late 1980s. Much of this growth has been encouraged by the development and expansion of the World Wide Web since its development in the early 1990s. Originally created to facilitate information exchange among scientists, the web has become part of everyday computer use for millions of people. It hosts a range of science-specific learning resources, including science outreach pages describing current research; instructional resources for children, educators, and parents; "serious games," and simulations of scientific phenomena. Other relevant digital technologies that harness scientific knowledge and interface with the web to support science learning include cellular phones, personal digital assistants (PDAs), radio-frequency identification (RFID) chips, and sensory probes. These technologies are harnessed with the intention of enriching learners' interactions with scientists and peers about scientific inquiry, and relaying science news to vast audiences.

While the Internet is not yet universally accessible in homes, schools, and libraries, it is increasingly accessible. The Pew Internet and American

Life Project conducted a survey of a random sample of 2000 U.S. adults and reports that 20 percent of people in America said they use the Internet for most of their science news (second only to television at 41 percent), 49 percent of Internet users have visited websites that specialize in science content, and the majority of respondents said they would turn to the Internet first to find information on specific scientific topics. Furthermore, 87 percent of online users have used the Internet to conduct research on some aspect of science, and 80 percent of online users have used the Internet to verify the accuracy of scientific claims (Horrigan, 2006). It is important to note, however, that these figures do not account for the speed and quality of connection to the Internet. High-speed Internet service is still beyond reach for many people.

Brown and Duguid (2000) argue that society has only begun to realize the transformative power of web-based technologies, which ultimately will be on par with the generation and use of electricity, permeating and redefining society. There are important features of the web that may support science learning in ways that other media do not. Unlike print media, the web allows users to both receive and send information. Through user-selected and designed interfaces, the web can honor diverse ways of knowing and learning, so that users can interact with content and with one another in ways that they deem valuable. As an expansive network of users and resources, individuals can leverage resources to communicate with huge numbers of people. Furthermore, these characteristics of the web—dialogic structure, user direction and organization, expansive networking of people and resources, and increasingly user created media—resonate with learning science and informal environments.

Is there evidence that high levels of Internet use in general result in positive science learning gains? The answer to this question is not yet in, although there is some evidence. The Pew project reports correlations between Internet-based science information-seeking and individuals' interest in science. For example, Horrigan (2006) notes that those seeking scientific information on the Internet are more likely to believe that science has a positive impact on society. They are also more likely to report having greater understandings of science, new scientific discoveries, and what it means to study something scientifically. Prior training in science plays a role in these perceptions, as people with college degrees who have taken science courses self-report higher interest in and knowledge of science. It is unclear whether use of the Internet for science learning promotes interest in science, or whether interest in science promotes the use of such tools. Is this correlation a function of selection bias or an outcome of Internet use? The answer to this question will be critical in establishing the impact of the Internet on science learning.

Online gaming and participation in virtual worlds occupy the time and attention of a significant and growing population of children and adults, and these environments are increasingly being pitched and analyzed as settings for science learning. Two popular virtual worlds, *World of Warcraft* and *Second Life*, report participation of 8.5 and 6.5 million users, respectively. Participants in *Sims Online* number in the hundreds of thousands (Bainbridge, 2007; Squire and Steinkuehler, 2001). Americans spent $8.2 billion on game software and accessories in 2004 and $10.5 billion in 2005 (Crandall and Sidak, 2006). Video games generate more money than Hollywood films, and they have also become objects for scholarly critique, much as literature, cinema, and other works of art are reviewed and evaluated.

The term "serious games" has been used to refer to recent collaborations between educators and game designers to create computer and video games that educate as well as entertain (e.g., Aldrich, 2005; Gee, 2003; Prensky, 2000; Shaffer, 2006; Squire, 2003). One of the virtues of these games is that people use them of their own volition, investing hours in play on a regular basis. Even early video games—which did not offer the rich social potential and startling graphics of today's virtual worlds—were notably compelling to users, who were intrinsically motivated to pursue rewards embedded in the play experience (Bowman, 1982). Many serious games build on this motivation to create large simulation environments that could take 40 or more hours to master and complete (Squire, 2006). For example, games like *Civilization* require players to dedicate long periods of time creating imaginary nations while allowing them to simulate and envision possible historical outcomes. The time expenditure could lead learners to deeper understandings of the complex social, economic, and political issues that underlie the success and failure of fictional and real nations (Squire and Barab, 2004).

The educational potential for these environments should be understood in light of the tasks they pose and enable users to work on. Although computer-based games used in homes and schools around the country have often intended to teach basic literacy, mathematical, or problem-solving skills (e.g., a child playing *Lemonade Stand* can learn some basic principles of economics), current and future virtual worlds have the potential to support science learning across the strands.

Success in gaming environments hinges on integrating a broad range of knowledge and skill. For example, *River City* is a multiuser virtual environment designed for use in middle grade science classrooms (Dede, Ketelhut, and Ruess, 2002). Students conduct scientific investigations around an illness that is spreading through a virtual city based on realistic historical, sociological, and geographical conditions. The authenticity of the virtual world allows learners to engage in practices that resemble those of real scientists (Strand 5). In order to "win" the game, players must form hypotheses, test these by creating and running controlled experiments, and interpret their data to make recommendations about possible courses of action. Being successful requires understanding data provided by characters, books, and scientific tools in the virtual world (Strand 2) and developing skills to transform these data into hypotheses that can be tested (Strand 3).

An additional quality of online gaming environments is the high degree of networking which enables participants to draw on the distributed cognitive resources of individuals within the room or around the globe to solve them. For example, Whyville.net is a virtual community of 1.2 million users, many of whom are children and teenagers (Feldon and Gilmore, 2006). Once a year, the Whyville designers unleash a virtual epidemic, Whypox, into the online community. When players are infected with the virus, their graphical avatars appear with rashes, and their chat room messages are randomly interrupted by the word "achoo" to represent sneezing. The virtual virus becomes an opportunity for players to track the spread of the disease, generate hypotheses about its cause and transmission, and predict when the epidemic will end. Resources in Whyville.net also allow players to learn about viral transmission and simulate portions of the epidemic.

While we uncovered no clear analysis of learning in this environment, Foley and La Torre (2004) provide some sense of the potential for learning. They observed that over 1,000 members of Whyville became immersed in the first outbreak of Whypox, exploring various resources and participating in discussions to prevent the spread of the virtual disease (Strand 5). The number of science-related comments in Whyville bulletin boards increased dramatically, although science was still a small part of the overall communications. Neulight and colleagues (2007) reported that player discussions involved comparing their Whypox experiences with existing understandings of disease transfer, but the overall experience did not significantly increase knowledge of the biological processes underlying infectious diseases (Strand 2).

This leads to questions about the forms of learning that serious games can facilitate. For example, the Federation of American Sciences is developing *Immune Attack*, a strategy game that simulates travel inside the human body to learn immunological principles. The anticipated learning outcomes for students who play include (1) increased interest and enthusiasm for immunology in particular and for science in general (Strand 1), (2) increased interest in biotechnology related careers (Strand 1), (3) increased understanding of scientific practices (Strand 4), (4) more frequent engagement in scientific practice (Strand 3), and (5) improved knowledge of the immune system (Strand 2). The Whyville studies suggest that interest and motivation are likely to occur (Strand 1), but knowledge improvements may be more difficult to achieve (Strand 2). Rather, it may be possible for these games to increase knowledge of facts and terminology (Strand 2), but it may be more difficult to help players conduct rigorous experimentation, generate causal explanations, and other activities that are typically associated with scientific practice (Strand 3).

Nonetheless, virtual worlds and gaming environments may be uniquely rich settings for identity development (Strand 6). As discussed in previous chapters, the notion of "third places" or "third spaces" may provide a useful way to think about this. As neither home nor work, third places are insulated from the strong influence of the real world and provide a unique potential for the development of identity where new resources and constraints evolve in the social milieu of the virtual space. The third place of the chat room or game, rather than the local community center or bar, can become a primary vehicle for identity construction. Whereas geographic, cultural, and technical boundaries have historically constrained cultural exchange among groups and individuals, virtual environments can facilitate transactions across these barriers, opening up new intersections of people, tools, and traditions to support identity development. The previously noted trends in participation are clear, suggesting that although this research is emergent, there is a clear trend in participation in third spaces.

Some have argued that the same qualities that make virtual environments rich sites for identity development also make them rich sites for social scientific research on the nature of identity. Although there is little evidence yet that virtual museums can drive powerful identity-building experiences, one can envision the enormous possibilities of an entirely new type of virtual museum, science center, or zoological or botanical collection.

## MEDIA IN VENUES AND CONFIGURATIONS

Thus far we have summarized studies of science learning through particular media. Next, we explore the role of media in particular venues and configurations for science learning. We consider in turn each of the venues discussed previously: everyday settings, designed settings, and programs.

### Everyday and Family Learning

How media shape people's relationship to science and science learning in their daily lives is not yet clear. On one hand, the connectedness that digital technologies afford is enticing. The promise of digital media for enhancing learning—linking learners to experts and knowledgeable peers, building communities around common interests, and even building new knowledge bases—is real and exciting. On the other hand, there are considerable concerns about the quality and reliability of media-based accounts of science. Creators and providers of scientific information (and information that is claimed to be scientific information) are multiplying. The proverbial "man on the street" is no longer a figment of political rhetoric, but a potential contributor to public discourse who can readily broadcast his opinions on stem cells, evolution, and science curriculum. The traditional, authoritative sources of scientific information—museums, disciplinary communities, even the mainstream news media—find themselves competing with political and ideological interest groups to convey science to the public. The results can be quite confusing. A Google search of "Is evolution real?" can sometimes bring up as the very first item "Some real scientists reject evolution." The reality of the vast and expanding world of digital media, which expands authorship dramatically, makes it even more important for individuals to develop the critical capacity to evaluate claims.

One crude measure of change in people's relation to information—if not science—is clear: More people are using computers and digital technologies to communicate, conduct research, and solve practical problems (Fox, 2006; Horrigan, 2006; Madden and Fox, 2006). However, reports of vast and broadening access should be interpreted in light of the kind of access that individuals have: dial-up or broadband, at home, at work, or at a community center or library. It is too early to say what stable patterns will emerge. Yet given the documented changes in behavior of the past 15 years, it seems wise to assume that future use of digital media in everyday life will make that life look very different from what it is today.

Not only are people using digital media, but they tend to enjoy it. Learners appear to prefer using digital technology for research over hard copy resources, such as books. This is particularly true for children. Children are drawn in by the quantity of information; they value the multimedia character of web resources and the ease of access (Large and Beheshti, 2000; Fidel, 1999; Ng and Gunstone, 2002; Watson, 1998). Their positive attitude about the Internet as a research tool is particularly interesting in light of their frequent failures to find what they are looking for in searches (Kuiper, Volman, and Terwel, 2005). Older adults—despite broadly held beliefs to the contrary—also value information technology and are interested in learning how to use it. Unlike children, however, their engagement with new media is contingent on seeing a particular added value. They will ask, "Why use a webpage when I could pick up an encyclopedia?" (Lindberg, Carstensen, and Carstensen, 2007).

Given the scale and trajectory of digital media use, it is important to assess how people use digital media. Some scholars have raised concerns that learners will be overwhelmed by the sheer volume of information that is available, forcing them to disengage or to rely on inaccurate information. For example, Agosto (2002) conducted a qualitative interview-based study of 22 ninth- and tenth-grade girls' use of Internet searches and found that study participants frequently experienced information overload. Meanwhile, the girls expressed great satisfaction when the parameters of the search task were reduced and at completion of a search.

Another area of considerable interest is determining how skillful people are at searching and sorting reliable and unreliable information. These questions are relevant to any source of scientific information—trade books, textbooks, lectures, etc.—and they have special salience for digital media. Of the basic Internet search strategies—using keywords, browsing, entering

URLs, following links—which do people grasp readily and use of their own volition?

Much of the research on Internet searching comes from small-scale studies that analyze K-12 students' efforts to approach an assigned topic with limited instructional guidance (e.g., Fidel, 1999; Wallace, Kupperman, Krajcik, and Soloway, 2000). Thus, the findings should be understood as somewhat distinct from what we have described as everyday learning. Students' motivations for searching in school may be distinct from those in nonschool settings. However, with these caveats in mind, the literature coalesces around several interesting findings.

Children are confident in their Internet search skills in ways that overestimate their actual abilities. Their searches tend to be intuitive rather than systematic. As they sort through websites and search engine hit lists, they tend to focus narrowly, looking for specific word combinations or sentences and rarely reading beyond headers and topic sentences. Rather than assembling a range of information resources, synthesizing these, and developing their own ideas, children go quickly to the resource that appears to be right and use it with little or no interpretation.

It is unclear what this tendency to focus narrowly and search out particular phrases means. Is it a reflection of the institutional setting, or is it a reflection of children's knowledge and skill? Jones (2001-2002), in their experimental study of 100 students searching the web under variably structured conditions, observed this pattern and interpreted it as a reflection of a broader pattern of teacher-student interactions. In this case, students were convinced that their job was to sort through the finite set of answers to identify the correct answer and have it corroborated by a teacher. Under this interpretation, the finding may be less relevant to everyday settings, in which learners select their own topics, set the pace, and define their own expectations. Bilal (2002a, 2002b) conducted a series of small-scale studies on the information-seeking behavior of children. For example, in an exploratory, noncomparative analysis of 22 middle school science students' searching behavior using the Yahooligans! search engine, she observed that the students were more motivated and more likely to complete web searches when they determined the topics than when topics were assigned (Bilal, 2002b). They were also equally effective in identifying relevant resources under both self-selected and assigned conditions.

As research matures on the matters of information overload and the search capabilities of users, two areas of work seem particularly important. Given the limited skill of novice searchers, especially young children, it would be helpful to understand what topics and techniques they explore when they are successful so that these may be leveraged for other areas of inquiry and educational practice. It would also be helpful to consider whether there are productive ways to constrain and focus digital tools (informational resources, search engines) to enhance the quality of searches performed by novices and to aid their efforts to synthesize and interpret results.

## Designed Settings

Museums, science centers, zoos, and aquariums can employ media to support science learning in several ways. As discussed above, informal institutions for learning may offer access to media tools, like networked computers, libraries, and digital databases, to support users' self-defined agenda or deploy prepackaged media products, like giant screen films and television programs in their exhibit spaces. In this section, we focus on ways in which media components are built into floor-based exhibitions and alternative virtual spaces that extend visitors' experiences and serve other visitors who do not visit physical locations.

Ucko and Ellenbogen (2008, p. 245) indicate

> interactive exhibits offer visitors the opportunity to explore real (and sometimes simulated) scientific phenomena, as well as aspects of historic and state-of-the-art technology. Interactives based on classical physics (e.g., force and motion) tend to be the most widespread because the phenomena lend themselves readily to direct visitor manipulation, although exhibits based on biology (e.g., Colson, 2005) and chemistry (e.g., Ucko, Schreiner, and Shakhashiri, 1986) have also been developed. Because the term "interactive" encompasses an extremely wide range of experiences, from simple tasks explored by individuals to complex tasks requiring multiple collaborators (Heath and vom Lehn, 2008), it is difficult to develop generalized findings about how they support or contribute to learning.

Science centers have begun to explore the use of newer technologies to create augmented and virtual environments (Roussou et al., 1999). Research on their impact, like other areas of technology in informal environments, is dominated by usability studies and has little to say about learning outcomes or the specific qualities of digitally augmented environments. An exception is a series of exploratory case studies by Roussou and colleagues (e.g., Roussou et al., 1999) that have begun to explore how collaboration in virtual reality (VR) environments can support learning of scientific design concepts (Strand 2).

The worlds of virtual reality (a computer-simulated environment) and augmented reality (an environment that is a combination of real-world and computer-generated information) pose particular problems for integrating interactivity. How can people who visit designed spaces in a group together share the same VR experience? What distinctions are made between mere navigational interaction and control over the VR environment? These two problems weaken many of the existing efforts to measure learning in VR environments. Participants report a high level of engagement and enjoyment, but few qualitative or quantitative measures have demonstrated conceptual learning. Further work is needed to demonstrate that this type of interactivity effectively mediates learning, along with serving as an attractor.

Technology holds great potential to support inquiry practices in designed spaces (Ansbacher, 1997). It has proven to be an effective scaffolding tool that helps learners engage in domain-specific inquiry (Strand 3) (Linn, Bell, and Davis, 2004). While no clear understanding of its contribution to learning is currently evident, novel technology may have a special appeal and unique potential. Visitors tend to use technology-based exhibits more frequently and for longer periods of time than traditional exhibits (Serrell and Raphling, 1992; Sandifer, 2003), a result that has been attributed to "technological novelty" (Sandifer, 2003). In an analysis of 47 visitors and 61 instances of visits to interactive exhibits, Sandifer sought to determine the characteristics of exhibits that sustained visitors' attention the longest. In a regression analysis, "technological novelty"—the presence of visible state-of-the-art devices or illustrating, through the use of technology, phenomena that would otherwise be impossible or laborious for visitors to explore on their own—was a significant factor in the variance of visitor holding time.

"A concern sometimes raised is that technology-based exhibits may reduce visitors' interactions with other exhibits or objects in the museum, or worse, replace authentic experiences" (Ucko and Ellenbogen, 2008, p. 246). However, studies suggest that well-designed technological tools can help people plan visits, instigate new interests, and stimulate them to seek out specific objects or experiences (Moussouri and Falk, 2002). There are also concerns that technology may decrease the social interaction that is a hallmark of learning in informal environments. The interfaces on technology-based exhibits, such as touch screens or joysticks, are often designed for one person (Flagg, 1991). Unless social interaction is prioritized in the design of technology-based exhibits, people will continue to be hampered in their efforts to use technology-based exhibits in social groups (Heath, vom Lehn, and Osborne, 2005).

A review of the literature on virtual museum visitors (Haley Goldman and Shaller, 2004) characterized the most common motivations for website visits as gathering information for an upcoming visit to the physical site, engaging in very casual browsing, self-motivated research for specific content information, and assigned research (such as a school or job assignment) for specific content information. Ucko and Ellenbogen (2008, p. 250) note that

> preliminary research suggests that the motivations for visits to museum Web sites of designed spaces differ significantly from motivations for visits to physical science museums (Haley Goldman and Shaller, 2004). Typical motivations for science museum visits include entertainment or recreation, social activity, education, a life-cycle event ("My mother always took me here, so now I take my children"), place ("We have to go to the Smithsonian

while we're in Washington, DC"), content interest, and practical reasons ("It's too cold to take the children to the park") (Moussouri, 1997; Rosenfeld, 1980).

As technology advances to provide users with new levels of control and authorship, significant evolution in motivations (Strand 1) for participation in virtual museums can be expected. Web 2.0 technologies, for example, enable users to create and modify their own media. One can imagine that motivations for accessing and using virtual collections will change as visitors have more control over the goals of their engagement and a wider variety of tools to use to provide input. Although there is little or no research on virtual designed spaces comparable to the identity-related motivation and psychological research in physical museums, research is growing on identity in virtual space, both on the Internet in general (Curtis, 1992; Donath and Boyd, 2004; Stone, 1996; Turkle, 1995, 2005) and, as previously discussed, in online gaming. Given the evidence that identity-related motivations strongly influence the learning and behavior of visitors to physical designed spaces, we think that an analogous situation may apply to the use of virtual designed spaces.

Designed spaces offer opportunities for overcoming some of the methodological problems involved in studying science learning outcomes associated with media. A preponderance of the research on outcomes is based on contrived circumstances in which the "dosage" of media is controlled. For example, researchers ask participants to agree to watch a television program for a predetermined period of time and conduct controlled testing to evaluate learning outcomes. While providing a starting point valuable, these studies do not get to the important question of what media learners will select of their own volition.

Because designed spaces have always employed media—whether an exhibition case, a diorama, a video, or an interactive—they provide a natural laboratory for seeing how learners select media. Virtual reality spaces, augmented-reality museum experiences, museum blogs, and podcasting art installations are just a few of the new media now appearing in science museums. Recent advances in research on the experiences of visitors' to designed environments provide a growing understanding of how and why visitors utilize these resources.

As discussed throughout this report, visitor motivations are particularly important. As media become further embedded in designed environments in both physical and virtual contexts, researchers must explore why visitors actually choose a particular medium and with what science-specific goals for learning. One component of this research must be to understand whether and how media satisfy people's identity-related motivations, especially in virtual learning environments which many believe offer unique opportunities for learners to explore or "try on" novel identities.

Ultimately, the goal of introducing new media technologies into designed science learning environments is not only to modernize the experience and space, but to significantly improve the quality of the visitor experience, including enhancing learning outcomes. Research on motivation has shown that both the quality of the visitor experience and the extent and breadth of learning outcomes are directly related to people's entering motivations. The better one understands the relationship between motivations and learning and how media-based experiences support these motivations (Strand 1), the more successful these efforts will be.

There are methodological obstacles to conducting research on "noncaptive" audiences, whether they are designed spaces visitors, television viewers, or web users. Studies often focus disproportionately on concerns related to usability, such as navigation. This emphasis can contribute to the ease with which users can access learning resources, but it obscures larger, more critical issues. For example, understanding how, why, and to what end people use science museum websites would help designers better select, organize, and present learning resources and activities. Understanding the impact of such experiences can also provide insights into how best to position this virtual resource in relation to the physical science museum, other museum websites, and complementary aspects of the learning infrastructure (e.g., books, magazines, television).

Designed spaces are also good sites for exploring the effects of new approaches to using media for creating, distributing, and incorporating content into informal settings. Hand-held personal data assistants and data probes, for example, have been used for years to extend science learning beyond the classroom. These devices are ideal for just-in-time learning (National Research Council, 2000) and field research (e.g., Gay, Reiger, and Bennington, 2002; Soloway et al., 1999). Many of these projects occupy an overlapping space between informal and formal environments, holding potential for linking these environments across the educational infrastructure.

Mobile devices such as cellular phones, PDAs, portable audio players, and radio frequency identification tags or transponders have the potential to enhance visitors experiences in designed spaces by supporting self-directed and customized learning at any time in any place. Ucko and Ellenbogen (2008) review the use and impact of mobile technology in designed paces. They note that the "challenges of integrating mobile technology into the museum experience are being addressed in numerous projects that are keeping pace with new developments in hardware and software (e.g., cell phone capabilities), as well as new uses for technology (e.g., podcasting)" (p. 248-249). Further research is needed on the potential for the devices to support learning.

## Programs for Science Learning

Several fundamental challenges common to after-school science programs can be alleviated or addressed through the integration of media programming. After-school programs are usually run on very limited budgets. In addition, recent shifts in the policy landscape have increased the emphasis on measurable academic and social growth for participants. Although many budgets have increased to support more focused programming, practitioners do not typically feel well trained to address the increasing academic demands and cite limited training and materials as limitations to their effectiveness (Nee, Howe, Schmidt, and Cole, 2006). Furthermore, many programs are committed to serving poor students, resulting in additional concerns for keeping costs of programs low.

A number of programs are integrating media to support children's and adolescents' academic and leisure-time engagement with science. After-school science programs have harnessed broadcast media (e.g., *Bill Nye the Science Guy*, *Cyberchase*, *Design Squad*) and digital media (e.g., *Kinetic City*). The Fifth Dimension Program (described in Chapter 6), though not science specific, offers an interesting example of an interactive, moderated virtual world.

Science learning media can support science and academic learning outcomes with relatively modest investments. For example, the Rockman Et Al (1996) evaluation of *Bill Nye the Science Guy* found that the popular television series was being used in after-school settings. The positive cognitive (Strand 2) and science practice gains (Strand 4) associated with participation were discussed earlier in this chapter. In addition, field researchers observed that children chose to watch the program over unmoderated free play activities with peers (Strand 1). They also observed upper elementary grade children holding sustained conversations among themselves about the program during and after viewing it with little facilitation from adults (Strand 5).

Boys were more likely to view *Bill Nye* regularly than girls. This observed gender disparity is perhaps one consequence of minimal instructional support and training. While leaning heavily on a standalone television program with little facilitation is not the optimal design for rich learning, the Rockman Et Al (1996) findings are interesting and suggest the possibility of developing strong, easily implemented after-school science learning programming.

We did not find studies exploring the use of media in adult programming.

# KEY THEMES

We have identified key ideas in particular segments of the literature on media and learning science in informal environments. Five cross-cutting themes or issues are raised by this literature:

1. Who uses media to learn science in informal environments?
2. The role of media in creating science identities.
3. Does format matter?
4. Science as a process.
5. The need for longitudinal and cross-media studies.

## Who Uses Media to Learn Science in Informal Environments?

Access is a universal challenge for educators, programs, and institutions concerned with science education. There are considerable inequities and related concerns in access to science learning media. Those who access science in informal settings are generally already interested in science, as participation in these environments is voluntary rather than mandatory (Crane, 1994).

The patterns of who participates are quite stark in broadcast media. Evidence suggests that individuals using media for science learning are privileged and highly educated. For example, PBS viewers "are 44 percent more likely than the average Joe to have a household income over $150,000; 39 percent more likely to have a graduate degree; and 177 percent more likely to have investments of $150,000 and up." And NPR listeners "are 152 percent more likely to own a home valued at $500,000 or more; 194 percent more likely to travel to France; and 326 percent more likely to read the *New Yorker*" (Schulz, 2005). It appears that both networks have a large, highly educated audience. While the evidence is not in, many people hold out hope that new media and new approaches to blending technologies will open up science learning to a more diverse population. The Internet provides unique ways to access and participate in scientific discussions that may suit the interests of groups that don't attend, for example to broadcast science media. The National Aeronautics and Space Administration, the National Geographic Society, *Scientific American*, and other reliable science sources regularly produce and distribute podcasts, audio/video recordings that can be listened to on personal computers and music devices. Some universities (e.g., University of California at Berkeley, Massachusetts Institute of Technology, Pennsylvania State University) record course lectures and make them available to the public. And there are science podcasts created by regular citizens who have knowledge to share with others (Strand 5).

This latter category of knowledgeable people sharing information over the Internet is an important trend as the act of producing ideas may invite new participants into science learning media. A website like the CommonCraft Show (http://www.commoncraft.com) presents opportunities for people to understand new computing trends through their explanatory video clips. Sites like dnatube.com and myjove.com hold collections of videotaped experiments for scientists and nonscientists to explore and learn from. Even the popular YouTube.com website, best known for its novelty videos, contains clips of real science content and experiments, often produced by individuals motivated to share what they know with the world.

We need to consider how to develop science media for informal settings that attract and retain more diverse audiences, who may be the people who benefit most from science content presented outside formal educational settings. Designers of informal media seem to understand how to attract audiences with preexisting interests in science. Better ways are needed to create and advertise programming that appeals to those who would typically avoid science for whatever reason.

## Questions of Identity

As designed environments become further embedded with media in both physical and virtual contexts, it is critical to align those technologies with an understanding of why visitors actually choose to use those resources with science-specific goals for learning. Too often, technologies are embraced because of an interest in advancing an institution's missions and agendas, without questioning how well those technologies actually serve the needs and interests of the museum's audiences. In other words, the success of new technologies in the museum context will partially be a consequence of its physical attributes, such as adaptability, availability, and usability, but of equal if not more importance will be whether the technology satisfies the situated, identity-related motivations of users. One clear need is to conduct research on visitors' identity-related motivations in virtual learning environments. Such research in the physical realm has begun to produce some very useful and provocative findings; parallel research in the virtual realm promises to do likewise.

## Does Format Matter?

The same science content can be presented in different media, leading to questions about whether the form and structure of a medium influences learning outcomes. For example, Richard Clark suggested that performance or efficiency gains attributed to particular media could be the result of instructional methods or novelty effects rather than anything unique about the media format (Clark, 1983). In other words, there may be reasons to choose a particular medium for content delivery, but it is the content itself that influences learner achievement. This led Clark to recommend that "researchers refrain from producing additional studies exploring the relationship between media and learning unless a novel theory is suggested" (Clark, 1983, p. 457).

A related body of research explores how different components in a multimedia environment interrelate to support learning (Mayer and Moreno,

2003; Moreno, 2006; Moreno and Mayer, 1999). Although Mayer and colleagues' cognitive model of multimedia learning has not been expressly tested and developed in informal environments, it may be of interest to the field. Researchers have developed this work through a series of laboratory-based studies in which they use computer technology to test the influence of (1) modality (text versus audio) and (2) spatial and temporal contiguity of particular media elements (e.g., the spatial and temporal proximity of text and a related video animation in a computer environment) on cognition. Through a series of controlled laboratory experiments primarily with college student volunteers, researchers have established that, when words and images are represented contiguously in time and/or space, the effectiveness of multimedia instruction increases, influencing recall (Mayer, 1989; Mayer, Steinhoff, Bower, and Mars, 1995) and transfer to novel problems (Mayer, 1997). Mayer and Anderson (1991, 1992) have also established that, in a laboratory setting, concurrent presentation of textual and graphic information (e.g., explanation of how a tire pump works and relevant imagery) is more conducive to recall and transfer than presentation of the same information in a series (e.g., text then graphic or graphics then text). They have also found that students presented with auditory verbal materials plus animations recalled more, solved problems better, and were better able to match the visual and verbal elements than those who learn with on-screen text plus animations. These ideas may be fruitful for design and further testing in informal environments for science learning.

There is also evidence that elements of certain media may help to focus learners on important issues. This feature can be useful for designed settings, in which opportunities to engage participants are often narrow and fleeting. For example, motion pictures use various cinematic techniques, such as panning and zooming, to help learners attend to filmed details that might otherwise go unnoticed during casual viewing (Salomon, 1994). Similar claims have been made about the interactive properties that computational media afford. For example, well-designed computer games and simulations have been touted by several researchers (Baillie and Percoco, 2001; Gee, 2007; Greenfield, 1984; Papert, 1980) as ideal spaces for learning science for a number of reasons: they enable learners to customize the learning environment; they situate learning in a more authentic context; they provide direct experiences and interaction with intangible, abstract, ideal, complex, or otherwise unavailable scientific phenomena; and they engage users in collaborative, active, and problem-based learning.

Perhaps more important is Robert Kozma's rebuttal to Clark's argument, especially in light of the strands discussed throughout this volume. Kozma (1991) argued that one could study learning occurred when the same message was presented in different media formats. But he emphasized the importance of theories of distributed cognition (Salomon, 1993) that describe how individual cognition is developed through interactions with peers and artifacts. For example, talking with expert scientists, teachers, and knowledgeable peers (Strand 5) can lead to greater knowledge of science. This land of learning can also take place when working with various objects, such as calculators, computers, and video clips.

Kozma's viewpoint accounts for the social contexts in which media are used. An episode of a children's television program can be used in multiple venues (e.g., classrooms, homes, after-school centers) for different purposes and lead to different learning outcomes. While some studies suggest that media formats have an impact on student learning, it is also worth considering if these studies have examined the possible effects of interactions with peers and the media objects.

## Science as a Process

Recent reforms in science education have been concerned with bringing rigorous scientific content into classrooms as well as introducing learners to the practices of scientific inquiry (American Association for the Advancement of Science, 1990; National Research Council, 1996, 2007). While traditional science learning is often thought of as acquiring concepts and terminology (Strand 1), inquiry reforms emphasize the need for students to perform tasks similar to those encountered in scientific practice (Strand 3): posing questions, generating and interpreting data, and developing conclusions based on their investigations (Linn, diSessa, Pea, and Songer, 1994). Developing deep understandings of science requires understanding the nature of scientific explanations, models, and theories as well as the practices used to generate these products (Strand 4). In other words, students should learn how to plan and conduct investigations of phenomena while also grounding these activities in specific theoretical frameworks related to particular scientific disciplines.

Despite these goals and recommendations, many science-related informal media focus on providing information about facts and phenomena. It may be easier to develop these types of programs than create materials that engage students in *doing* science. For example, science documentaries can present information about earthquakes and tsunamis, but the narrative flow of these programs might be compromised by including experiments for viewers to conduct. Furthermore, it would be difficult to know if viewers were actually performing these experiments in classrooms and homes.

As discussed above, few studies speak to the impact of science-related programs on the audience's understanding of science. For example, Hall, Esty, and Fisch (1990) investigated the impact of a television program, *Square One TV*, on children's problem-solving heuristics when working with complex mathematical problems. They found program viewers using more problem-solving actions and being more mathematically rigorous in post-test problems than nonviewers. Evaluations of *Bill Nye the Science Guy* have also been conducted to demonstrate differences between viewers and nonviewers' abilities to make observations and comparisons (Rockman Et Al, 1996). Studies like these suggest ways to assess whether informal media can model scientific processes for learners.

Some media are clearly well suited for engaging learners in the process of science (Strand 3). Computer- and web-based environments can present facts along with simulations that allow people to generate and test hypotheses. In some cases, learners become immersed in science and engineering by designing their own materials. For example, Resnick, Berg, and Eisenberg's (2000) *Beyond Black Boxes* project used small, programmable computers called Crickets to allow students to develop monitoring instruments to study scientific problems that interested them (e.g., how many birds visit a backyard bird feeder each day). A growing number of game environments or engines allow users to customize their gaming experiences by building and expanding game behavior. Some studies have found a range of skills that can be learned through this customization, including computer programming, software engineering, and mathematics (Harel, 1991; Hooper, 1998; Kafai, 1994; Seif El-Nasr and Smith, 2006; Seif El-Nasr et al., 2007).

Museum exhibits that incorporate media can also be created that focus on the process of science (Strand 3) rather than its findings, such as the *Mysteries of Çatalhöyük* exhibition at the Science Museum of Minnesota (Pohlman, 2004). Many of the media associated with programs, such as citizen science programs, are also focused on process rather than scientific fact (Bonney, 2004).

## Longitudinal and Cross-Media Studies

Many studies of learning science in informal environments look at single or a small number of exposures to a medium. These studies provide information about the effects of particular media instances, but there is a lack of research dealing with repeated exposures to programs over long periods of time. For example, how do people come to appreciate science after watching *NOVA* for 1, 3, or 12 months?

Little is known about how people learn about a single content or domain area across different media formats. To illustrate this, consider a child reading a book about dinosaurs at age 3. She may like the book and ask to read it many times. Sensing her excitement for dinosaurs, her parents may take her to a museum to see an exhibit on her fourth birthday. The parents may have also bought her several dinosaur models from a local toy store during that period. A television program on dinosaurs may air after the museum visit, providing more information. And, in the era of networked computing, the family may seek dinosaur information together on the Internet.

Crowley and Jacobs refer to the repeated exposure to a single topic across multiple media as an island of expertise (Crowley and Jacobs, 2002).

These islands begin to form with initial interests and ultimately develop into deep, rich knowledge about a particular domain. Chi and Koeske's (1983) study of a young dinosaur expert demonstrates how children can develop categorization and recall skills around a domain through repeated readings of books. Crowley and Jacobs build on this by suggesting that islands of expertise can also develop over time with exposure to different media formats and conversations with knowledgeable mentors and peers.

Researchers of learning science in informal environments need to consider the effects of long-term exposure to single, specific media exemplars (e.g., the cumulative effects of watching *Bill Nye the Science Guy* for a year) as well as multiple media formats presenting the same content in different ways (e.g., books, films, museum exhibits on dinosaurs). These studies are very difficult to envision and carry out. There are methodological obstacles to conducting research on "noncaptive" audiences, whether nature center visitors, television viewers, or web users. Exploring the repeated interaction of multiple media and venues would provide insights into how best to position virtual and physical resources for science learning, including better understanding of the relationship between designed spaces, websites, book, magazines, television, and digital entertainment.

## CONCLUSION

Science-related media are likely to continue to play a major role in the ways that people learn about science informally. The public often cites broadcast, print, and digital media as their major sources of scientific information. Media producers seek large audiences, and they have developed techniques to present scientific content in entertaining and engaging ways. These modes of engagement are aligned with Strand 1, helping learners develop initial interests in science. Studies of science media have also demonstrated effects on people's perceptions of science and scientists (Strand 4). In the best cases, they can portray science as an interesting practice, scientists as a diverse group of individuals who lead normal lives, and demonstrate the realities of scientific investigation.

## REFERENCES

Agosto, D. (2002). Bounded rationality and satisficing in young people's web-based decision making. *Journal of the American Society for Information Science and Technology, 53*(1), 16-27.

Aldrich, C. (2005). *Learning by doing: A comprehensive guide to simulations, computer games, and pedagogy in e-learning and other educational experiences.* San Francisco: Wiley.

American Association for the Advancement of Science. (1990). *Science for all Americans: Project 2061.* New York: Oxford University Press.

Ansbacher, T. (1997). If technology is the answer, what was the question? Technology and experience-based learning. *Hand to Hand, 11*(3), 3-6.

Baillie, C., and Percoco, G. (2001). A study of present use and usefulness of computer-based learning at a technical university. *European Journal of Engineering Education, 25*(1), 33-43.

Bainbridge, W.S. (2007). The scientific research potential of virtual worlds. *Science, 371*(5837), 472-476.

Bauer, M.W., and Gaskell, G. (Eds.). (2002). *Biotechnology: The making of a global controversy.* New York: Cambridge University Press.

Bauer, M.W., Durant, J., Ragnarsdottir, A., and Rudolfsdottir, A. (1995). *Science and technology in the British press, 1946-1990: The media monitor project, vols. 1-4* (Technical Report). London: Science Museum and Wellcome Trust for the History of Medicine.

Bauer, M.W., Petkova, K., Boyadjieva, P., and Gornev, G. (2006). Long-term trends in the public representation of science across the "Iron Curtain": 1946-1995. *Social Studies of Science, 36*(1), 99-131.

Bilal, D. (2002a). Children's use of the Yahooligans! web search engine: III. Cognitive and physical behaviors on fully self-generated search tasks. *Journal of the American Society for Information Science and Technology, 53*(13), 1170-1183.

Bilal, D. (2002b). Perspectives on children's navigation of the World Wide Web. *Online Information Review, 26*(2), 108-117.

Bonney, R. (2004). Understanding the process of research. In D. Chittenden, G. Farmelo, and B. Lewenstein (Eds.), *Creating connections: Museums and the public understanding of current research* (pp. 199-210). Walnut Creek, CA: AltaMira Press.

Bowman, R.F. (1982). A Pac-Man theory of motivation: Tactical implications for classroom instruction. *Educational Technology, 22*(9), 14-17.

Bransford, J.D., Franks, J.J., Vye, N.J., and Sherwood, R.D. (1989). New approaches to instruction: Because wisdom can't be told. In S. Vosniadou and A. Ortony (Eds.), *Similarity and analogical reasoning* (pp. 470-497). New York: Cambridge University Press.

Broks, P. (2006). *Understanding popular science.* London: Open University Press.

Brossard, D., and Shanahan, J. (2003). Do they want to have their say? Media, agricultural biotechnology, and authoritarian views of democratic processes in science. *Mass Communication and Society, 6*(3), 291-312.

Brossard, D., Scheufele, D., Kim, E., and Lewenstein, B.V. (2008). Religiosity as a perceptual filter: Examining processes of opinion formation about nanotechnology. Submitted to *Public Understanding of Science.*

Brown, J.S., and Duguid, P. (2000). *The social life of information.* Boston: Harvard Business School Press.

Burnham, J. (1987). *How superstition won and science lost: Popularizing science and health in the United States.* New Brunswick, NJ: Rutgers University Press.

Callanan, M., and Jipson, J.L. (2001). Explanatory conversations and young children's developing scientific literacy. In K. Crowley, C.D. Schunn, and T. Okeda (Eds.), *Designing for science: Implications from everyday, classroom, and professional settings* (pp. 21-49). Mahwah, NJ: Lawrence Erlbaum Associates.

Callanan, M.A., Jipson, J., and Soennichsen, M. (2002). Maps, globes, and videos: Parent-child conversations about representational objects. In S. Paris (Ed.), *Perspectives on object-centered learning in museums* (pp. 261-283). Mahwah, NJ: Lawrence Erlbaum Associates.

Chambers, D.W. (1983). Stereotypic images of the scientist: The draw-a-scientist-test. *Science Education, 67*(2), 255-265.

Chi, M.T., and Koeske, R.D. (1983). Network representations of a child's dinosaur knowledge. *Developmental Psychology, 19*, 29-39.

Clark, R.E. (1983). Reconsidering research on learning from media. *Review of Educational Research, 53*(4), 445-459.

Clifford, B.R., Gunter, B., and McAleer, J. (1995). *Television and children: Program evaluation, comprehension, and impact*. Mahwah, NJ: Lawrence Erlbaum Associates.

Colson, C. (2005). Accessing the microscopic world. *PLoS Biology, 3*(1), 27-29.

Crandall, R.W., and Sidak, J.G. (2006). *Video games: Serious business for America's economy*. Washington, DC: Entertainment Software Association.

Crane, V. (1994). An introduction to informal science learning and research. In V. Crane, H. Nicholson, M. Chen, and S. Bitgood (Eds.), *Informal science learning: What the research says about television, science museums, and community-based projects* (pp. 1-14). Dedham, MA: Research Communications.

Crowley, K., and Callanan, M.A. (1998). Identifying and supporting shared scientific reasoning in parent-child interactions. *Journal of Museum Education, 23*, 12-17.

Crowley, K., and Jacobs, M. (2002). Islands of expertise and the development of family scientific literacy. In G. Leinhardt, K. Crowley, and K. Knutson (Eds.), *Learning conversations in museums* (pp. 333-356). Mahwah, NJ: Lawrence Erlbaum Associates.

Curtis, P. (1992). Mudding: Social phenomena in text-based virtual realities. Part of the *Proceedings of Directions and Implications of Advanced Computing* (DIAC 92) Symposium, Berkeley, CA. Available: http://citeseerx.ist.psu.edu/viewdoc/summary?doi=10.1.1.51.6586 [accessed October 2008].

Dede, C., Ketelhut, D., and Ruess, K. (2002). Motivation, usability, and learning outcomes in a prototype museum-based multi-user virtual environment. In P. Bell, R. Stevens, and T. Satwicz (Eds.), *Proceedings of the fifth international conference of learning sciences* (pp. 406-408). Mahwah, NJ: Lawrence Erlbaum Associates.

Donath, J., and Boyd, D. (2004). Public displays of connection. *BT Technology Journal, 22*(4), 71-82.

Dornan, C. (1989). Science and scientism in the media. *Science as Culture, 7*, 101-121.

Dowmunt, T. (1993). Introduction. In T. Dowmunt (Ed.), *Channels of resistance: Global television and local empowerment* (pp. 1-15). London: BFI.

Efthimiou, C.J., and Llewellyn, R.A. (2006). Avatars of Hollywood in physical science. *Physics Teacher, 44*, 28-32.

Efthimiou, C.J., and Llewellyn, R.A. (2007). Cinema, Fermi problems, and general education. *Physics Education, 42*, 253-261.

Federal Communications Commission. (1996). *Policies and rules concerning children's television programming: Revision of programming policies for television broadcast stations* (MM Docket No. 93-48). Washington, DC: Author.

Feldon, D.F., and Gilmore, J. (2006). Patterns in children's online behavior and scientific problem-solving: A large-N microgenetic study. In G. Clarebout and J. Elen (Eds.), *Advances in studying and designing (computer-based) powerful learning environments* (pp. 117-125). Rotterdam, The Netherlands: Sense.

Fidel, R. (1999). A visit to the information mall: Web searching behavior of high school students. *Journal of the American Society for Information Science, 50*(1), 24-37.

Fisch, S.M. (2004). *Children's learning from educational television: Sesame Street and beyond.* Mahwah, NJ: Lawrence Erlbaum Associates.

Fisch, S.M. (2006). *Informal science education: The role of educational TV.* Presentation to the National Research Council Learning Science in Informal Environments Committee. Available: http://www7.nationalacademies.org/bose/LSIE%201_Meeting_Presentation_Fisch.pdf [accessed February 2009].

Fisch, S.M., Yotive, W., Brown, S.K.M., Garner, M.S., and Chen, L. (1997). Science on Saturday morning: Children's perceptions of science in educational and non-educational cartoons. *Journal of Educational Media, 23*(2/3), 157-168.

Fisher, M. (1997). The effect of humor on learning in a planetarium. *Science Education, 81*(6), 703-713 (Informal Science Education Special Issue).

Flagg, B.N. (1991). Visitors in front of the small screen. *Association of Science-Technology Centers Newsletter, 19*(6), 9-10.

Flagg, B.N. (2000). Impact of *Science Friday* on public radio member listeners. *The Informal Learning Review, 44*(Sept./Oct.), 6-7.

Flagg, B.N. (2002, June). *Earth and sky summative evaluation, study 1.* Multimedia Research Group unpublished report, San Jose, CA.

Flagg, B.N. (2005a). Beyond entertainment: Educational impact of films and companion materials. *The Big Frame, 22*(2), 50-56, 66.

Flagg, B.N. (2005b). Can 90 seconds of science make a difference? *Informal Learning Review, 75*(2-3), 2-22.

Foley, B.J., and La Torre, D. (2004). Who has why-pox: A case study of informal science education on the net. In Y.B. Kafai, W.A. Sandoval, and N. Enyedy (Eds.), *Proceedings of the sixth international conference on the learning sciences* (p. 598). Mahwah, NJ: Lawrence Erlbaum Associates.

Fox, S. (2006). *Online health search.* Washington, DC: Pew Internet and American Life Project.

Gamson, W.A., and Modigliani, A. (1989). Media discourse and public opinion on nuclear power: A constructionist approach. *American Journal of Sociology, 95*(1), 1-37.

Gaskell, G., and Bauer, M.W. (Eds.). (2001). *Biotechnology 1996-2000: The years of controversy.* London: Science Museum.

Gaskell, G., Bauer, M.W., Durant, J., and Allum, N.C. (1999). Worlds apart? The reception of genetically modified foods in Europe and the U.S. *Science, 285*(5426), 384-387.

Gay, G., Reiger, R., and Bennington, T. (2002). Using mobile computing to enhance field study. In R. Hall, T. Koschmann, and N. Miyake (Eds.), *CSCL 2, Carrying forward the conversation* (pp. 507-528). Mahwah, NJ: Lawrence Erlbaum Associates.

Gee, J.P. (2003). *What video games have to teach us about learning and literacy.* New York: Palgrave Macmillan.

Gee, J.P. (2007). *Good video games and good learning: Collected essays on video games, learning, and literacy*. New York: Peter Lang.

Gerbner, G. (1987). Science on television: How it affects public conceptions. *Issues in Science and Technology, 3*(3), 109-115.

Gleason, M.E., and Schauble, L. (2000). Parents' assistance of their children's scientific knowledge. *Cognition and Instruction, 17*(4), 343-378.

Greenfield, P.M. (1984). *Mind and media: The effects of television, video games, and computers*. Cambridge, MA: Harvard University Press.

Gunter, B., and McAleer, J. (1997). *Children and television* (2nd ed.). London: Routledge.

Haefner, M.J., and Wartella, E.A. (1987). Effects of sibling coviewing on children's interpretations of television programming. *Journal of Broadcasting and Electronic Media, 31*(2), 153-168.

Haley Goldman, K., and Shaller, D. (2004). Exploring motivational factors and visitor satisfaction in on-line museum visits. In D. Bearman and J. Trant (Eds.), *Museums and the web 2004*. Toronto, ON: Archives and Museum Informatics.

Hall, E.R., Esty, E.T., and Fisch, S.M. (1990). Television and children's problem-solving behavior: A synopsis of an evaluation of the effects of Square One TV. *Journal of Mathematical Behavior, 9*, 161-174.

Hansen, A., and Dickinson, R. (1992). Science coverage in the British mass media: Media output and source input. *Communication, 17*, 365-377.

Harel, I. (1991). *Children designers: Interdisciplinary constructions for learning and knowing mathematics in a computer-rich school*. Norwood, NJ: Ablex.

Hartley, J. (1999). *Uses of television*. London: Routledge.

Haynes, R.D. (1994). *From Faust to Strangelove: Representations of the scientist in western literature*. Baltimore: Johns Hopkins University Press.

Heath, C., and vom Lehn, D. (2008). Construing interactivity: Enhancing engagement and new technologies in science centres and museums. *Social Studies of Science, 38*(1), 63-91.

Heath, C., von Lehm, D., and Osborne, J. (2005). Interaction and interactives: Collaboration and participation with computer-based exhibits. *Public Understanding of Science, 14*(1), 91-101.

Hendershot, C. (1997). The atomic scientist, science fiction films, and paranoia: The Day the Earth Stood Still, This Island Earth, and Killers from Space. *Journal of American Culture, 20*(1), 31-41.

Hooper, P.K. (1998). *They have their own thoughts: Children's learning of computational ideas from a cultural constructionist perspective*. Unpublished doctoral dissertation, Massachusetts Institute of Technology.

Horrigan, J. (2006). *The Internet as a resource for news and information about science*. Washington, DC: Pew Internet and American Life Project.

Houck, M.M. (2006). *CSI: Reality. Scientific American, 295*(1), 84-89.

Huston, A.C., Zuckerman, D., Wilcox, B.L., Donnerstein, E., Fairchild, H., Feshbach, N.D., Katz, P.A., Murray, J.P., and Rubinstein, E.A. (1992). *Big world, small screen: The role of television in American society*. Lincoln: University of Nebraska Press.

Jones, B. (2001-2002). Recommendations for implementing internet inquiry projects. *Journal of Educational Technology Systems, 30*(3), 271-291.

Jones, R.A. (1997). The boffin: A stereotype of scientists in post-war British films (1945-1970). *Public Understanding of Science, 6*(1), 31-48.

Jones, R.A. (2001). "Why can't you scientists leave things alone?" Science questioned in British films of the post-war period (1945-1970). *Public Understanding of Science, 10*, 365-382.

Jones, R., and Bangert, A. (2006). The *CSI* effect: Changing the face of science. *Science Scope, 30*(3), 38-42.

Kafai, Y.B. (1994). *Minds in play: Computer game design as a context for children's learning.* Mahwah, NJ: Lawrence Erlbaum Associates.

Kane, T.T., Gueye, M., Speizer, I., Pacque-Margolis, S., and Baron, D. (1998). The impact of a family planning multimedia campaign in Bamako, Mali. *Studies in Family Planning, 29*(3), 309-323.

Karlyn, A.S. (2001). The impact of a targeted radio campaign to prevent STIs and HIV/AIDS in Mozambique. *AIDS Education and Prevention, 13*(5), 438-451.

Kirby, D.A. (2003a). Scientists on the set: Science consultants and communication of science in visual fiction. *Public Understanding of Science, 12*(3), 261-278.

Kirby, D.A. (2003b). Science consultants, fictional films and scientific practice. *Social Studies of Science, 33*(2), 1-38.

Kirshner, J. (2001). Subverting the cold war in the 1960s: Dr. Strangelove, The Manchurian Candidate, and The Planet of the Apes. *Film and History, 31*(2), 40-44.

Kozma, R. (1991). Learning with media. *Review of Educational Research, 61*(2), 179-211.

Kubey, R., and Csikszentmihalyi, M. (1990). *Television and the quality of life: How viewing shapes everyday experience.* Mahwah, NJ: Lawrence Erlbaum Associates.

Kuiper, E., Volman, M., and Terwel, J. (2005). The web as an information resource in K-12 education: Strategies for supporting students in searching and processing information. *Review of Educational Research, 75*(3), 258-328.

LaFollette, M.C. (1990). *Making science our own: Public images of science 1910-1955.* Chicago: University of Chicago Press.

LaFollette, M.C. (in press). Scientific and technical publishing in the United States, 1880-1950. In *A history of the book in America (1880-1950)* (vol. 4). Chapel Hill: University of North Carolina Press.

Large, J.A., and Beheshti, J. (2000). The web as a classroom resource: Reactions from the users. *Journal of the American Society for Information Science, 51*(12), 1069-1080.

Lewenstein, B.V. (in press). Science books since World War II. In D.P. Nord, M. Schudson, and J. Rubin (Eds.), *The enduring book: Publishing in post-war America.* Chapel Hill: University of North Carolina Press.

Lindberg, C.M., Carstensen, E.L., and Carstensen, L.L. (2007). *Lifelong learning and technology.* Background paper for the Committee on Learning Science in Informal Environments. Available: http://www7.nationalacademies.org/bose/Lindberg_et%20al_Commissioned_Paper.pdf [accessed October 2008].

Linn, M.C., Bell, P., and Davis, E.A. (2004). Specific design principles: Elaborating the scaffolded knowledge integration framework. In M.C. Linn, E.A. Davis, and P. Bell (Eds.), *Internet environments for science education* (pp. 315-340). Mahwah, NJ: Lawrence Erlbaum Associates.

Linn, M.C., diSessa, A., Pea, R.D., and Songer, N.B. (1994). Can research on science learning and instruction inform standards for science education? *Journal of Science Education and Technology, 3*(1), 7-15.

Madden, M., and Fox, S. (2006). *Finding answers online in sickness and in health.* Washington, DC: Pew Internet and American Life Project.

Mayer, R.E. (1989). Systematic thinking fostered by illustrations in scientific text. *Journal of Educational Psychology, 81*(2), 240-246.

Mayer, R.E. (1997). Multimedia learning: Are we asking the right questions? *Educational Psychologist, 32*(1), 1-19.

Mayer, R.E., and Anderson, R.B. (1991). Animations need narrations: An experimental test of a dual-dual coding hypothesis. *Journal of Educational Psychology, 83*(4), 484-490.

Mayer, R.E., and Anderson, R.B. (1992). The instructive animation: Helping students build connections between words and pictures in multimedia learning. *Journal of Educational Psychology, 84*(4), 444-452.

Mayer, R.E., and Moreno, R. (2003). Nine ways to reduce cognitive load in multimedia learning. In R. Bruning, C.A. Horn, and L.M. PytlikZillig (Eds.), *Web-based learning: What do we know? Where do we go?* (pp. 23-44). Greenwich, CT: Information Age.

Mayer, R.E., Steinhoff, K., Bower, G., and Mars, R. (1995). A generative theory of textbook design: Using annotated illustrations to foster meaningful learning of science text. *Educational Technology Research and Development, 43*(1), 31-43.

Moreno, R. (2006). Does the modality principle hold for different media? A test of the method affects learning hypothesis. *Journal of Computer Assisted Learning, 22*(3), 149-158.

Moreno, R., and Mayer, R.E. (1999). Cognitive principles of multimedia learning: The role of modality and contiguity. *Journal of Educational Psychology, 91*, 358-368.

Moussouri, T., (1997). *Family agendas and family learning in hands-on museums.* Unpublished doctoral dissertation, University of Leicester.

Moussouri, T. (2003). Negotiated agendas: Families in science and technology museums. *International Journal of Technology Management, 25*(5), 477-489.

National Research Council. (1996). *National science education standards.* National Committee on Science Education Standards and Assessment. Washington, DC: National Academy Press.

National Research Council. (2000). *How people learn: Brain, mind, experience, and school* (expanded ed.). Committee on Developments in the Science of Learning. J.D. Bransford, A.L. Brown, and R.R. Cocking (Eds.). Washington, DC: National Academy Press.

National Research Council. (2007). *Taking science to school: Learning and teaching science in grades K-8.* Committee on Science Learning, Kindergarten Through Eighth Grade. R.A. Duschl, H.A. Schweingruber, and A.W. Shouse (Eds.). Board on Science Education, Center for Education, Division of Behavioral and Social Sciences and Education. Washington, DC: The National Academies Press.

National Science Board. (2008). Science and technology: Public attitudes and public understanding. In *Science and Engineering Indicators—2008.* Washington, DC: Author.

Nee, J., Howe, P., Schmidt, C., and Cole, P. (2006). *Understanding the afterschool workforce: Opportunities and challenges for an emerging profession.* Report prepared by the National Afterschool Association. Houston: Cornerstones for Kids. Available: http://nextgencoalition.org/files/NAA_PDF_rw111506.pdf [accessed October 2008].

Nelkin, D. (1987). *Selling science: How the press covers science and technology.* New York: Freeman.

Neulight, N., Kafai, Y.B., Kao, L., Foley, B., and Galas, C. (2007). Children's participation in a virtual epidemic in the science classroom: Making connections to natural infectious diseases. *Journal of Science Education and Technology, 16*(1), 47-58.

Newsom, C. (Ed.) (1952). *A television policy for education: Proceedings of the Television Programs Institute held under the auspices of the American Council on Education at Pennsylvania State College.* Washington, DC: American Council on Education.

Ng, W., and Gunstone, R. (2002). Students' perceptions of the effectiveness of the World Wide Web as a research and teaching tool in science learning. *Research in Science Education, 32,* 489-510.

Nisbet, M.C., and Goidel, R.K. (2007). Understanding citizen perceptions of science controversy: Bridging the ethnographic-survey research divide. *Public Understanding of Science, 16*(4), 421-440.

Nisbet, M.C., and Huge, M. (2006). Attention cycles and frames in the plant biotechnology debate: Managing power participation through the press/policy connection. *Harvard International Journal of Press/Politics, 11*(2), 3-40.

Nisbet, M.C., Brossard, D., and Kroepsch, A. (2003). Framing science: The stem cell controversy in an age of press/politics. *Harvard International Journal of Press/Politics, 8*(2), 36-70.

Papert, S. (1980). *Mindstorms: Children, computers, and powerful ideas.* New York: Basic Books.

Piotrow, P.T., Kincaid, D.L., Rimon, J.G., and Rinehart, W. (1997). *Health communication: Lessons from family planning and reproductive health.* Westport, CT: Praeger.

Piotrow, P., Rimon, J., Winnard, K., Kincaid, D.L., Huntington, D., and Convisser, J. (1990). Mass media family planning promotion in three Nigerian cities. *Studies in Family Planning, 21*(5), 265-274. Available: http://www.jstor.org/stable/1966506?seq=1 [accessed October 2008].

Podlas, K. (2006, July). *Investigating the CSI effect's impact on jurors.* Paper presented at the Law and Society Association annual meeting. Available: http://www.allacademic.com/meta/p95911_index.html [accessed October 2008].

Pohlman, D. (2004). Catching science in the act: Mysteries of Çatalhöyük. In D. Chittenden, G. Farmelo, and B.V. Lewenstein (Eds.), *Creating connections: Museums and the public understanding of research* (pp. 267-275). Walnut Creek: AltaMira Press.

Prensky, M. (2000). *Digital game-based learning.* New York: McGraw-Hill.

Priest, S.H. (2001). Misplaced faith: Communication variables as predictors of encouragement for biotechnology development. *Science Communication, 23*(2), 97-110.

Reiser, R.A., Tessmer, M.A., and Phelps, P.C. (1984). Adult-child interaction in children's learning from Sesame Street. *Educational Communications and Technology, 32*(4), 217-233.

Reiser, R.A., Williamson, N., and Suzuki, K. (1988). Using Sesame Street to facilitate children's recognition of letters and numbers. *Educational Communication and Technology Journal, 36*(1), 15-21.

Resnick, M., Berg, R., and Eisenberg, M. (2000). Beyond black boxes: Bringing transparency and aesthetics back to scientific investigation. *Journal of the Learning Sciences, 9*(1), 7-30.

Rockman Et Al. (1996). *Evaluation of Bill Nye the Science Guy: Television series and outreach.* San Francisco: Author. Available: http://www.rockman.com/projects/124. kcts.billNye/BN96.pdf [accessed October 2008].

Rockman Et Al. (2007). *Media-based learning science in informal environments.* Background paper for the Learning Science in Informal Environments Committee of the National Research Council. Available: http://www7.nationalacademies.org/ bose/Rockman_et%20al_Commissioned_Paper.pdf [accessed October 2008].

Rose, C. (2003). How to teach biology using the movie science of cloning people, resurrecting the dead, and combining flies and humans. *Public Understanding of Science, 12*(3), 289-296.

Rosenfeld, S. (1980). *Informal education in zoos: Naturalistic studies of family groups.* Unpublished doctoral dissertation, University of California, Berkeley.

Roussou, M., Johnson, A., Moher, T., Leigh, J., Vasilakis, C., and Barnes, C. (1999). Learning and building together in an immersive virtual world. *Presence: Teleoperators and Virtual Environments Journal, 8*(3), 247-263.

Salomon, G. (1977). Effects of encouraging Israeli mothers to co-observe Sesame Street with their five year-olds. *Child Development, 48*(3), 1146-1151.

Salomon, G. (Ed.). (1993). *Distributed cognitions: Psychological and educational considerations.* New York: Cambridge University Press.

Salomon, G. (1994). *Interaction of media, cognition, and learning: An exploration of how symbolic forms cultivate mental skills and affect knowledge acquisition* (2nd ed.). Mahwah, NJ: Lawrence Erlbaum Associates.

Sandifer, C. (2003). Technological novelty and open-endedness: Two characteristics of interactive exhibits that contribute to the holding of visitor attention in a science museum. *Journal of Research in Science Teaching, 40*(2), 121-137.

Scanlon, E., Jones, A., and Waycott, J. (2005). Mobile technologies: Prospects for their use in learning in informal science settings. *Journal of Interactive Media in Education, 21*(5), 1-17.

Scheufele, D.A., and Lewenstein, B.V. (2005). The public and nanotechnology: How citizens make sense of emerging technologies. *Journal of Nanoparticle Research, 7*(6), 659-667.

Schulz, W. (2005). *The "assault" on public broadcasting.* Available: http://www.cpb. org/ombudsmen/display.php?id=7 [accessed October 2008].

Seif El-Nasr, M., and Smith, B.K. (2006). Learning through game modding. *ACM Computers in Entertainment, 4*(1), Article 3B.

Seif El-Nasr, M., Yucel, I., Zupko, J., Tapia, A., and Smith, B.K. (2007). Middle-to-high school girls as game designers—What are the learning implications? In I. Parberry (Ed.), *Proceedings of the 2nd annual microsoft academic days conference on game development in computer science education* (pp. 54-58), Available: http://www.eng.unt.edu/ian/Cruise2007/madgdcse2007.pdf [accessed April 2009].

Serrell, B., and Raphling, B. (1992). Computers on the exhibit floor. *Curator, 35*(3), 181-189.

Shaffer, D.W. (2006). *How computer games help children learn*. New York: Palgrave Macmillan.

Sherry, J.L. (1997). *Do violent video games cause aggression? A meta-analytic review*. Top student paper, Instructional and Developmental Communication Division, International Communication Association Annual Convention, Montreal, Quebec.

Singer, E., and Endreny, P.M. (1993). *Reporting on risk*. New York: Russell Sage Foundation.

Singhal, A., and Rogers, E. (1989). *India's information revolution*. New Delhi: Sage/India.

Singhal, A., and Rogers, E. (1999). *Entertainment-education: A communication strategy for social change*. Mahwah, NJ: Lawrence Erlbaum Associates.

Sobchak, V. (2004). *Screening space: The American science fiction film*. New Brunswick, NJ: Rutgers University Press.

Soloway, E., Grant, W., Tinker, R., Roschelle, J., Mills, M., Resnick, M., Berg, R., and Eisenberg, M. (1999). Science in the palms of their hands. *Communications of the ACM, 42*(8), 21-26.

Squire, K. (2003). Video games in education. *International Journal of Intelligent Games and Simulation, 2*(1), 49-62.

Squire, K. (2006). From content to context: Videogames as designed experience. *Educational Researcher, 35*(8), 19-29.

Squire, K.D., and Barab, S.A. (2004). Replaying history. In Y. Kafai, W.A. Sandoval, N. Enyedy, A. Dixon, and F. Herrera (Eds.), *Proceedings of the 2004 international conference of the learning sciences* (pp. 505-512). Mahwah, NJ: Lawrence Erlbaum Associates.

Squire, K.D., and Steinkuehler, C.A. (2001). Generating cyberculture/s: The case of star wars galaxies. In D. Gibbs and K.L. Krause (Eds.), *Cyberlines: Languages and cultures of the Internet* (2nd ed.). Albert Park, Australia: James Nicholas.

Steinke, J. (1997). Portrait of a woman as a scientist: Breaking down barriers created by gender-role stereotypes. *Public Understanding of Science, 6*(4), 409-428.

Steinke, J. (1999). Women scientist role models on screen: A case study of contact. *Science Communication, 21*(2), 111-136.

Steinke, J. (2005). Cultural representations of gender and science: Portrayals of female scientists and engineers in popular films. *Science Communication, 27*(1), 27-63.

Steinke, J., and Long, M. (1996). A lab of her own? Portrayals of female characters on children's educational science programs. *Science Communication, 18*(2), 91-115.

Stone, A.R. (1996). *The war of desire and technology at the close of the mechanical age*. Cambridge, MA: MIT Press.

Storksdieck, M. (2005). *Field trips in environmental education*. Berlin: Berliner Wissenschafts-Verlag.

Ten Eyck, T.A. (1999). Shaping a food safety debate: Control efforts of newspaper reporters and sources in the food irradiation controversy. *Science Communication, 20*(4), 426-447.

Ten Eyck, T.A. (2005). The media and public opinion on genetics and biotechnology: Mirrors, windows, or walls? *Public Understanding of Science, 14*(3), 305-316.

Ten Eyck, T.A., and Williment, M. (2003). The national media and things genetic: Coverage in the *New York Times* (1971-2001) and *The Washington Post* (1977-2001). *Science Communication, 25*, 129-152.

Turkle, S. (1995). *Life on the screen: Identity in the age of the Internet*. New York: Simon and Schuster.

Turkle, S. (2005). *The second self: Computers and the human spirit*. Cambridge, MA: MIT Press.

Turow, J. (1989). *Playing doctor: Television, storytelling and medical power*. New York: Oxford University Press.

Turow, J., and Gans, R. (2002). *As seen on TV: Health policy issues on TV's medical dramas*: Report to the Kaiser Family Foundation. Available: http://www.kff.org/entmedia/John_Q_Report.pdf [accessed October 2008].

Ucko, D.A., and Ellenbogen, K.M. (2008). Impact of technology on informal science learning. In D.W. Sunal, E. Wright, and C. Sundberg (Eds.), *The impact of the laboratory and technology on learning and teaching science K-16* (Ch. 9, pp. 239-266). Charlotte, NC: Information Age.

Ucko, D.A., Schreiner, R., and Shakhashiri, B.Z. (1986). An exhibit on everyday chemistry: Communicating chemistry to the public. *Journal of Chemical Education, 63*, 1081.

U.S. Congress. (1990). *Children's Television Act of 1990* (P.L. no. 101-437, 104 Stat. 996-1000).

Valente, T.W., and Saba, W. (1998). Mass media and interpersonal influence in a reproductive health communication campaign in Bolivia. *Communication Research, 25*, 96-124.

Valente, T.W., Kim, Y.M., Lettenmaier, C., Glass, W., and Dibba, Y. (1994). Radio and the promotion of family planning in the Gambia. *International Family Planning Perspectives, 20*(3), 96-100.

Valente, T.W., Poppe, P.R., and Merritt, A.P. (1996) Mass media-generated interpersonal communication as sources of information about family planning. *Journal of Health Communication, 1*, 247-265.

Valente, T.W., Watkins, S., Jato, M.N., Van der Straten, A., and Tsitsol, L.M. (1997). Social network associations with contraceptive use among Cameroonian women in voluntary associations. *Social Science and Medicine, 45*, 677-687.

Vieth, E. (2001). *Screening science: Contexts, texts, and science in fifties science fiction film*. Lanham, MD: Scarecrow Press.

Wallace, R.M., Kupperman, J., Krajcik, J., and Soloway, E. (2000). Science on the web: Students online in a sixth-grade classroom. *Journal of the Learning Sciences, 9*(1), 75-104.

Walters, L.M., Wilkins, L., and Walters, T. (Eds.). (1989). *Bad tidings: Communication and catastrophe*. Mahwah, NJ: Lawrence Erlbaum Associates.

Watson, J. (1998). If you don't have it, you can't find it: A close look at students' perceptions of using technology. *Journal of the American Society for Information Science, 49*(11), 1024-1036.

Weingart, P., and Pansegrau, P. (2003). Introduction: Perception and representation of science in literature and fiction film. *Public Understanding of Science, 12*(3), 227-228.

Weiss, C.H., and Singer, E. (1988). *Reporting of social science in the national media.* New York: Russell Sage Foundation.

Wilkins, L., and Patterson, P. (Eds.). (1990). *Risky business: Communicating issues of risk, science, and public policy.* Westport, CT: Greenwood Press.

Yu, K.C. (2005). Digital full domes: The future of virtual astronomy education. *Planetarian, 34*(3), 6-11.

# Part IV

# Conclusions, Recommendations, and Future Directions

# 9

# Conclusions and Recommendations

Learning science in informal environments is a vast and expanding area of study and practice that supports a broad range of learning experiences. Informal environments for science learning include not only science centers and museums but also a much broader array of settings, ranging from family discussions at home to everyday activities like using the Internet, watching television, gardening, participation in organizations like the Girl Scouts and Boy Scouts, and recreational activities like hiking and fishing.

Each year tens of millions of Americans, young and old, choose to visit informal science learning institutions, participate in programs, and use media outlets to pursue their interest in science. Thousands of organizations dedicate themselves to building, documenting, and improving informal science learning for learners *of all ages and backgrounds*. They include informal learning and community-based organizations, think tanks, institutions of higher education, private companies, government agencies, and philanthropic foundations. And through after-school programs and field trips, schools facilitate science learning in informal environments on a broad scale.

Virtually all people of all ages and backgrounds engage in informal science learning in the course of daily life. Informal environments can stimulate science interest, build learners' scientific knowledge and skill, and—perhaps most importantly—help people learn to be more comfortable and confident in their relationship with science. Researchers and educators interested in informal settings are typically committed to open participation in science: building and understanding science learning experiences that render science accessible to a broad range of learners. There is increasing interest in understanding cultural variability among learners and its implications: how learners participate in science and the intersections of values, attitudes, histories, and practices that are evident in learner and scientific communities. Accordingly, two notions of the culture of science underlie the committee's conclusions and the recommendations that follow.

In one sense, there is a culture of science in that science involves specialized practices for exploring questions through evidence (e.g., the use of statistical tests, mathematical modeling, instrumentation) which people must acquire if they wish to enter the formal domains of science. This first sense of the culture of science also includes social practices such as peer review, publication, and debate. In a second sense, science reflects the cultural values of those who engage in it—in terms of choices about what is worthy of attention, differing perspectives on how to approach various problems, and so on. From this latter perspective, as is the case with any cultural endeavor, differences in norms and practices within and across fields reflect not only the varying subject matters of interest but also the identities and values of the participants. The recognition that science is a cultured enterprise implies that there is no cultureless or neutral perspective on science, nor on learning science—any more than a photograph or painting can be without perspective. Thus, diversity of perspectives is beneficial both to science and to the understanding of learning. It also stands as a potential resource for the design of informal environments for science learning.

This chapter presents the committee's conclusions and recommendations for research and practice. We begin with conclusions drawn from the research reviewed by the committee, beginning with evidence about learners and learning, and then move on to informal learning settings and how to broaden participation in science learning. Finally, we outline our recommendations for practice and research that flow from our conclusions.

## LEARNERS AND LEARNING

**Conclusion 1: Across the life span, from infancy to late adulthood, individuals learn about the natural world and develop important skills for science learning.**

As the committee discussed in Chapter 4, a vast literature documents young children's learning about the natural world. Even infants observe regularities in the world and build tacit understandings that help them reliably anticipate physical phenomena and create order in their experience. Very young children learn a great deal about the natural world in the first few years. They notice changes in the world around them (flowers blooming, the moon changing shape, snow melting, airplanes flying overhead), they learn the names of objects and processes, and they engage in learning conversations with other people about these events. Children extend these early experiences by engaging with science-related media, asking spontaneous questions of adults and peers, making predictions, evaluating evidence, and building explanations of changes in the physical world. Throughout the school years, children and adolescents, with support, can piece together school-based and informal learning experiences to build scientific understanding of the natural world.

The drive to understand and explain the world continues into adulthood. As discussed in "Who Learns in Everyday Settings" (Chapter 4), over the life span, additional motivations stimulate science learning: pragmatic needs (e.g., dealing with health care issues, academic tasks, local environmental concerns), science-rich hobbies, and workplace tasks. Like children, adults (including scientists) pursue questions of personal interest and assemble evidence from their everyday experience to develop their understanding of the world. Increased memory capacity, reasoning, and metacognitive skills that come with maturation enable adult learners to explore science in new ways, summarized in Chapter 6 in the section "Programs for Older Learners." Senior citizens retain many of these capabilities, and as they mature their interests change. Informal environments are of fundamental importance for supporting science learning by adults, particularly because they thrive in environments that acknowledge their needs and life experiences.

**Conclusion 2: A great deal of science learning, often unacknowledged, takes place outside school in informal environments—including everyday activity, designed spaces, and programs—as individuals navigate across a range of social settings.**

Most people routinely circulate through a range of social settings that can support science learning. The committee found abundant evidence of learning in everyday life experiences, designed educational settings, and programs.

As discussed in Chapter 4, as individuals interact with the natural world and participate in family and community life, they develop knowledge about nature and about science-relevant interests and skills. Long-term, sophisticated science learning can occur through the individual and social processes (e.g., mentorship, reading scientific texts, watching educational television) associated with science-related elective pursuits and hobbies—for example, amateur astronomy clubs, robot-building leagues, and conservation groups.

Designed settings—including museums, science centers, zoos, aquariums, and nature centers—can also support science learning. Rich with educationally framed real-world phenomena, these are places where people can pursue and develop science interests, engage in science inquiry, and reflect on their experiences through conversations. There has been very little synthesis of this research to date. However, the committee compiled and reviewed extensive evidence from visitor studies, program evaluations, and design studies (Chapter 5) that sketch out the empirical evidence and the promise of designed settings for science learning.

Programs for science learning are offered through informal learning institutions, schools, community-based organizations, and private companies. As discussed in Chapter 6, there is mounting evidence that such experiences can stimulate and enhance the science-specific interests of adults and children. There is also some evidence that participation in informal programs for science learning, such as those involving networks of science volunteers in data collection (e.g., citizen science programs for tracking migrations, environmental monitoring and clean-up), can promote informed civic engagement on science-related issues, such as local environmental concerns and policies.

Science is also receiving more emphasis in out-of-school-time programs (clubs, after-school and summer programs, scouts) as part of an increased focus on academic subjects for school-age learners during nonschool hours (see Conclusion 12). With increased public and private funding, existing programs are adopting a science focus, and new science initiatives are being developed. In our review of this literature in Chapter 6, we found that the current evidence base of science-specific learning in these programs is limited to data from individual program evaluations. These studies suggest that science programs can make important contributions to students' understanding of scientific and mathematical concepts, their ability to think scientifically, and their use of scientific language and tools. They also can be effective in improving students' attitudes toward science and toward themselves as science learners.

> **Conclusion 3: Learning science in informal environments involves developing positive science-related attitudes, emotions, and identities; learning science practices; appreciating the social and historical context of science; and cognition. Informal environments can be particularly important for developing and validating learners' positive science-specific interests, skills, emotions, and identities.**

The committee outlined six strands of science learning that encompass a broad, interrelated network of knowledge and capabilities that learners can develop in these environments. In Chapters 4, 5, and 6 we use the strands to organize our review of the literature in order to illustrate the ways in which research supports these particular learning outcomes. The strands are statements about what learners do when they learn science, reflecting the practical as well as the more abstract, conceptual, and reflective aspects of science learning. Learners in informal environments:

Strand 1: Experience excitement, interest, and motivation to learn about phenomena in the natural and physical world.

Strand 2: Come to generate, understand, remember, and use concepts, explanations, arguments, models and facts related to science.

Strand 3: Manipulate, test, explore, predict, question, observe, and make sense of the natural and physical world.

Strand 4: Reflect on science as a way of knowing; on processes, concepts, and institutions of science; and on their own process of learning about phenomena.

Strand 5: Participate in scientific activities and learning practices with others, using scientific language and tools.

Strand 6: Think about themselves as science learners and develop an identity as someone who knows about, uses, and sometimes contributes to science.

The strands are distinct from, but necessarily overlap with, the science-specific knowledge, skills, attitudes, and dispositions that can be developed in schools. Specifically, a previous National Research Council report (2007) on K-8 science learning, *Taking Science to School*, proposed a four-strand framework from which the current six-strand model evolved. By building on that four-strand framework, we underscore that the goals of schools and informal, nonschool settings are both overlapping and complementary. The two additional strands—Strands 1 and 6—are prominent and of special value in informal learning environments. Strands 2 through 5 are explained in greater detail in Chapter 3.

Strand 1, which focuses on the development of interest and motivation to learn through interaction with phenomena in the natural and designed world, is fundamental. Strand 1 emphasizes the importance of building on prior interests and motivations by allowing learners choice and agency in their learning. Strand 1 is particularly relevant to informal environments that are rich with phenomena—a local stream, backyard insects, a museum exhibit illustrating Newtonian physics, watching pigeons downtown, ranger-led national park tours. Such phenomena often inspire scientific inquiry for scientists and nonscientists alike. They often serve as an "on ramp" to help the learner build familiarity with the natural and designed world and to establish the experience base, motivation, and knowledge that fuel and inform later science learning experiences.

Strand 6 is another strand that is particularly important to informal environments, addressing how learners view themselves with respect to science—their "science learner identity." This strand speaks to the process by which some individuals come to view themselves and come to be socially recognized as comfortable with, knowledgeable about, interested in, and capable of engaging in science. Learning in this strand is the consequence of multiple science learning experiences across settings over significant time scales (i.e., weeks, months, years), reflecting multiple opportunities for learners to participate in science. It is important to note, however, that one's identity as a science learner is also shaped by factors that may be external to or beyond experiences with science, such as social expectations and stereotypes. Informal environments have the potential to promote nondiscriminatory expectations for learners and nonstereotyped views of participant groups and their capabilities in science, to support identity development.

The strands may also facilitate developing shared enterprises between informal learning environments and schools. For example, there is currently extensive work being undertaken to develop and test learning progressions (National Research Council, 2007). A learning progression organizes science learning so that learners revisit important science concepts and practices over multiple years. Rooted in a few major scientific ideas (e.g., evolution, matter) and starting with children's early capabilities, learning progressions increase in depth and complexity over the months and years of instruction. At each phase, learners draw on and develop relevant capabilities across the strands.

Although this is a relatively new and developing area of work, informal settings could play a complementary role in supporting learning progressions despite the episodic nature of informal learning experiences. For example, informal settings could be designed with the explicit intent of supporting learning progressions in a manner tightly aligned with K-12 science curriculum goals. Alternatively, informal environments could differentiate themselves from the K-12 agenda. If schools were to go "deep" with a commitment to a small number of learning progressions, this could invite informal settings to go "broad," focusing on incorporating other scientific issues that may not be evident in learning progressions.

**Conclusion 4: Members of cultural groups develop systematic knowledge of the natural world through participation in informal learning experiences and forms of exploration that are shaped by their cultural-historical backgrounds and the demands of particular environments and settings. Such knowledge and ways of approaching nature reflect a diversity of perspectives that should be recognized in designing science learning experiences.**

Although there are examples of culturally valued knowledge and practices being at odds with science (including spiritual and mystical thought, folk narratives, and various accounts of creation), a growing body of research documents that some knowledge and many skills developed in varied cultures and contexts serve as valid and consistent interpretations of the natural world that can form the basis for further science learning. This literature is reviewed in Chapter 7 and includes evidence from cultural psychology, anthropology, and educational research. The committee thinks that the diverse skills and orientations that members of different cultural communities bring to formal and informal science learning contexts are assets to be built on. For example, researchers have documented that children reared in rural agricultural communities who have more intense and regular interactions with plants and animals develop more sophisticated understanding of ecology and biological species than urban and suburban children of the same age. Others have identified connections between children's culturally based story-telling and argumentation and science inquiry, and they have documented pedagogical means of leveraging these connections to support students' science learning. The research synthesized in this volume demonstrates the importance of enlisting, embracing, and enlarging diversity as a means of enhancing learning about science and the natural world.

**Conclusion 5: Learners' prior knowledge, interest, and identity—long understood as integral to the learning process—are especially important in informal environments.**

The committee urges that researchers, practitioners, and policy makers pay special attention not only to the long-established importance of prior knowledge (National Research Council, 2000), but also to the broader array of learners' prior capabilities and interests reflected in the six strands and discussed throughout this report (see especially Chapters 3 through 6). The committee underscores the idea that prior interest and identity are as important as prior knowledge for understanding and promoting learning.

Prior knowledge, experience, and interests are especially important in informal learning environments, where opportunities to learn can be fleeting, episodic, and strongly learner-driven. At any point in the life span, learners have knowledge and interests, which they can tap into for further science learning. This includes their comfort and familiarity with science. Although learners' knowledge may remain tacit and may not always be scientifically accurate, it can serve as the basis for more sophisticated learning over time. Educators can support learners of all ages by intentionally querying, drawing on, and extending their interests, ideas about self, and knowledge.

# INFORMAL ENVIRONMENTS

**Conclusion 6: Informal science learning, although composed of multiple communities of practice, shares common commitments to science learning environments that:**

- engage participants in multiple ways, including physically, emotionally, and cognitively;
- encourage participants' direct interactions with phenomena of the natural and designed physical world largely in learner-directed ways;
- provide multifaceted and dynamic portrayals of science; and
- build on learners' prior knowledge and interests.

Direct access to phenomena of the natural and designed physical world—both familiar and foreign—is fundamental to informal environments. In informal environments, basic aspects of daily life are frequently framed in light of associated scientific ideas (e.g., draining a bath tub, swinging on a rope, throwing a baseball pitcher's curveball, and setting off the chain reaction of dominos falling can all be examined from the standpoint of physical mechanics). Informal environments may also provide access to phenomena and experiences that are difficult or impossible for learners to access otherwise, such as extreme micro- and macro-scale phenomena (e.g., views of earth from space, the merging of a human sperm and egg), cutting-edge science (e.g., nanotechnology), and historical and contemporary tools of scientific inquiry.

Hallmarks of learning in informal environments include interactivity driven by learner choice, an emphasis on the emotional responses of individual participants, and group experiences. At its best, informal science learning builds on both long-term and momentary or situated interests and motivations of learners. These hallmarks are evident in research and evaluation and in the practices, tools, and institutions of informal science learning.

Informal science education portrays science as multifaceted, highlighting that the knowledge and processes for building knowledge vary across fields. For example, much of physics and cognitive science is experimental. Many fields—astronomy, geology, anthropology, evolutionary biology—also draw on observational and historical reconstruction methods. Also, the values and practices of science reflect the diverse cultural values of practicing scientists as well as their shared professional commitments.

Although science is fundamentally evidence-based and draws its predictive power from scientists collectively testing theoretical models against evidence in the natural world, there is sociological and historical evidence that its accomplishments are shaped by who participates in science and how it is carried out (see Chapter 7). The influence of diverse perspectives on science is most evident in research in which a dominant view is ultimately overturned or challenged. For example, in making the case for increasing the participation of women in science, numerous examples have been identified that show how a scientist's gender can shape the questions asked and influence the interpretation of data. One of the most powerful examples is the involvement of a critical mass of female scientists in biology, which has been extremely influential in challenging assumptions about female health issues based on findings historically drawn from the data of only male subjects.

Science as portrayed in informal settings reflects a growing understanding that it is a dynamic enterprise in terms of its tools, practices, and knowledge. For example, until the past few decades, biology relied primarily on observed phenotypic features of species to determine the story of their evolution. Now powerful tools help scientists sequence DNA to provide precise and sometimes very surprising insights into speciation, such as the close relation of birds and dinosaurs. Some relationships once accepted on the strength of phenotypic similarities have been dashed in light of new genetic information, and new relationships are being established. Informal science learning environments seek to provide insight into how the creative tension between stable and changing information and perspectives creates reliable knowledge.

> **Conclusion 7: Broadcast, print, and digital media can play an important role in facilitating science learning across settings. The evidence base, however, is uneven. Although there is strong evidence for the impact of educational television on science learning, there is substantially less evidence regarding the impact of other media—newspapers, magazines, digital media, gaming, radio—on science learning.**

Educational programming, "serious games," entertainment media, and science journalism provide a rich and varied set of resources for learning science. Through technologies such as radio, television, print, the Internet, and personal digital devices, science information is increasingly available to people in their daily lives. For most people, television is the single most widely referenced source of scientific information, though it may be losing ground to the Internet. Media can support learning by expanding its reach to larger and more varied audiences. They can also be used in combination with designed spaces or particular educational programs to enhance learners' access to natural and scientific phenomena, scientific practices (e.g., data visualization, communication, systematic observation), and scientific norms (e.g., through media-based depictions of scientific practice). Interactive media have the potential to customize portrayals of science, for example, by allowing learners to select developmentally appropriate material and culturally familiar portrayals (e.g., choosing the language of a narrative, the setting of a virtual investigation).

Media offer tools that can be used well or poorly and may or may not influence science learning in desirable ways. While many learning experiences may be enhanced with media, evidence suggests that educators must carefully consider learning goals, learners' experience bases and interests, and the trade-offs associated with a particular form of media and means of delivery. Although there are not yet enough data to generalize, some studies have shown the power of media to support science learning in informal environments. For example, as summarized in Chapter 8, several studies (both experimental and quasi-experimental) have examined science television as an informal learning medium for children and youth and have shown that educational science programming can support science learning. In particular, researchers have documented concept development (Strand 2), some evidence that television can support learning science inquiry skills (e.g., Strand 3), and that it can positively influence interest in scientific topics (Strand 1).

Digital media, including user-developed media, are expanding rapidly. As discussed in Chapter 8, there are strong theoretical reasons to believe that deeply immersive digital gaming environments and simulations designed for informal educational purposes may enable learners to test out new identities and develop a sense of science and science careers. However, to date empirical evidence in this new research area is limited to a handful of studies of gaming and simulations, most of which are not science-specific.

**Conclusion 8: Designers and educators can make science more accessible to learners when they portray science as a social, lived experience, when they portray science in contexts that are relevant to learners, and when they are mindful of diverse learners' existing relationships with science and institutions of science learning.**

While it is the case that, as a group, scientists have the goal of being objective and place a premium on replicable empirical results, the very presence of debates in science reveals it to be a social activity in which competing background assumptions and judgments come strongly into play. The committee views science learning, science instruction, and the practice of science itself as forms of sociocultural activity. The practices and epistemological assumptions of science reflect the culture, cultural practices, and cultural values of scientists and others involved in the scientific endeavor more broadly.

Learning to communicate in and with a culture of science is a much broader undertaking than mastering a body of discrete conceptual or procedural knowledge (see, e.g., Strands 4 and 5, Conclusion 2). Aikenhead (1996), for example, describes the process of science education as one in which students must engage in "border crossings" from their own everyday-world culture into the subculture of science. The subculture of science is in part distinct from other cultural activities and in part a reflection of the cultural backgrounds of scientists themselves. As we have argued throughout this report and in particular in Chapter 7, by developing and supporting experiences that engage learners in a broad range of science practices, educators can increase the ways in which diverse learners can identify with and make meaning from their informal science learning experiences.

**Conclusion 9: Informal environments can have a significant impact on science learning outcomes for individuals from nondominant groups who are historically underrepresented in science.**

Several studies suggest that informal environments for science learning may be particularly effective for youth from historically nondominant groups—groups with limited sociopolitical status in society, who are often marginalized because of their cultural, language, and behavioral differences. For example, as discussed in Chapter 6, evaluations of museum-based and after-school programs suggest that these experiences can support academic gains for children and youth from nondominant groups. These successes often draw on local issues and the prior interests of participants (e.g., integration of science learning and service to the community, projects that involve participants' own backyard or local community). Several case studies of community science programs targeting participation of youth from historically nondominant groups document participants' sustained, sophisticated engagement with science and sustained influence on school science course selection and career choices. In these programs, children and youth play an active role in shaping the subject and process of inquiry, which may include local health or environmental issues about which they subsequently educate the community.

**Conclusion 10: Partnerships between science-rich institutions and local communities show great promise for fostering inclusive science learning. Developing productive partnerships requires considerable time and energy.**

Many designers in informal science learning are making efforts to address inequity and wish to partner with members of diverse communities. Effective strategies for organizing partnerships include identifying shared goals; designing experiences around local issues of local relevance; supporting participants' patterns of participation (e.g., family structure, modes of discourse); and designing experiences that satisfy the values and norms and reflect the practices of all partners.

Community-based programs that involve diverse learners in locally defined science inquiry, such as identifying and studying local health and environmental concerns, show promise for developing sustained, meaningful engagement (see Chapter 6, "Citizen Science and Volunteer Monitoring Programs"). Specific cultural resources can also be harnessed in program design (see Chapter 7, "Science Learning Is Cultural"). Many cultural groups spend leisure time in extended, multigenerational families, and partnerships have explicitly built cross-generational educational experiences in efforts to capitalize on this social configuration. These efforts merit replication and further study, including analysis of how science-rich institutions can collaborate with and serve community-based organizations and how these programs support and sustain participants' engagement.

## PROMOTING LEARNING

**Conclusion 11: Parents, adult caregivers, peers, educators, facilitators, and mentors play critical roles in supporting science learning.**

Just as informal settings for learning vary tremendously, so do the practices in which facilitators, educators, and parents engage to support it. Even in everyday settings, facilitators can enhance learning. For example, a child's cause-seeking "why" questions are an expression of an everyday, intense curiosity about the world. Parents, older peers, facilitators, and teachers can and often do support these natural expressions of curiosity and sense-making. Evidence indicates that the more they do, the greater the possibility that children will learn in these moments. Recognizing expressions of curiosity and sense-making supports and encourages learning as productive and signals this value to learners (e.g., by listening to learners, helping them inquire into and answer their own questions, and involving them in regular activities that place learners into contact with natural and designed phenomena and scientific concepts).

Their means of supporting learning range from simple, discrete acts of assistance to long-term, sustained relationships, collaborations, and apprenticeships. For example, just by interacting with children in everyday routine activities (e.g., preparing dinner, gardening, watching television, making health decisions), parents, caretakers, and educators are often helping them learn about science. In addition, family and social group activities often involve learning and the application of science as part of daily routines. For example, agricultural communities regularly analyze environmental conditions and botanical issues.

Even facilitators who are not experts in science (e.g., in after-school and community-based programs) can serve as intermediaries to informal science learning experiences. For example, the choice of pursuing a science badge in Girl Scouts may rest on the enthusiasm and assistance of a facilitating adult (see Chapter 5, "Doing and Seeing" and "Meaning-Making").

Cognitive apprenticeships are a specialized form of informal science learning in which learners enter into relationships with more knowledgeable others who help them refine their science understanding and skill deliberately over sustained time periods. For example, seasoned science enthusiasts may serve as de facto mentors for newcomers in hobby groups (e.g., amateur astronomy, gardening).

Productive science learning relationships frequently involve sustained individual inquiry but also intensive social practices with affinity interest groups and in apprenticeship relationships. Distributed and varied expertise in groups allows less knowledgeable individuals to interact with more knowledgeable peers and mentors. Frequently the roles of expert and novice shift back and forth over time, based on specific aspects of the inquiry in question.

> **Conclusion 12: Programs for school-age children and youth (including after school) are a significant, widespread, and growing phenomenon in which an increasing emphasis is placed on science.**

Programs, especially during out-of-school time, afford a special opportunity to expand science learning experiences for millions of children. These programs, many of which are based in schools, are increasingly focused on disciplinary content, but by means of informal education. Out-of-school-time programs allow sustained experiences with science and reach a large audience, including a significant population of individuals from nondominant groups. Ensuring that the principles of informal science learning (e.g., learner choice, low-stakes assessments for learners) are sustained as out-of-school-time programs grow will require careful attention to professional development, curricula, and best practices.

## INFORMAL ENVIRONMENTS AND K-12 SCHOOLS

> **Conclusion 13: Currently there are not good outcome measures for assessing the science learning goals of informal settings. Conventional academic achievement measures (e.g., standardized tests of science achievement) are too narrow and not well aligned to the goals of informal providers.**

One of the more noteworthy features of informal learning settings is the absence of tests, grades, class rankings, and other familiar approaches to documenting achievement that are characteristic of schools. The informal science community has nonetheless recognized the need to assess the impact of informal learning experiences in order to understand how everyday, after-school, museum, and other types of settings contribute to the development of scientific knowledge and capabilities. Everyday interactions about science frequently involve embedded informal evaluation and assessment of activity and reasoning.

In Chapter 3 we outline three criteria that need to be satisfied in order to develop the types of assessments that are most useful for science learning in informal environments. First, the assessments should not be limited to factual recall or other narrow cognitive measures of learning, but should address the range of relevant capabilities (depicted in the six strands) that informal environments are designed to promote. Second, the assessments used should be valid, providing authentic evidence of participants' learning and competencies. Third, assessments of informal science learning should fit with the experiences that make these environments attractive and engaging; that is, any assessment activities undertaken in informal settings should not undermine the very features that make for effective engagement, such as learner choice, voluntary participation, and pursuit of science-related interests.

> **Conclusion 14: Learning experiences across informal environments may positively influence children's science learning in school, their attitudes toward science, and the likelihood that they will consider science-related occupations or engage in lifelong science learning through hobbies and other everyday pursuits.**

Although, as discussed in Conclusion 13, the committee has serious reservations about using academic measures to assess learning in informal settings, we did find evidence that these settings may support improvements in student achievement, attainment, and career choices (see, for example, discussion of Strand 2 in Chapter 6). These outcomes reflect a degree of overlap between academic and informal settings. However, informal environments may particularly foster capacities that are unlikely to register traceable effects on conventional academic measures, notably around interest and motivation (Strand 1) and identity (Strand 6).

## TOWARD A COMMON FIELD

> **Conclusion 15: The literature on learning science in informal environments is vast, but the quality of the research is uneven, at least in part due to limited publication outlets (i.e., dedicated journals and special editions) and a lack of incentives to publish for many researchers and evaluators in nonacademic positions.**

Although there is a tremendous body of evidence relevant to learning science in informal environments, there is a limited (but growing) number of peer-reviewed outlets for publication devoted to it. While many scholars publish in a variety of peer-reviewed journals in education, psychology, and museum studies, others are not in academic positions and hence receive few rewards for publication. At present, much of the literature that informs the science learning in informal environments has not undergone rigorous, systematic peer review. In fact, the committee observed enormous variety in norms with respect to evidence, warrants, publication, and peer-review practices.

**Conclusion 16: Evaluation reports on particular programs provide an important source of evidence that can inform practice and theory more generally. Other kinds of research and data are needed, however, to build and empirically shape a shared knowledge base.**

Evaluations can be designed to support improvement during the design and implementation phase (formative) and to measure final impact of educational practice (summative). Although a substantial body of high-quality evaluation reports informs the knowledge base on learning in informal settings, important findings are not always widely disseminated and reports can be difficult to obtain. Also, evaluation is typically carried out by external evaluation consultants hired by the science learning institution. This arrangement can result in uncomfortable conflicts between the material interests of the institution to document successes and the interests of obtaining even-handed and theoretically oriented analysis. Many opportunities to learn from evaluations are lost as reports of outcomes are often not accompanied with careful description of practice or relevant comparisons to prior efforts and findings.

**Conclusion 17: There is an interdisciplinary community of scholars and educators who share an interest in developing coherent theory and practice of learning science in informal environments. However, more widely shared language, values, assumptions, learning theories, and standards of evidence are needed to build a more cohesive and instructive body of knowledge and practice.**

The literature reviewed in this report is derived from widely varied traditions, including researchers in different academic disciplines, evaluators and communities of inquiry and practice associated with informal learning institutions. Although their disciplinary and organizational affiliations vary, these scholars and educators share common interests in understanding, building, and supporting science learning in informal environments. Further development of common frameworks, standards of evidence, language and values will require new ways to share knowledge and expertise. Several leading thinkers have recognized this need. Journal special issues, the new Center for Advancement of Informal Science Education, and new guidelines from the National Science Foundation on evaluating the impact of informal science education have initiated and furthered this work, with the goal of contributing to better knowledge integration.

**Conclusion 18: Ecological perspectives on informal environments can facilitate important insights about science learning experiences across venues.**

The committee stresses the broad theoretical relevance of an ecological perspective on learning science in informal environments. An ecological learning perspective makes learners' activity and learning the organizing element in educational research. Rather than focusing on discrete moments of learning (e.g., as in a short-term, pre-post assessment), an ecological perspective strives to understand learning across settings: exploring, for example, how learning experiences in one setting prepare learners to participate in other settings. Working from an ecology of learning perspective, educators and researchers focus on learning experiences as they occur in specific settings and cultural communities and on the continuity of a learner's experiences across science learning environments—from classrooms to science centers to community sites.

# RECOMMENDATIONS FOR PRACTICE AND RESEARCH

The committee has developed a view of informal science education that takes learning seriously while maintaining a clear focus on personal engagement and enjoyment of science. In other words, we use the term "learning" in a broad sense that incorporates motivation and identity (see the six strands). Advancing the research and practice in ways that reflect this view of learning more fully will require careful consideration of goals, alignment of goals with learning experiences, and design of experiences that are informed by the values and interests of learners.

Our recommendations flow from the conclusions presented in this chapter and focus on improving both science learning experiences and research on learning science in informal environments. Given the nature of the evidence base, the recommendations for improving informal learning environments should be understood as promising ideas for further development that require additional validation through research and evaluation. These recommendations reflect practices that have been developed in some settings and may have been replicated; however, they have not been adopted widely.

These recommendations are relevant for a range of actors involved in science learning in informal settings. We consider three major groups: exhibit and program designers, front-line educators (e.g., scout leaders, club organizers, docents, parents and other care providers) who facilitate these experiences, and researchers and evaluators. These actors shape the educational experiences of learners in important ways collectively and individually. Through their collective actions they convey important messages about what science is, how science can benefit learners and society, and how learners can and should engage with science.

While their roles, actions, and goals overlap in important ways, particular actors have varying levels of control over different aspects of the learning environment. Here we organize our recommendations around the responsibilities of the three major groups.

## Exhibit and Program Designers

Exhibit and program designers play an important role in determining what aspects of science are reflected in learning experiences, how learners engage with science and with one another, and the type and quality of educational materials that learners use.

**Recommendation 1:** Exhibit and program designers should create informal environments for science learning according to the following principles. Informal environments should

- be designed with specific learning goals in mind (e.g., the strands of science learning)
- be interactive
- provide multiple ways for learners to engage with concepts, practices, and phenomena within a particular setting
- facilitate science learning across multiple settings
- prompt and support participants to interpret their learning experiences in light of relevant prior knowledge, experiences, and interests
- support and encourage learners to extend their learning over time

Learners are diverse and may be driven by a range of motivations, including nonscience ones (e.g., entertainment, socializing with family and friends). To increase the likelihood of engaging diverse learners with science, experiences should be multifaceted and interactive and developed in light of science-specific learning goals.

Designers should also be cognizant of the fact that learning experiences in informal settings can be sporadic and that, without support, learners may not find ways to sustain their engagement with science or a given topic. To support productive learning experiences and promote sustained engagement, designers should draw on learners' prior experience and knowledge and illustrate for learners both immediate and distal pathways for engagement and learning.

**Recommendation 2:** From their inception, informal environments for science learning should be developed through community-educator partnerships and whenever possible should be rooted in scientific problems and ideas that are consequential for community members.

Local community members and individuals from nondominant groups, including culturally diverse groups, older adults, and people with disabilities, should play an active role in the development, design, and implementation of science learning experiences—serving as designers, advisers, front-line educators, and evaluators of such efforts. The questions, materials, and contexts that constitute science learning experiences should be infused with the interests, knowledge, local activities, and concerns of partnering communities and diverse groups.

**Recommendation 3:** Educational tools and materials should be developed through iterative processes involving learners, educators, designers, and experts in science, including the sciences of human learning and development.

The relevant knowledge and skill needed to design state-of-the-art learning experiences reside among a constellation of actors. Ideally the design of science learning experiences in informal environments would begin with such diverse teams, who would work collaboratively over time in the development process. Over time repeated observation of participants' experiences and learning outcomes should inform efforts to improve educational tools and materials.

## Front-Line Educators

Front-line educators include the professional and volunteer staff of institutions and programs that offer and support science learning experiences. Front-line educators influence learning experiences in a number of ways. They may model desirable science learning behaviors and help learners develop and expand scientific explanations and practice, in turn shaping how learners interact with science, with one another, and with educational materials. They may also work directly with science teachers and other education professionals, who themselves are responsible for educating others. Given the diversity of community members who do (or could) participate in informal environments, front-line educators should embrace diversity and work thoughtfully across diverse groups.

In important ways, even parents and other care providers who interact with learners in these settings are front-line educators. They organize group visits, facilitate interactions among learners, and even convene pre- and postvisit activities. Thus, while parents and other care providers are not trained education professionals, they shape learning experiences and can be supported to do so more effectively.

**Recommendation 4:** Front-line staff should actively integrate questions, everyday language, ideas, concerns, worldviews, and histories, both their own and those of diverse learners. To do so they will need support opportunities to develop cultural competence, and to learn with and about the groups they want to serve.

In order to serve the goal of broadening participation in science, front-line staff should have the disposition and repertoire of practices and tools at their disposal to help learners expand on their everyday knowledge and skill to learn science. They need well-honed questions, dialogue prompts, and multiple examples of how the science that learners encounter in informal environments can be related to everyday experiences.

Professionals on the front line also should embrace diversity, so that they can empower learners by drawing on their knowledge, skills, and language to promote science learning. For example, youth are deeply interested in peer relations and so may benefit from intentional efforts to foster and sustain peer networks for science learning. Peers may be particularly important for encouraging underrepresented groups to participate in science learning. Front-line staff can also encourage and support cross-generational dialogue for multigenerational family groups. In order to accomplish this, practitioners need professional development to support their efforts.

## Researchers and Evaluators

Improving the quality of evidence on learning science in informal environments is a paramount challenge. Research and evaluation efforts rely on partnerships among curators, designers, administrators, evaluators, researchers, educators, and other stakeholders whose varied interests, expertise, and resources support and sustain inquiry. Accordingly our recommendations address investigators and the broader community that collaborates with investigators and consumes research and evaluation results.

**Recommendation 5:** Researchers, evaluators, and other leaders in informal education should broaden opportunities for publication of peer-reviewed research and evaluation, and provide incentives for investigators in nonacademic positions to publish their work in these outlets.

**Recommendation 6:** Researchers and evaluators should integrate bodies of research on learning science in informal environments by developing theory that spans venues and links cognitive, affective, and sociocultural accounts of learning.

Building and testing theoretical frameworks is a central goal of scientific inquiry, which drives the development of research questions, methods, tools, and integration of previous research findings across fields. Several bodies of work on nonschool learning are well established, although they often exist in isolation from other areas and could be integrated more broadly. Research on informal environments for science learning could enhance the community-wide development of theoretical frameworks by (1) making their theoretical frameworks and influences explicit in research and evaluation reports and presentations, (2) further testing common theoretical frameworks in science learning activities and analyses, and (3) exploring how theoretical frameworks in other social science fields can inform science learning in informal environments.

**Recommendation 7:** Researchers and evaluators should use assessment methods that do not violate participants' expectations about learning in informal settings. Methods should address the science strands, provide valid evidence across topics and venues, and be designed in ways that allow educators and learners alike to reflect on the learning taking place in these environments.

One of the main challenges at present is the development of means for assessing participants' learning across the range of experiences. Currently, studies that measure similar constructs often include unique measures, scales, or observation protocols. For example, research on media and learning tends to take different methodological approaches depending on the type of media in question (e.g., television, radio, digital environments). While some of this diversity reflects responsiveness to real differences inherent in the learning characteristics of such media, the lack of coherence hinders synthesis of research findings and the development of reliable measures. Rigorous, shared measures and methods for understanding and assessing learning need to be developed, especially if researchers are to attempt assessment of cumulative learning across different episodes and in different settings.

At the same time, the focus of assessment must be not only on cognitive outcomes, but also on the range of intellectual, attitudinal, behavioral, sociocultural, and participatory dispositions and capabilities that informal environments can effectively promote (i.e., the strands). They must also be sensitive to participants' motivation for engaging in informal learning experiences, and, when the experience is designed, assessments should be sensitive to the goals of designers.

## AREAS FOR FUTURE RESEARCH

Informal environments can be powerful environments for learning. They can be organized to allow people to create and follow their own learning agenda and can provide opportunities for rich social interactions. While this potential is often only partially fulfilled, research has illustrated that experience in informal environments can lead to gains in scientific knowledge or increased interest in science. However, further exploration is needed to provide a more detailed understanding of not only what is learned, but also of how the distinct features of informal environments contribute to their broad and long-term impact on learners. The committee outlines below the areas in which further research is particularly needed.

### Tools and Practices That Contribute to Learning

Additional research is needed to explore what physical, social, and symbolic tools best support science learning in informal environments. Researchers should build on the current research findings from studies of science learning in informal settings and draw more on approaches from across related fields (educational research, cognition, anthropology) to identify and adapt methods of discourse and conversational analysis, as well as observation techniques, that have been effective in describing settings in such a way that they can be compared, measured, and analyzed for change over time (e.g., Waxman, Tharp, and Hilberg, 2004).

### Learning Strands

The committee's six-strand framework represents a broader view of science learning than is typically found in the research. This view includes aspects of science learning that are supported in informal environments as well as in schools. It also aligns with the commitment informal education to participant engagement and development of interest and identity (Strands 1 and 6).

Evidence to support the impact of experiences in informal learning on Strand 6 is emergent. It is commonly believed that even participants who do not demonstrate increased knowledge as measured in pre-post assessment designs take away the potential to learn later. Do participants whose interest is sparked go on to learn more in the months that follow? Do they seek out other, related learning experiences? Does their relationship to science and science learning fundamentally shift? There is a need for studies to investigate how interest, future learning, and identity develop through informal science learning experiences over long time spans (e.g., weeks, months, and years). In order to better understand participants' perceptions of opportunity and how deeper forms of knowledge and enjoyment can be supported, it would be useful to further explore the relationship between interest and other motivational factors in an informal learning context. The inquiry-based, free-choice nature of these experiences offers the possibility of examining how motivation and interest relate to future science learning across a range of venues. In addition, an exploration of how interest development influences learner identity may help create a better understanding of science learning not only in informal environments, but also in more formal learning environments.

### Cumulative Effects

Science learning, and informal science learning more specifically, is a cumulative process. The impact of informal learning is not only the result of what happens at the time of the experience, but also the product of events happening before and after an experience. And interest in and knowledge of science is supported by experiences in informal environments and in schools. Although it is important to understand the impact of informal environments, a more important question may be how science learning occurs across the range of formal and informal environments. The science learning literatures and fields are segmented (e.g., into school learning, informal education) in ways that are at odds with how people routinely traverse settings and engage in learning activities. Thus, research should attempt to explore learners' longer term, cross-cutting experiences. Further work should increase understanding of the connections or barriers in learning between more formal and more informal science learning environments.

The committee calls for additional efforts to explore science learning in longer term increments of time, tracking learners (rather than exhibits, tools, programs) across school and informal environments. Such research would allow researchers to examine the influence of experiences in different settings and over time and to explore how these experiences build on or connect to each other. It will require developing and refining research methods for tracking individuals over time and solving other problems pertaining to security of participants' personal information and attrition.

### Learning by Groups, Organizations, and Communities

One of the more difficult but important research challenges that the science education in informal settings community faces is developing the means to study learning, growth, and change at the level of a group, organization, or community. How do social groups learn science through dinner table conversations, visits to the zoo, science laboratory meetings, hobbyist interest groups, civic engagement, and other everyday activities? Such interactions influence not only the individuals who participate in them, but also the group itself. What are the relevant changes at the group level? The literatures are turning toward exploring the learning that takes place through social interactions, yet it remains unclear what factors (e.g., participant's behaviors, attitudes, intrinsic interest) are responsible for the impact of these social exchanges and how they play out at the group or organizational level.

### *Supporting Learning for Diverse Groups*

Informal environments for science learning may be particularly important for science learning for diverse groups. Research exists on how different groups participate in various venues, but questions remain about how to best empower science learning for diverse groups through informal learning environments. Research has documented that participation in many venues (e.g., designed informal settings, science media) is skewed toward the dominant cultural group and those most interested in science, although there are several important exceptions. School group visits to designed settings, community-based organizations and after-school programs, and exhibitions designed around local scientific or health issues have all been observed to serve a more diverse audience, in part because they are often designed with underserved populations in mind (as described in the committee's conclusions). Yet there is variability in the success of these environments in attracting and engaging their diverse audiences. A better understanding of the naturally occurring science learning in nondominant and dominant cultures is needed to inform basic theory and to design learning experiences that meaningfully attend to the cultural practices of diverse groups.

### *Media*

Media, in particular television and Internet resources, are the most sought-out tool for learning about science. Meanwhile, through media, the nature of learners' interactions with science has changed. Many people now have at their fingertips immersive, interactive platforms that allow them to pursue their interest in science. Through various forms of digital media—blogs, virtual spaces, wikis, serious games, RSS feeds, etc.—access to scientific ideas and information and knowledgeable others has become, if not pervasive, at least widespread.

It is unclear whether more frequent use of media is the by-product of engagement in and enjoyment of science learning experiences, or vice versa. Existing studies, with the exception of extensive research on television, are primarily correlational in nature, indicating that there is a relationship between enjoyment of science learning and frequency of use of media, but these studies do not indicate whether one factor causes the other or if there is a complex dynamic of interacting influence. Further studies are needed to determine whether the use of tools, such as media for science learning, promotes interests in science, whether interest in science inspires the use of such tools, or both in specific ways.

Arguments about the transformative power of media for informal science learning are based on very modest evidence and warrant further investigation. Many emergent media forms allow users to receive and send information, leverage resources to communicate with huge numbers of learners, and honor diverse ways of knowing and learning (through user-selected and designed interfaces), so that users can interact with content and with one another in ways that they deem valuable. And these characteristics of new digital technologies—dialogic structure, user-direction and organization, and expansive networking of learners and resources—resonate with the values and research findings of the informal science learning community. Research on the impact of media is needed to understand how the unique features of media can support different aspects of science learning (e.g., the six strands).

Another related area worthy of further research is exploration of how learners evaluate the validity of science information from emergent media-based sources. Technologies have made it possible for almost anyone to author information about science and to make that content accessible to very broad audiences. This leaves the learner with the difficult task of deciphering the validity of information and discerning the likely sources of bias for any resource. With ever-increasing user-generated information spaces, it will be important for researchers to continue studying how learner characteristics influence their judgment of information presented through these media.

## REFERENCES

Aikenhead, G.S. (1996). Science education: Border crossing into the subculture of science. *Studies in Science Education, 26,* 1-52.

National Research Council. (2000). *How people learn: Brain, mind, experience, and school.* Committee on Developments in the Science of Learning, J.D. Bransford, A.L. Brown, and R.R. Cocking (Eds.), and Committee on Learning Research and Educational Practice, M.S. Donovan, J.D. Bransford, and J.W. Pellegrino (Eds.). Commission on Behavioral and Social Sciences and Education. Washington, DC: National Academy Press.

National Research Council. (2007). *Taking science to school: Learning and teaching science in grades K-8.* Committee on Science Learning, K-8. R.A. Duschl, H.A. Schweingruber, and A.W. Shouse (Eds.). Washington, DC: The National Academies Press.

Waxman, H., Tharp, R.G., and Hilberg, R.S. (Eds.). (2004). *Observational research in U.S. classrooms: New approaches for understanding cultural and linguistic diversity.* New York: Cambridge University Press.

# Appendix A

# Biographical Sketches of Committee Members and Staff

**Philip Bell** (*Cochair*) is associate professor of learning sciences in education at the University of Washington, Seattle. He directs the ethnographic and design-based research of the Everyday Science and Technology Group (http://everydaycognition.org). As a learning scientist, he has studied everyday cognition and expertise in science, children's argumentation, the use of digital technologies in youth culture, the design and use of novel learning technologies, and new approaches to inquiry instruction in science. Bell is a coleader of informal learning research for the Learning in Informal and Formal Environments Center (http://life-slc.org/) and a coprincipal investigator of COSEE-Ocean Learning Communities (http://cosee-olc.org/). He is a board member of the International Society of the Learning Sciences and the Board on Science Education of the National Research Council. He has a background in human cognition and development, science education, electrical engineering, and computer science. He has a B.S. from the University of Colorado, Boulder, and M.A. and Ph.D. degrees from the University of California, Berkeley.

**Bruce Lewenstein** (*Cochair*) is professor of science communication at Cornell University, with appointments in the Departments of Communication and Science and Technology Studies at Cornell. At heart, he is a historian (his current favorite personal project is a history of science books in the years since World War II), but overall his work focuses on both historical and contemporary issues involving the public understanding of science. He is a former editor of the journal *Public Understanding of Science* (1998-2003) and an elected fellow of the American Association for the Advancement of Science (AAAS). He is a past member of the AAAS Committee on Public

Understanding of Science and Technology and is on the advisory board to the Sciencenter, an interactive science museum in Ithaca. His work has been supported by the National Science Foundation, the U.S. Department of Agriculture, the U.S. Department of Energy, the National Aeronautics and Space Administration, and the Chemical Heritage Foundation. He has a B.A. from the University of Chicago and an M.A. and a Ph.D. from the University of Pennsylvania, in the history and sociology of science and in science and technology policy.

**Sue Allen** is director of visitor research and evaluation at the Exploratorium in San Francisco, where she oversees all aspects of visitor studies, education research, and evaluation on all projects involving the museum's public space. She was the in-house evaluation coordinator for the California Framework Project, which explored the roles that a science museum can play in assisting science education reform in the schools. She and her colleagues study visitors' learning in the museum's public space and work collaboratively with practitioners in the design of their research and evaluation agendas. Her current research interests include methods for assessing learning, exhibit design, personal meaning-making, and scientific inquiry. She has lectured in the Department of Museum Studies at the John F. Kennedy University. She teaches a graduate-level, action-oriented course on thinking and learning in science in the School of Education at the University of California, Berkeley. She has contributed numerous articles and book chapters to the informal science field. She is a member of many professional associations, including the Visitor Studies Association, the Museum Education Roundtable, and Cultural Connections. She has a Ph.D. in science education from the University of California, Berkeley.

**B. Bradford Brown** is professor of human development and former chair of the Department of Educational Psychology at the University of Wisconsin–Madison. His research has focused on adolescent peer relations, especially teenage peer groups and peer pressure and their influence on school achievement, social interaction patterns, and social adjustment. He is the former editor of the *Journal of Research on Adolescence* and a past member of the Executive Council of the Society for Research on Adolescence (SRA). Currently, he serves as chair of the SRA Study Group on Parental Involvement in Adolescent Peer Relations. He is the coeditor or coauthor of five books, including *The Development of Romantic Relationships in Adolescence*, *The World's Youth: Adolescence in 8 Regions of the Globe*, and *Linking Parents and Family to Adolescent Peer Relations: Ethnic and Cultural Considerations*. He has served as a consultant for numerous groups, including the Carnegie Council on Adolescent Development, the National Campaign to Prevent Teen Pregnancy, and the Blue Ribbon Schools Program of the U.S. Department of Education. He has an A.B. in sociology from Princeton University and a Ph.D. in human development from the University of Chicago.

**Maureen Callanan** is professor and chair of the Department of Psychology at the University of California, Santa Cruz. Her research focuses on cognitive and language development in young children, exploring how they come to understand the world through everyday conversations with their parents. One particular focus is on how children's intuitive theories about the world (e.g., how heat makes things melt, what makes people sad) develop in parent-child conversations. Children's "why" questions and parents' explanations are studied through parents' diary reports of children's questions and through videotapes of parent-child activities, such as reading books, baking muffins, and visiting children's museums. The research explores how children and parents construct shared understandings of concepts and of causal theories about particular domains, with special attention to scientific domains. Callanan has also focused on how children learn word meanings and understand multiple names for the same objects. Her studies on examining how parents and children name objects in everyday conversations have demonstrated important links between parents' language and children's interpretations of new words. She has an A.B. from Mount Holyoke College and a Ph.D. from Stanford University.

**Angela C. Cristini** is executive director of special programs at the Ramapo College of New Jersey. She also directs the Meadowlands Environment Center, in which over 20,000 children, educators, families, adults, and senior citizens per year participate in informal science programming. She designed and directs the Master of Science in Educational Technology Program, which is directed at the needs and concerns of the professional education community. Previously she was president of the American Littoral Society, a national, not-for-profit, membership organization, dedicated to the environmental well-being of coastal habitats. She was assistant director of environmental research in the Division of Science and Research of the New Jersey Department of Environmental Protection, where she was responsible for setting the research agenda for state environmental issues. She was a recipient of competitive grant funding from the National Science Foundation and the New Jersey State Department of Education for informal science, teacher enhancement, and curriculum development projects in science and technology. Her research interests include comparative physiology and the ecology of invertebrates, the effects of pollutants on marine organisms, and the endocrinology and mechanisms of ionic regulation of crustaceans. She has a B.A. from Northeastern University and a Ph.D. from the City University of New York.

**Kirsten Ellenbogen** is the director of evaluation and research in learning at the Science Museum of Minnesota. Her research focuses on fostering science discourse in out-of-school learning environments and positioning museums in systemic education reform. Previously, as project director of the Center for Informal Learning and Schools at King's College London, she collaborated on the development and management of a training and research program designed to address the need for research degrees in informal science education. Most recently, she was a senior associate at the Institute for Learning Innovation, where she developed an initiative to coalesce the last decade of research on learning in museums into frameworks for practitioners. She began her work in museums as a demonstrator, and her award-winning exhibition development work has focused on inquiry experiences, multimedia interactives, and dual-purpose spaces appropriate for both school and family groups. She has a Ph.D. in science education from Vanderbilt University.

**Michael A. Feder** (*Senior Program Officer*) is on the staff of the Board on Science Education. At the National Research Council, he serves as a program officer for the Committee on Learning Science in Informal Environments and the Committee on Understanding and Improving K-12 Engineering Education. Previously, he worked as an education program evaluator, contributing to such projects as a review of interventions for English-language learners and the evaluation of the Math and Science Partnership Programs in New Jersey and Ohio. His research includes the effects of subsidized non-Head Start day care on the academic achievement of Hispanic children and the psychological and academic adjustment of refugee children exposed to wartime trauma. He has a Ph.D. in applied developmental psychology from George Mason University.

**Cecilia Garibay** is principal of Garibay Group, where she leads audience research and evaluation. Her research focuses on exhibits and programs in informal learning environments, particularly those aimed at reaching underrepresented audiences. Some of her recent efforts have involved research with Latino and immigrant communities—particularly regarding leisure values and informal learning, conceptions of museums, and perceptions of science. She regularly consults with institutions on audience development and community inclusion and to develop frameworks and strategic plans for making exhibitions and programming accessible to multiple and diverse audiences. She brings a bicultural/bilingual perspective to her work and specializes in culturally responsive and contextually relevant research and evaluation approaches. She has consulted with a wide range of free-choice learning organizations, including the Association of Science-Technology Centers, Children's Museum of Houston, Cornell Lab of Ornithology, the Exploratorium, Smithsonian (National Museum of American History), Monterey Bay Aquarium, and the Oregon Museum of Science and Industry. Her 20 years of research and evaluation experience also include working with nonprofit organizations, foundations, and corporations. She is currently conducting research on issues of audience diversity and organizational change in museums.

**Laura Martin** is the director of science interpretation at the Arizona Science Center and a visiting professor in the College of Education at Arizona State University. Previously she held positions at the Phoenix Zoo; the Center for Teaching and Learning and the Center for Informal Learning and Schools, both at the Exploratorium in San Francisco; and the Children's Television Workshop. She has authored numerous publications and book chapters on learning science and mathematics in a variety of settings, as well as research reports for the Arizona Science Center. She has also written curricula and developed demonstration projects in science. She is a member of many associations and is the organizer as well as a participant of the Informal Learning Opportunities Network. She has a B.A. in history from the University of Chicago, an M.S. in education from the Bank Street College of Education, and an M.A. and a Ph.D., the latter in psychology, cognition, and development, from the University of California, San Diego.

**Dale McCreedy** is director of the Gender and Family Learning Programs Department at The Franklin Institute. She is also the program director of multiple National Science Foundation–funded collaborations, two with Girl Scouts of the U.S.A., one with the School District of Philadelphia, one with the Free Library of Philadelphia, and one with the Institute for Learning Innovation. She has led the development of program structures and resources, as well as collaborations with local and national partner organizations. Dr. McCreedy is an advocate for lifelong learning, with a focus on girls and women's science learning, and the cultivation of family and after-school support in science. She participates on numerous advisory boards and girl-based organizations, and makes frequent presentations about girls, women, and families in science. She was awarded a lifetime membership in Girl Scouting in 1996, and was the 2002 recipient of the Maria Mitchell Women in Science Award. She received a Ph.D. in education from University of Pennsylvania.

**Douglas L. Medin** is the Louis W. Menk Chair in psychology at Northwestern University. He previously taught at the University of Illinois and the University of Michigan. Best known for his research on concepts and categorization, his recent research interests have extended to decision making; cross-cultural studies of reasoning and categorization; models of similarity, culture, and cognition; and cognitive dimensions of moral reasoning. He is an editor for *Cognition* and for the *Journal of Experimental Psychology: Human Learning and Memory*. He also serves as a consulting editor for *Cognitive Psychology* and was past editor of *Psychology of Learning Motivation*. He has conducted research on cognition and learning among both indigenous and majority culture populations in Guatemala, Brazil, Mexico, and the United States. He is a member of the American Academy of Arts and Sciences and a recent recipient of the American Psychological Association's Distinguished Scientific Contribution Award. He was elected to the National Academy of Sciences in 2005. He has a B.A. from Moorhead State College and an M.A. and a Ph.D. from the University of South Dakota.

**Vera Michalchik** is a researcher social scientist at SRI International. Trained in both anthropology and cognitive science, she focuses on the social and cultural aspects of learning. Her recent research includes case studies of technology use and learning in informal settings, analyses of science discourse and communication, and studies of representational technology in learning science and mathematics. She is especially interested in how various types of social interactions and modes of communication support or detract from students' participation in learning activities and how this affects issues of equity. Her work on such projects as ChemSense, the Community Technology Centers evaluation, and NetCalc reflect this interest. She has extensive experience in conducting case studies, video analyses of learning environments, and ethnographic fieldwork. She has a B.A. in film studies from the University of California, Berkeley (1985), an M.Ed. in human development from Harvard University (1986), and a Ph.D. in educational psychology from Stanford University (2000).

**Gil G. Noam** is founder and director of the Program in Education, Afterschool and Resilience (PEAR), and an associate professor at Harvard Medical School and McLean Hospital. Trained as a clinical and developmental psychologist and psychoanalyst in both Europe and the United States, he has a strong interest in supporting resilience in youth, especially in educational settings. He served as the director of the Risk and Prevention Program and is the founder of the RALLY Prevention Program, a Boston-based intervention that bridges social and academic support in school, after-school, and community settings. He has also followed a large group of high-risk children into adulthood in a longitudinal study that explores clinical, educational, and occupational outcomes. PEAR is actively engaged in research on after-school topics and is also working with Boston after-school programs, in partnership with Achieve Boston, to develop an after-school training and technical assistance infrastructure. Noam is also involved in a private-public partnership, which includes the Institute for Educational Science, the Piper Trust, and the Haan Foundation, in conducting a randomized control study of a reading and resilience intervention for young struggling readers in after-school settings. He has published numerous papers, articles, and books in the areas of child and adolescent development as well as risk and resiliency in clinical, school, and after-school settings. He is the editor-in-chief of the journal *New Directions in Youth Development: Theory, Practice and Research*, which has a strong focus on out-of-school time.

**Andrew W. Shouse** (*Senior Program Officer*) is a staff member of the Board on Science Education. He codirected the study that produced the 2007 report *Taking Science to School: Learning and Teaching Science in Grades K-8*. He also serves as study director for the Learning Science K-8 Practitioner Volume, a "translation" of *Taking Science to School*. He is an education researcher and policy analyst whose interests include teacher development, science education in formal and informal settings, and communication of education research to policy and practice audiences. Prior to joining the National Research Council, he worked as an education research and evaluation consultant, science center administrator, and elementary and middle grade teacher. He has a Ph.D. in curriculum, teaching, and educational policy from Michigan State University.

**Brian K. Smith** is associate professor of information sciences and technology and education at the Pennsylvania State University. He studies the use of computation to support and augment human performance and learning. Examples of his work include video annotation systems for biology education, cameras equipped with global positioning systems and image databases for history education, and interventions around photography and computer visualizations to promote awareness of personal health practices. Current projects are under way to explore information design for informal, everyday decision making. Previously he was on the faculty of the Media Laboratory of the Massachusetts Institute of Technology, where he led a research consortium of 20 collaborating corporations to define new methods for information description, design, and dissemination. He is currently the principal investigator of the medical informatics research initiative in the College of Information Sciences and Technology at Penn State and the research director of its involvement in Apple Computer's Digital Campus Initiative. He received a Faculty Career Development Award from the National Science Foundation in 2000 to begin a research agenda around visual learning. He received the Jan Hawkins Award for early career contributions to humanistic research and scholarship in learning technologies from the American Education Research Association in 2004. Apple Computer also named him an Apple Distinguished Educator in 2004. He has a B.S. in computer science and engineering from the University of California, Los Angeles, and a Ph.D. in learning sciences from Northwestern University.

# APPENDIX B

# Some Technical Considerations in Assessment

## *Interviewing Groups Versus Individuals*

Learners in informal environments, such as museums, generally participate in multigenerational groups rather than as individuals, and these groups often move loosely through the environment, splitting and reforming as members make new discoveries and share what they are experiencing with one another. This makes interviewing particularly challenging. It is difficult to craft interview questions that are suitable for a broad range of people. Also, the learning experience varies frequently between an individual and group focus.

There are trade-offs between interviewing individuals and groups:

(a) Interviewing individuals has the advantage that participants do not influence each other's opinions, that the resulting data are amenable to statistical methods with equal weighting for each person, and that the time taken to conduct an interview is relatively short. However, such interviews require selecting an individual interviewee (and often individuals prefer to self-nominate rather than accept a random sampling method), locating the rest of the group to explain what is happening, ensuring that minors have appropriate child care, and finding a nearby yet quiet location to conduct the interview. Also, parents often respond to questions by reporting on what they think their children learned rather than what they themselves learned, because their children's experience is often their framing reason for attending. Unless parents' interpretations of children's experience are the focus of the study or the child is too young to be interviewed, this approach is problematic because it relies on indirect inferences rather than self-report.

(b) Interviewing groups has the advantage of not separating group members, so families are more likely to agree to participate. Also, the responses they give, as a group, reflect the actual learning when the group members were jointly engaged in activity. One disadvantage of group interviews is that one member may dominate (typically an adult), and group members will often fall into agreement with each other's opinions. A way to reduce this tendency is to question members individually but in inverse order of status. However, some researchers feel that unequal power dynamics are likely to be representative of the learning dynamic, and therefore interviews with asymmetrical participation are authentic. Group interviews also present the problem of how to quantify data from groups of different sizes, particularly if the study attempts to characterize frequencies of responses. Some researchers code response frequencies into 3 categories: "1," "2," and "many." Finally, although group interviews are more relaxing for participants, there is rarely time to ask all questions equitably before the group becomes restless, so most persons in the group typically do not complete the interview. Alternatively, interviews conducted from a more qualitative or naturalistic perspective may allow for a much looser participation structure by the group members, but they require extended and careful analysis by the researcher afterward.

### Control Groups

Because informal environments emphasize learning by choice, using random assignment of learners to treatment and control groups may sometimes be logistically impossible, upsetting to the learners, threatening to the study validity, or all of the above. In such cases, it may be desirable to reference a comparison group that is not a strict control but that provides some sense of plausible baseline behavior (data from visitors to other museums or exhibitions, literature that cites common knowledge, behaviors, or attitudes to a topic, etc.).

### Video- and Audiotaping

With increasing interest in such process-based outcomes as engagement, conversations, and actions, research in informal environments has made increasing use of recording systems, such as audio- and videotape. These raise technical and ethical issues. Technically, the main challenge is often to obtain audio of sufficiently high quality to hear what people are saying above the ambient noise. Attempts at solution include using a Dictaphone (Borun, Chambers, and Cleghorn, 1996), wearing of cordless microphones (e.g., Leinhardt and Knutson, 2004), or placement of microphones on individual exhibits (e.g., Gutwill, 2003). The ethical issues, namely the need to have visitors give informed consent to being recorded, have been addressed by posting signs, augmenting posting-signs (Gutwill, 2003), asking for consent when visitors arrive and placing a sticker on their clothing to alert the videographer (Crowley and Callanan, 1998), or getting explicit consent as visitors enter a space. Such methods are generally compromises, and researchers should always refer to their local institutional review board for approval of their specific data collection method.

### Time as a Measure of Learning

In environments such as museums, botanical gardens, and zoos, where learners move freely through a physical space of options, time spent ("holding time" or "dwell time") is a commonly used measure of impact in summative evaluations. At the same time, there is controversy about what exactly it assesses in relation to learning. There are various approaches to thinking about time, including:

(a) Some researchers regard it as a necessary but not sufficient condition for learning. In this view, learners need to pause and engage with objects, people, or activities in order to have a chance to learn from them, but learning is not necessarily linearly related to time spent. Some researchers have interpreted histograms of holding time as bimodal or multimodal, revealing different audience characteristics in terms of background or motivation (browsers, grazers, etc.), but these are controversial: most exhibitions show a single peak at the short end of the spectrum of time spent (Serrell, 1998, 2001).

(b) Some regard it as an indicator of learning, using the well-established principle that time on task is the most universal correlate with learning across contexts. However, the meaning of "on task" is particularly ambiguous in free-choice environments (Shettel, 1997), as is the definition of learning. A few studies have shown direct evidence that time spent in exhibitions correlates with learning, as measured by previsit questionnaires on the exhibit topic (Abler, 1968) or free recall of objects seen (Barnard and Loomis, 1994).

(c) Some regard time spent as a direct measure of learning, defined as engagement in socially sanctioned collaborative activity. From this sociocultural perspective, participants are learning throughout their engagement, although the exact nature of what they learn may be quite different from institutional expectations.

### Internet Surveys

Increasingly, the Internet is being used to conduct surveys of learners. These may be assessments of online resources or may ask about previous experiences in another setting (such as a museum visit, viewing of a TV series, etc.). They may be contained within emails, or, increasingly, be web-based. Compared with paper surveys, Internet surveys are inexpensive and generate quick responses, but often raise concerns about response rates and biased populations of respondents. For recent reviews of the literature on web surveys in informal science learning environments, including suggestions for effective design and usage, see Parsons (2007), Yalowitz and Ferguson (2007), and Storksdieck (2007).

## REFERENCES

Abler, T.S. (1968). Traffic patterns and exhibit design: A study of learning in the museum. In S.F. DeBorhegyi and I. Hanson (Eds.), *The museum visitor* (vol. 3, pp. 103-141). Milwaukee, WI: Milwaukee Public Museum.

Barnard, W.A., and Loomis, R.J. (1994). The museum exhibit as a visual learning museum. *Visitor Behavior, 9*(2), 14-17.

Borun, M., Chambers, M., and Cleghorn, A. (1996). Families are learning in science museums. *Curator, 39*(2), 123-138.

Crowley, K., and Callanan, M. (1998). Describing and supporting collaborative scientific thinking in parent-child interactions. *Journal of Museum Education, 17*(1), 12-17.

Gutwill, J. (2003). Gaining visitor consent for research II: Improving the posted sign method. *Curator, 46*(2), 228-235.

Leinhardt, G., and Knutson, K. (2004). *Listening in on museum conversations.* Walnut Creek, CA: AltaMira Press.

Parsons, C. (2007). Web-based surveys: Best practices based on the research literature. *Visitor Studies, 10*(1), 13-33.

Serrell, B. (1998). *Paying attention: Visitors and museum exhibitions.* Washington, DC: American Association of Museums.

Serrell, B. (2001). In search of the elusive bimodal distribution. *Visitor Studies Today, 4*(2), 4-9.

Shettel, H.H. (1997). Time—is it really of the essence? *Curator, 40*, 246-249.

Storksdieck, M. (2007). Using web surveys in early front-end evaluations with open populations: A case study of amateur astronomers. *Visitor Studies, 10*(1), 47-54.

Yalowitz, S., and Ferguson, A. (2007). Using web surveys in summative evaluations: A case study at the Monterey Bay Aquarium. *Visitor Studies, 10*(1), 34-46.

# Index